The geomorphology of North-west England

edited by

R. H. Johnson

Manchester University Press

Published by
Manchester University Press
Oxford Road, Manchester M13 9PL, UK
and 51 Washington Street, Dover
New Hampshire 03820, USA

British Library cataloguing in publication data
The Geomorphology of north-west England.
  1. Landforms – England, Northern
  I. Johnson, R. H.
  551.4′09427   GB436.G7

Library of Congress cataloging in publication data

The geomorphology of north-west England.
  Bibliography: p. 372
  Includes index.
  1. Geomorphology – England, Northern. I. Johnson, R. H. (Richard Hugh)
GB436.G7G46   1985   551.4′09427   85-275

ISBN 0 7190 1745 9 *cased*
       0 7190 1790 4 *paper*

Photoset by
Elliott Brothers & Yeoman Limited, Woodend Avenue, Speke, Liverpool L24 9JL
Printed in Great Britain by Bell and Bain Ltd., Glasgow

Contents

Preface

This book is not an excursion guide nor is it a regional textbook. It has been compiled in the hope that it will generate interest in a number of geomorphological problems currently being studied in areas near Manchester. It is also hoped that it will provide a useful reference text not only for those geomorphologists able to attend the First International Geomorphological Conference (Manchester: September 1985), but also for others studying Geomorphology in local universities, colleges and schools. The selection of essay topics has been somewhat arbitrary for although most relate to areas that are either drained by streams flowing to the Irish Sea or covered by Late Devensian ice during the Pleistocene Epoch and thus have a regional connotation, others have been included which possess neither of these elements. Their inclusion is justified because they cover topics of special interest in areas which are readily accessible from Manchester and will be visited by Conference visitors or student field parties. For this reason then the Peak District of north Derbyshire is included whereas other areas, such as north Cumbria, are excluded even though on morphological grounds they could be considered to be parts of the north-west regional province.

In a volume as wide ranging as this there are of course constraints affecting length of essays and content, but no attempt has been made to reach a single general view about the origin, age or mode of formation of features described. The book is arranged in three main sections. The first provides a general regional physiographic survey relating to geology, climate, denudation history, soils, rivers and the coast. The second records the legacy of past ice-sheet incursions into North-west England and finally several studies are included in the third section which examine post-glacial and historic changes in localised areas and notably within the Manchester conurbation itself.

List of contributors

Stephen Belbin, Department of Geography, University of Manchester
Dr F. M. Broadhurst, Department of Geology, University of Manchester
Professor Ian Douglas, Department of Geography, University of Manchester
Dr Stephen J. Gale, College of Arts and Technology, Cambridge
Dr John Gunn, Department of Environmental and Geographical Studies, Manchester Polytechnic
C. Harrison and Dr J. R. Petch, Department of Geography, University of Salford
Dr A. M. Harvey, Department of Geography, University of Liverpool
Dr R. H. Johnson, Department of Geography, University of Manchester
B. S. Kear, Department of Geography, University of Manchester
A. Knowles, Department of Geography and Recreational Studies, North Staffordshire Polytechnic, Stoke-on-Trent
David Longworth, School of Languages and Humanities, Lancashire Polytechnic, Preston
Leslie F. Musk, Department of Geography, University of Manchester
Dr D. W. Shimwell, Department of Geography, University of Manchester
Dr John H. Tallis, Department of Botany, University of Manchester
Dr G. S. P. Thomas, Department of Geography, University of Liverpool
Dr Michael J. Tooley, Department of Geography, University of Durham
Dr Peter Vincent, Department of Geography, University of Lancaster
Professor Peter Worsley, Department of Geography, University of Nottingham

Acknowledgements

The Editor and Manchester University Press wish to express their appreciation to the Council of the Manchester Geographical Society who generously gave a grant to cover the production costs for the colour illustrations.

The following acknowledgements are made:
To Professor Ian Douglas and the staff of Manchester University Press for much editorial advice and help at all stages in the production of this book. To Messrs N. Scarle and G. Bowden for their editorial work on the illustrations and for their advice on the many cartographical problems that have arisen during publication.

Stephen Belbin, Ian Douglas, Richard Johnson, Brian Kear, Leslie Musk and David Shimwell wish to express their thanks to the secretarial staff, past and present, in the Geography department for their work in preparing the several chapters for publication. They would also wish to thank Mr N. Scarle, Mr G. Bowden and Miss A. Lowcock for their work in drawing the text figures and diagrams.

Permission for the reproduction of figures, maps and other diagrams has been given by many copyright owners and their names are listed under the names of each author.

Copyright permission.
F. M. Broadhurst: The Editors of Proc. Yorks. Geol. Soc. for permission to use Figures 2:2 & 3; The Editors of Lethaia for Fig. 2.5 and The Geological Journal for Figures 2.6 and 7. *L. F. Musk*: Professor H. H. Lamb for permission to use Figures 14.1, 15.22, and 16.11 from his book 'Climate, Present, Past and Future'. *B. S. Kear*: The Soil Survey of England and Wales for permission to use information from official sources for the construction of Figures 5.1 and 5.2. *M. J. Tooley*: The Director of the British Geological Survey for permission to reproduce Figures 6.2 and 6.3. *R. H. Johnson*: Gebrüder Borntrager, (West Germany) for permission to use Figure 1.1 from J. Büdel's book 'Climatic Geomorphology'; The Director of the British Geological Survey for permission to use information from official sources for the construction of Figures 13.2–9; The Editors of The Geological Journal for permission to use data for Figure 13.9; J. Gunn: Figure 14.2 is reproduced by permission of Dr T. D. Ford and Geo Abstracts Ltd.

The drawing of text figures and other technical assistance.
The undermentioned authors gratefully acknowledge help from the following people.

M. J. Tooley: Dr M. A. Geyh (Niedersächsisches Landesamt für Bodenforschung, Hanover) is thanked for providing the radio-carbon assays; J. Telford, Hazel Evans and D. Hudspeth (technical assistance); A. Corner and G. Brown (text figures). *A. M. Harvey* and *G. S. P. Thomas*: The Drawing Office Staff, the University of Liverpool (text figures and photographs). *P. Vincent*: P. Mingins (text figures). *D. Longworth*: The Drawing Office Staff, Preston Polytechnic (text figures). *A. Knowles*: Jane Williams (text figures). *J. Gunn*: Catherine Gunn (text figures). *D. M. Shimwell*: M. McKigney (technical assistance). *J. Petch and C. Harrison*: G. Dobryznsky (text figures and photographs).

The preparation of typescripts and other manuscript material.
M. J. Tooley: Sheila Shippam. *J. Gunn*: Marie Cunningham. *D. M. Shimwell*: Eileen Wynne.

Useful discussion.
The undermentioned authors wish to express their thanks to the following for helpful discussion.
S. Belbin: Professor Ian Douglas and Dr R. H. Johnson. *M. J. Tooley*: Mrs Margaret McAllister, Drs I. Shennan, R. T. R. Wingfield and H. Pantin. *P. Worsley*: Drs J. Rose and D. Holyoak. *R. H. Johnson*: The Geological Section of the Greater Manchester Council Planning Department, (especially Mr G. Edwards), and also Mr P. N. J. Robinson (Manstock Geotechnical Services) concerning the use of bore hole data and records and Professor Ian Douglas. *J. Gunn*: Drs A. C. Waltham and T. D. Ford. *S. Gale*: C. O. Hunt. *C. Harrison and J. Petch*: The Engineering Departments of Salford and Bury Metropolitan Borough Councils and Greater Manchester Council.

1 *R. H. Johnson*

The geomorphology of the regions around Manchester: an introductory review

Although it is scarcely 50 km from one side of the Lake District to the other, there is in that small space, a variety of scenery and geology almost unequalled in Britain (Moseley, 1978).

Few provinces of Britain contain such profound yet simple contrasts in the character of their physical landscapes as Lancastria, a region divided neatly and clearly between morphological units of radically different types (Freeman, Rodgers and Kinvig, 1966).

Introduction

If it is true that almost every part of the world has its academic and literary advocates who have extolled in their writings its scenic qualities, then, as the above quotations indicate, the north-west part of England is no exception. The Lake District and Cumbrian landscapes are some of the best known in Britain, having been described, painted, photographed, and intensively studied by many of Britain's greatest writers, artists and scientists during the last two hundred years, but its dramatic qualities have perhaps tended to direct interest away from its neighbouring region to the south, Lancastria. This is a region of considerable topographic diversity ranging from deeply dissected, plateau uplands and tabular hills through undulating terrain drained for the most part by westward-flowing Pennine rivers to flat, lowland till plains, wetlands and coastal dune systems. Historically Lancastria has long been regarded by geographers as a discrete, unified region which complements the more rugged terrain of the other main area in North-west England. Together, these two regions, Cumbria and Lancastria, with geographical extensions to include the Pennine karst areas of north Derbyshire and west Yorkshire, the floor of the Irish Sea, and the Isle of Man form the focus of study for this book.

In this introductory review the overall geomorphic character and history of the regions are outlined for the benefit of those not already familiar with the area. The review concerns itself with the distinction between the uplands and lowlands through a discussion of the nature of the transition zone between them and then considers the physique and structure, the erosion history and the glacial modification of each major terrain unit. A final section reviews the changes in the regions in the Late and Post-glacial periods. To avoid needless duplication, maps have not been included in this chapter, and the reader is asked to refer to figures in subsequent chapters where appropriate. Reference tables for Britain and northern Europe during the later Pleistocene and Holocene periods are also included, but these must be considered as extremely tentative and unconfirmed.

The physical regions of Lancastria and Cumbria

The Pennine upland/lowland transition zone (Figs. 2.1, 7.1 and 13.1)

In the north and east of the region the Cumbrian mountains and Pennine moorlands have been eroded from Palaeozoic rocks which have been strongly folded and sometimes faulted during times of uplift. In contrast, the same up-arching of the rocks caused rift-valley subsidence in the western and southern parts of the region, leading to the development of graben structures in which late Palaeozoic, Mesozoic and Cainozoic sediments were accumulated in geosynclinal basins, located in the area now consisting of the Irish Sea, Cheshire and Lancashire. Although strongly fractured and faulted, the strata were not greatly deformed by any fold movements, but underwent intermittent erosion at various times, until eventually they were covered by glacial (Pleistocene) and Recent (Flandrian) and other deposits.

Between these two major morphologic/tectonic units there is a relatively narrow zone, 2–3 km wide, of fault-disturbed rocks which mark the structural discontinuity between the two regions. The faults themselves are not manifest as a major topographic feature, but a strong relief feature is aligned parallel to them which has a relative relief of some 200 m. This 'scarp' extends from Newcastle-under-Lyme northwards along the edge of the Pennines to the Rossendale hills, and is continuous except where streams leave the uplands by way of narrow water-gaps. The high 'scarp' continues along the west margin of the Rossendale and Bowland Forests, and into the Lake District. Here it becomes merged into the high mountain relief, but geological studies of the tectonic features (Moseley, 1972), indicate that the fault zone is an extremely important influence on the topography of southern Lakeland. An escarpment feature similar to that found at the edge of the Pennines occurs both at the margin of the Lake District where it overlooks the west Cumbrian lowland and in the Eden valley, where it marks the western edge of the North Pennines.

Despite its domination of the lowlands which lie to the west, the 'scarp' has no distinct topographic name. This is because it is not a single feature, but is formed by a line of hills on the western edge of the Pennines whose present morphology no longer suggests that the feature is one of great geological antiquity. Nevertheless this erosional fault-line scarp is possibly the most distinctive feature in Lancastria, perhaps having originated as far back as the Carboniferous Period (King, 1976). Douglas (1984) has suggested that it forms the eastern margin of a rift depression extending from west Cumbria, along the Pennine edge, southwards to Worcester and the West Midlands (see later). Some 30–60 million years ago further rifting occurred with fault movements taking place along both old and new fracture lines. These subsequent displacements once again emphasised the relief discontinuity between the upland and lowland, and today the fault-line scarps form either a single, or a multiple series of west-facing escarpments, which lie well to the east of the main fault-lines. The faults at Alderley, Macclesfield (Red Rock), Horwich, Chorley, Garstang, Lancaster and at Barrow and Annaside in west Cumbria are now much degraded, and partly buried by glacial debris, but they probably predate almost all the features of the uplands or lowlands adjacent to them.

The uplands – their physique and structure

The southern Pennines Within the South Pennines the high relief rises to well above 600 metres O.D. In many places the local relative relief is 300 to 400 metres. The dominant relief forms are extensive flat summit plateaux, isolated tabular hills, and deep valleys whose slopes to the floors below are broken by extensive benches or terraces formed by the differential erosion of the grits and shales. In the south-west Pennines however, these topographic features are replaced by a succession of cuestas and asymmetrical vales.

The more elevated areas are formed of coarse Namurian sandstones (gritstone) and mudstones. On these extensive high moorlands have developed which provide not only a 'peat covered, rain sodden wilderness', to use Rodgers's apt phrase, but also are areas of extreme climatic severity for this part of Britain. Present-day rainfall exceeds 1650 mm on the two highest hills near Manchester, Bleaklow and Kinderscout, and in mid-winter, temperatures on the summits may remain below zero for 8–13 hours each day (Garnett, 1956). Much of the peat cover has been eroded by gullying and mass movement (see Tallis, Chapter 17), to expose many large grit boulders and sandstone bedrock surfaces. Sometimes the boulders lie on, or are embedded in, a subsoil of rotted and cryoturbated '*grus*' (weathered rock), for periglacial processes have been extremely active in the South Pennines, etching out tors, and upland surfaces, creating blockfields and transporting large amounts of debris in solifluction lobes down the slopes. Beneath the plateau levels the valleys are deeply incised, with steep valley sides and narrow floors. Frequently in these 'dales', the higher slopes of the valley are formed by massive sandstone formations which provide a conspicuous scarp or 'edge' to the hillside. Since these rocks almost invariably overlie thick beds of incompetent, weak mudrocks, many of the slopes are inherently unstable and slope movements are almost ubiquitous. These latter features range from various forms of rockslides, most of which occurred by rock slump rotation, to modern soil slumps, terracettes and various soil creep features (Johnson, 1980, 1981).

All these strata have been involved in fold movements which began in the late Carboniferous (Coal Measures/Westphalian) period, and have been affected by upward movements of the crustal plate since those times. The dominant structure is a monocline best seen, in its simplest form, at Blackstone Edge (SD 972161) east of Rochdale. To the east of the axis are high plateaux with strong west-facing scarps developed on almost horizontal rocks, while to the west the smooth cuesta slopes are much disrupted by faulting. Here, the strata plunge steeply beneath the synclinal lowlands in which Burnley and Rochdale lie. Nearer Manchester, the upfold axis moves closer to the Pennine western margin where it becomes a more complicated structure. Here, in the extreme western parts of the uplands, the strata dip steeply to form a series of east-facing scarps and north–south trending vales where the anticline has become displaced by faulting. To the east of the anticline, the level bedded grits rise in a step succession of scarps which culminate in the Bleaklow (630 m) and Kinderscout (637 m) hills.

Further south, the main folding of the western Pennines was more intense and produced several fold structures which are arranged '*en échelon*' to each other, and whose axes are aligned north to south. Etched out by a complex drainage system they form a cuesta pattern which is 'Appalachian' in its degree of dissection and is probably the best example of its kind in Britain. In the Potteries the same type of topography is present, but the fold structures have been complicated by extensive faults and are fo.med of younger Coal Measure rocks.

In the Peak District the limestones are exposed in the core area of the upland and are sometimes differentiated as the White Peak area, in contrast to the surrounding, more elevated, Dark Peak grit moorlands described above. The limestones were uplifted into a broad dome structure whose north–south axis can be seen in the Wye valley just east of Buxton. These strata are interbedded with volcanic rocks, some of which are impermeable and provide water tables within the strata. At its northern edge the limestone core passes unconformably beneath the Millstone Grit rocks where its presence has been revealed in a borehole sited in Alport dale to the east of Kinderscout.

The gentle up-and-down swells of this plateau surface are coincident with the structures of the region, with upfolds where the limestones outcrop at their widest, and downfolds where they become narrower. The plateau itself is morphologically homogeneous, rising from about 300±20 m in the south to about 400±60 m some 40 km further north. It is dissected by many deep, deep-sided dry valleys and by large dales whose rivers receive most of their runoff from catchment areas located in the shale-grit terrain. The smaller tributary dry valleys often head in broad, shallow depressions with sinks on the broad valley floors, but lower down they become narrow and steep-sided with bare outcrops exposed in the upper cliff slopes; scree covers the lower flanks. Many of the upper tributary valleys lost their streams as a result of rapid downcutting of the main dales, and at such times, the extensive cavern systems were extensively enlarged and developed.

The Central Pennines The massive limestones again outcrop extensively in the Central Pennines, around Malham and in the Ribble valley, forming both the higher plateau of elevations of *c*.450 m O.D., and a series of low ridges within the dale itself. These two groups of limestone belong, however, to two distinct structural areas, for the high plateau are part of the Askrigg Block, a stable rigid massif, whereas the reef knolls at Cracoe and Clitheroe are a series of steep-sided, rounded hills developed in a subsidence zone and separated from the upland by the Craven Fault system. Much of this limestone scenery is very similar to that found further south in the Peak District, but the glacial imprint has been stronger and there are extensive limestone pavements, and in general the karst is better developed and the cavern systems larger.

Overlying these northern limestones there is an alternating sequence of sandstones, grits, shales and impure limestones which together constitute the lower part of the Millstone Grit (Namurian) succession. They form the main mass of the Bowland Forest but elsewhere are replaced by the higher Millstone Grit formations which consist of hard, cemented gritstone bands separated by more easily weathered mudrocks (shales, siltstones and mudstones). Some coals were also formed in thin seams and these became thicker higher up in the succession, where the Millstone Grits ultimately are replaced by Lower Coal Measures.

To the north of Manchester there are two broad salients of the Pennines which are known as the Rossendale and Bowland Forests. The Rossendale hills are remnants of a broad, flattened anticline of ENE–WSW trend which was broken and disturbed by a series of faults aligned NNW–SSE. The most prominent scarp former is the Rough Rock Grit which is a conspicuous sandstone member in the Upper Millstone Grit succession, but the faulting has also preserved some Middle Coal Measures sandstones and these form small mesa hills that rise above the level of the Rossendale plateau. In the Bowland hills the relief is more varied for here limestones, both reef and massive, are interbedded with shales and sandstones, and are exposed in the centre of a sharply dipping anticline. Although there is some adjustment by drainage to structure, and the more resistant grits still tend to dominate the skyline, which is formed by a series of gently undulating hill surfaces with structural and erosional benches being often prominent.

Between the two Forest regions, there is a belt of intensively folded strata which has also been badly affected by faulting. The whole succession between the Middle Coal Measures and Carboniferous Limestones is compressed into a series of tight flexures, whose axes are mostly WNW–ESE. South of Pendle Hill, and in the area of greatest subsidence, a syncline has been eroded by the tributaries of the Darwen and Lancashire Calder into a series of parallel cuestas and vales, but not all the Coal Measures have been removed. The northern edge of this basin culminates in the double ridge of Pendle Hill, a monoclinal structure, which rises to over 550 m O.D., but north of this escarpment there are several tightly compressed folds with breached anticlines occurring in the Ribble and Hodder valleys at Clitheroe and Easington respectively. The relief is generally low-lying with Longridge Hill being the most upstanding feature.

The Lake District (Cumbria) Geologically the Lake District is an elongated dome structure, in which the Lower Palaeozoic rocks outcrop within a ring of Carboniferous strata. It may be subdivided into three distinct physical and geological terrains which run west to east, but the Lake District also includes some granite and granophyre outcrops, in Eskdale, Shap and Ennerdale, that provide their own distinctive terrains. The most northerly of the belts is formed by the highly contorted and cleaved Skiddaw Slates of Lower Ordovician age that have been moulded into a mountain terrain of smooth rounded summits, deeply dissected by glaciated valleys. The central part is formed by rocks of the Borrowdale Volcanic Series, which are largely pyroclastics ranging from tuffs to coarse agglomerates, but also include many interbedded lavas of variable character. Most of these are andesites, but basaltic and rhyolitic flows are also present. This wide range of rocks has been sculptured by ice and associated processes into the rugged topography for which the Lake District is so renowned. Further south these rocks are replaced by other formations belonging to the Silurian Series. These include some (Coniston) limestones, but consist for the most part of shales and flags which grade into sandstones and mudstones in the upper part of the succession. The smooth slopes formed on these rocks typify not only the southern part of the Lakeland area, but also the adjacent Howgill Fells. Located east of the Lune valley the Howgills are less rugged than the Lakeland fells, but both areas owe their present elevation to recent uplifts in the Tertiary Period. Together with the rest of the Lake District, the Howgills were raised along an east–west axis, and form a monoclinal block which has gently dipping strata to the north. An unconformity separates the overlying Carboniferous limestones from the Silurian strata and if extrapolated this unconformity would pass only a short distance above the Howgill summits. There is thus a geological explanation for the more bevelled rounded summits in the Howgills, which perhaps mark a transition zone between the Lake District and the Pennine Askrigg Block where the topography reflects structure controls to a much greater degree.

Erosional history of the uplands

It has long been accepted that the British uplands have experienced many phases of uplift and subsequent erosion and that, as a consequence, their geological structure has been truncated or bevelled. In the course of such a history many planation surfaces and terrace features were eroded, and are now located well above or below the levels at which they were formed. How and when this bevelling took place is a matter of considerable speculation and dispute. The problem was comprehensively reviewed in the regional texts by King (1976) and Straw and Clayton (1979) and a further important contribution to the problem was also made by King (1977) in her study of the Quaternary landscapes and related Neotectonic matters. Much of the material presented here is based on her conclusions, but important advances in this field have been made recently and these are discussed here and also by Belbin (Chapter 3) and by other contributors (Chapters 9, 12 and 13).

The great thicknesses of sediments preserved in the Irish Sea and North-west England subsidence zones record not only a history of depositional conditions, but also reveal evidence of important erosional events. In Lancashire and Cheshire the basin sediments comprise basal Lias and older Permo-Triassic rocks and these also infill the graben depression that lies between the Isle of Man and the mainland. Further south, off the Welsh coast, Oligocene and Miocene sediments (600 m thick) are preserved and rest unconformably upon Lias beds that are themselves part of a post-Carboniferous sedimentary succession which is over

2135 m thick. The existence of deep basins between areas of uplift therefore constitutes a consistent feature in the structure of the British Isles and the adjacent seas and has largely resulted from Britain's position on a trailing edge coast of a continental plate, at a time when there was active splitting and spreading of the ocean floor (see Broadhurst, Chapter 2). This rupture, which began at the end of the Cretaceous Period some 70 million years ago, initiated major sea-level changes that were themselves dependent upon the size and configuration of the oceans and the growth of the Antarctic ice cap. Since these fluctuated through time, Britain's island status has varied, with Britain either becoming part of the European land mass, or alternatively, a much more fragmented archipelago. The present outline of Britain, however, was probably generally defined for much of the Tertiary, and there are indications that crustal tilting of the upland blocks was initiated at the end of the Cretaceous Period at least 70 million years ago, and that the present Irish Sea graben began to form some 30–40 million years later with much displacement of the strata along both old and new faults.

Today, many geomorphologists would accept King's hypothesis, that the Palaeozoic rocks of the North England uplands were already exposed at the beginning of the Eocene and that there must have been strong differential movements occurring between upland blocks and subsidence basins from mid-Oligocene onwards. It would appear that much of the drainage either, became adjusted to structure as this warping progressed, or was antecedent to any fold movements that may have occurred in the Cheshire basin, the English Midlands and the Solway area during the Miocene Period (King, 1963, 1976, 1977; Johnson, 1969a). The most important diversions appear to have taken place in the Lune and Derbyshire Derwent catchments where the exposure of the Palaeozoic rocks provided structural conditions which permitted those favoured streams to extend their systems at the expense of those which had originated on the warped surfaces uplifted in the early Tertiary epochs.

No deposits associated with these events have, however, been found anywhere in North-west England, excepting those at Brassington, Derbyshire, where they are fortuitously preserved in doline solution features, that now form part of an undulating plateau surface, whose limits lie between 300 and 350 m O.D. The sediments date from the Mio-Pliocene transition period, originating in a fluvial environment on a low gradient surface near to sea level. This formation has been involved in crustal movements which raised it some 450 m after its deposition, but as the result of sinkhole formation in dolomitic limestones, the deposits were subsequently lowered some 150–200 m to their present position (Walsh *et al.*, 1972).

In 1956 D. L. Linton recognised three major surfaces in the South Pennines; these he termed the Upland Summit (S), the Upland Plain (U) and the Upland Valley Surfaces (Fig. 1.1a). The Summit Surface is located only on the higher sandstone plateaux and is an undulating surface with an elevation between 480 and 600 metres. The middle surfaces (U) are much less extensive, forming benches at lower levels, but on the limestone plateau they become the dominant element in the morphology occurring between 280 and 350 m O.D. Below these plateau levels the valleys are deeply incised but contain within them suites of terraces and bench

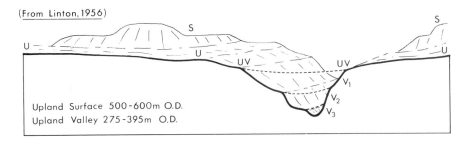

(From Linton, 1956)

Upland Surface 500-600m O.D.
Upland Valley 275-395m O.D.

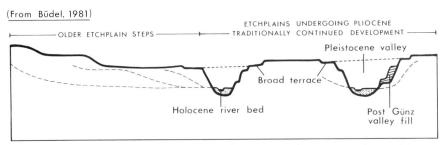

(From Büdel, 1981)

Fig. 1.1   Development of Upland Surfaces. (a) after Linton (1956), (b) after Büdel (1981).

'flats', eroded during the Pleistocene Epoch. Linton's model was based on the belief that subaerial processes would lower the landscape to one of low relief during prolonged stillstands and that uplift had occurred only intermittently during the Tertiary. He also believed that the residual hills on the Summit Surface marked the approximate position of a sub-Cretaceous unconformity that had once passed over most of upland Britain.

Such a hypothesis appears unfounded today, and equally unacceptable is that offered in 1974 by T. N. George who argued that all planation surfaces in upland parts of 'Britain were parts of an unbroken sequence which extended from 950 m down to 200 m O.D. He envisaged that these surfaces were eroded from an earlier peneplane feature which had been formed in the early Tertiary and subsequently warped to form an 'envelope' surface. The warping was either continuous or occurred intermittently from the Miocene onwards to at least the early Pleistocene. The several surfaces were developed by both marine and fluvial processes with the higher and older ones being eroded *de facto* in the Neogene (Miocene–Pliocene) Period. Included in such a sequence would be the broad summit plateau of the South Pennines, many of which show little adjustment to the underlying structures, and are at elevations between 300 and 650 m O.D., but this concept of a 'stairway' of surfaces is no longer acceptable, as it is linked to an assumption that at all times the surfaces were created in response to a falling base level provided by changes in sea level.

Linton's denudation model for the southern Pennines is, however, very similar in its form to the one developed by Büdel (1981), which was conceived for the middle Rhine valley (Fig. 1.1b). Like Linton, Büdel attributes valley incision to the work of Quaternary processes, and he recognised a series of broad terraces

('*Breitterrassen*') as the 'traditional development of the older surfaces along narrow ribbons of terrain which at the same time were the first forerunners of today's valley network'. These are the equivalent to what Linton termed Upland Valley Surfaces.

At higher levels above the benches, Büdel recognised composite surfaces, which were formed under quite different conditions to those envisaged by Linton and the other mid-twentieth century workers in Britain. Given warm, possibly sub-tropical climates in which deep weathering occurred, it is possible for two or more rock-eroded surfaces to form at the same time by the process of etchplanation. As the upper surface develops by the work of surface processes which wear back the slopes and create pediments, the lower is formed at the basal limit of weathering and beneath the subsoil, becoming only exhumed when the regolith is removed. When this occurs both surfaces become further degraded, but the rate at which they undergo change is reduced, and if raised above levels at which the main streams are working, the surfaces are likely to be preserved, even though strongly modified by periglacial processes in late Pleistocene times (McArthur, 1981).

It is thus possible for two surfaces to be developed concomitantly, yet be separated by an etchplain escarpment which has itself been affected by tectonic disturbance (Fig. 1.2).[1] Büdel's model therefore provides a better explanation for the evolution of the present landscape, for the surfaces have continued to evolve since their inception in the Neogene Period. Such a model, as Belbin will argue, does not preclude some parts of the land surface from being much older, having been exhumed from beneath a now vanished rock cover, but it is improbable that their present form retains any record of such events although some residual hills on the upland etchplain surfaces may lie very close to where once an unconformable rock cover passed over them.

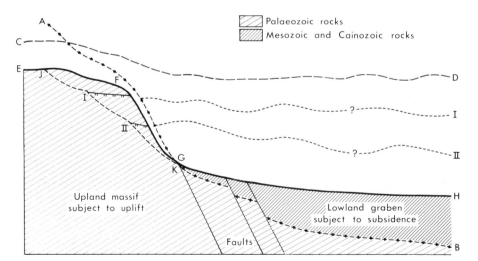

Fig. 1.2   The evolution of the West Pennine escarpment and the Cheshire Plain lowland[1]

Glacial modification of the upland landscape

During the Pleistocene epoch, Britain probably experienced at least eleven, and possibly sixteen, extremely severe cold climates during which land ice may have formed (Price, 1983), but like all other uplands in northern Britain, there is little evidence for any glacial stage other than that which took place within the last 30 000 years. Some patches of weathered and older till have been preserved in dolines on the limestones and elsewhere in the Peak District (see Chapters 13 and 14), and there is a pre-Devensian buried soil overlying older tills at Mosedale, in the Lake District. Apart from these isolated instances, the record of the earlier Pleistocene changes has to be inferred from cave earth materials (Table 1.1).[2]

The glacial history is described in full in the succeeding chapters, so only a brief comment is necessary here. Essentially, it is a history of Devensian events, and involves the growth, expansion and ultimate decay of an ice sheet from its feeder areas in Scotland and Northern Ireland southwards into the Irish Sea basin. The ice sheet extended as far as South Wales and south-east Ireland, but at its margin there was a sea-ice shelf which terminated along the north Cornish coast. Between Wales and the Pennines, the same ice sheet spread southwards to the West Midlands and to the vicinity of Worcester.

For our present purpose the upland regions can be divided into two categories dependent on whether ice was generated in sufficient quantity to migrate from its source area or not. Valley glaciers or ice sheets were formed in the Lake District and more northern Pennine valleys, and spread out from their source regions, but in the south-west Pennines, the Bowland and Rossendale Forests, all but the highest hills were affected by ice which had emanated from other areas. This fundamental difference resulted in the creation of different types of landforms and sediments which can be classified into distinct glacial landform and sediment assemblages. In the Lake District and Howgill areas the assemblages are closely associated with valley glaciers and indeed many textbooks have taken their examples of glaciated landforms from this region. Glacially eroded landforms are ubiquitous, but there is now some controversy concerning the efficacy of the Devensian ice, which appears to have been a very effective transporter of debris but may have only modified features that had been fashioned in earlier glacial periods.

Within the northern parts of North-west England glacial erosion has been extremely important with extensive corrie (cirque) formation, overdeepening of valley floors and the breaching of watersheds by both diffluent and transfluent ice. Such breachings were mostly caused when Irish Sea and Scottish ice flows prevented Cumbrian valley glaciers from escaping to the lowlands to the north and west of the mountains. Substantial amounts of ice were diverted southwards through cols at Dunmail Raise and Kirkstone but diffluent ice movements also occurred when the glaciers formed an ice dome covering all but the highest peaks.

In the eastern and more southerly parts of the Lake District, the trunk valleys are more 'Icelandic' than 'Alpine' in form, for although the lower plateau surfaces were scoured and highly mammillated by slow moving ice, there are a number of drumlins and rock-drumlins in the valleys whose forms show that rapidly moving

PLEISTOCENE CHRONOLOGY

YEAR B.P. (000's)	V28-238 STAGES	after BOWEN 1978		3*	after GASCOIGNE 1983
		NORTH EUROPE STAGES	BRITISH STAGES		BRITISH STAGES
10	1	Flandrian	Flandrian		Flandrian
			Windermere		Late Devensian
20	2		Late Devensian (Ice sheet)		Main Devensian (glaciation)
30					
40	3	Weichselian	Devensian		
50	4		Upton Warren I/s		Mid-Devensian (interstadial)
60	5a	Brørup	Chelford		
70					
80					Early Devensian (cold)
90	5b-d	Early Weichselian	Early Devensian		
100					Ipswichian
	5e	Eemian	Trafalgar Square		
		Warthe			
150	6	Eemian ?	Ipswichian Brundon		Wolstonian ?
200	7	Eemian ?	Hoxnian ?		warm Hoxnian ?
250	8	Saale ?	Wolstonian ?		
300	9	Eemian	Ipswichian Ilfordian		Anglian ?
	10	Saale	Anglian ?		
400	11	Holstein	Hoxnian	?	warm interglacial
	12	Elster	Anglian	?	cold extensive glaciation
500	13	Holstein	Hoxnian		

Table 1.1. The Pleistocene chronology.

glaciers were diverted by local relief across the lower watersheds, and formed a major piedmont glacier before joining with the main Irish Sea ice sheet on the Lancashire lowland. These large glaciers were capable of eroding a series of deep rock basins such as those that form the lakes Windermere and Coniston, but it is not clear if this overdeepening was achieved in only one, presumably the last, major glacial episode or not. In the central core of the Lake District many corries and other ice-eroded features were modified by glaciers during the short 'Loch Lomond' glacial stage, which terminated the Devensian cold climate period, but these glaciers failed to extend very far into the main trunk valleys.

Somewhat surprisingly, the English Lakeland contains few glacial depositional features of note. The 'Loch Lomond' stage glaciers have left a series of hummocky features in the upper catchment areas, but beyond the limits of this particular advance there are few morainic ridges and those that occur in Borrowdale, Langdale and downstream of the Windermere and Coniston lakes are probably 'cross valley moraines', owing their formation to glacier stagnation and downwasting. Many of the lower slopes are only thinly mantled with drift deposits and these are in part overlain by scree, fan talus and alluvial material. Following the removal of the ice many small lakes on the valley floors have become drained, and the streams are now choked with an over-abundant supply of outwash material. As a consequence flood prevention measures have had to be undertaken in all the larger valleys.

In the Pennines, the Lune and Ribble valleys were able to generate their own glaciers and in the Devensian stage these ice streams crossed the main England divide at several places. These glaciers were much less effective in their erosional work when compared to those in the Lake District, but many of the limestone plateau surfaces were overridden and heavily ice-scoured, and the main valleys were affected by ice plucking and smoothing. In the Bowland Forest, ice from the north and the upper Ribble covered the hill summits, and left a thick residue of tills and other sediments on the lower slopes. Further south, however, this Pennine ice stream was much less powerful than the main Irish Sea and Lake District ice sheet, and was diverted eastwards to abut and only partly encroach into the north Rossendale and Trawden uplands. Here it breached the watershed at Cliviger (SD 885275), and Gorple (SD 911314), but the ice sheet legacy is primarily one of deposits and meltwater erosion. An extensive drumlin field was formed in the Ribble valley to the north of the Pendle ridge and in lower Ribblesdale and the Calder valleys the sides were mantled with tills and considerable thickness of fluvio-glacial material. Meltwater channels were cut on the northern flanks of the Bowland, Trawden and Rossendale hills and several cols on the watersheds were deepened by meltwater erosion.

A similar denudational history also occurred within the Rossendales and the south-west Pennines with ice moving over the lowlands and encroaching onto the higher land. The Devensian ice limits for the most part do not extend up to the main English divide, but some ice and meltwaters did reach the upper reaches of the Churnet, a Trent headwater stream. It would appear that successive ice movements transported erratics from the Lake District, Scottish Southern Uplands and Irish Sea floor to the Pennine slopes, but most of the till matrix is composed of Permo-Triassic and Carboniferous materials. On the steeper slopes, meltwaters were the most effective erosional agent and many ice margin channels were eroded especially along the slopes that form the hill margin adjacent to the lowlands. Meltwaters were also important in creating substantial drainage derangements within the Pennine and Rossendale valleys. As the ice downwasted, many of the larger valleys became infilled with glacial sediments and the streams were deflected from the line of their old courses to cut deep and narrow gorges at the edge of the now buried valley. These gorges are steep and irregular in profile and contrast sharply with the wide open valley sections that lie adjacent to them. The history of such diversions is often quite complex in detail and in some locations the valleys were never infilled with drift, but many have been occupied by stagnating ice. This left little or no debris where it stagnated, but the ice itself was of sufficient thickness to divert the streams. Examples of these derangements will be given in Chapter 13, but such diversions are very common and can be found in most large tributary valleys in the Mersey and Ribble catchments.

In the southern Pennines however, periglacial processes played a very significant part in modifying the landscape. Substantial areas of the Peak District lay beyond the Devensian ice limits and the gritstone escarpments especially were affected by frost wedging and other frost action processes with many rock falls and topples occurring on the scarp edges. On the shale slopes and across the sandstone benches, strong soliflual movements of frost riven debris took place forming blockfields, mudslides and solifluction lobes. These are now stabilised, but even when they were forming the conditions were not so severe that all the vegetation cover was destroyed. Radiocarbon dates (BM 1807, 2126 and 2127), derived from bone samples found in cave earths, indicate that animals were present in the hills both during the waxing and waning phases of the ice sheet growth and even during the period of its maximum development. Reindeer and probably other animals were able to roam over the North Staffordshire hills at the time of the glacial maxima (*c.*20 000 years BP) and where the ice front was probably only a few kilometres away from their grazing grounds.

The physique of the Lowlands

The relief and geological structure

The Lancastrian lowland has been variously defined, but in general it has a triangular shape, having an apex north of Lancaster and its base south of Whitchurch, where it extends from Wales to the South Pennines. At the edge of the till plain and in the central part of the lowland there are a number of hills where Palaeozoic and Mesozoic rocks outcrop, and these rise above the general level of the lowland to provide a distinctive element in the topography. But for the most part the relief is formed by glacial deposits and only occasionally rises above 135 m O.D.

The Pleistocene sedimentary cover masks a landscape which is both irregular in form and of variable age. Most of the lowland is underlain by Triassic rocks, but older rocks outcrop at the margins of the graben-depression and subsidence probably began in the late Carboniferous (Coal Measures) and has been continuing intermittently ever since. As a result of this subsidence the unconformity underlying the Permo-Triassic rocks now lies some 1500 m below present sea level in both the Cheshire and Lancashire basins, but the movements which led to its subsidence also caused a saddle structure between Liverpool and the Rossendales, thereby separating the two main subsidence zones. Both in this saddle and along the margins of the basins intense faulting along NW–SE lines took place during the Cretaceous Period, and through much of the Tertiary Era. These movements produced a series of interrupted Palaeozoic outcrops, but the relief forms they created have been largely obliterated. They have been replaced by cuestas and plateaux which are erosional features having bevelled surfaces that are only thinly drift-covered. The western margin of the Rossendales too, is sharply defined by a fault-line scarp which has retreated eastwards from a fault that passes through Burscough and Bickerstaffe. At Parbold (SD 510120), Ashurst Beacon (SD 501076) and Billinge (SD 520021), the summits lie between 150 and 175 m O.D. and similar heights mark the summits of the plateaux in the Coal Measures on the south flank of the Rossendales between Bolton and Rochdale. West of Runcorn, Bunter Sandstones form the highest ground in Liverpool and the Wirral, having been downfaulted by more than 300 m in some instances: fault lines also determine the outline of the Dee estuary.

In the Permo-Triassic stratigraphical succession there is considerable lithological variation, ranging from the weak saliferous Keuper Marls through to the friable Bunter and hard Keuper sandstones. The latter were probably abraded by ice action but protrude through the drift mantle to form a series of cuestas and escarpments within the inner part of the Cheshire basin. They can be traced from Bowden and Lymm, where they overlook the Mersey, westwards to Frodsham and Helsby. From here the scarps turn southwards and become part of the western limb of the broad synclinal structure that underlies the Cheshire plain and whose axis is aligned NE–SW. As elsewhere, the west-facing Keuper scarps have been affected by faulting, but here it has emphasised the relief height relative to the Dee valley further west. The scarps are not continuous, but are particularly conspicuous at Delamere (SJ 510550), Peckforton (SJ 530570) and Bickerton (SJ 498530). Further south they lose height and only form high hills again when they outcrop in the Market Drayton area (see Knowles, Chapter 12). On the eastern limb of the structure the sandstones again form a strong north-facing fault-scarp at Alderley Edge (SJ 860780), but this is not continued northwards and only at Stockport is there a low, partly drift-covered escarpment to mark the limit of the basin.

The erosional history of the lowland areas

In the geomorphological literature for this region there are many references to the existence of planation surfaces lying at levels below 300 m O.D. Arguments were put forward that most of these were eroded by marine processes in the Pliocene or during the early Pleistocene period, but in the absence of datable deposits no firm conclusions were possible. Since most of the platforms occur well within the areas covered by ice, their exact form is impossible to determine and indeed in some limestone areas in the Lake District, previously recognised 'marine' cliff lines (Parry, 1960) are regarded by at least one author (Gale, 1981) to have originated solely by solutional and ice-plucking processes, and therefore cannot be used to interpret past landscape history.

North of the Mersey, planation features were mapped along the lower slopes of the Bowlands and in the area between Liverpool and Skelmersdale (Moseley, 1961; Gresswell, 1964). Related to baselevels at heights ranging between 47 and 120 m O.D., they were considered to be marine-abraded platforms of probably early Pleistocene age (King, 1976), but as further data has become available the surfaces appear to have a single sloping gradient beneath the drift, and a stepped platform sequence is now thought to be improbable (Longworth, Chapter 10).

Within the Manchester embayment, the rockhead surface is very subdued, being for the most part between present sea level and 60 m O.D. It is dissected by a number of deep trenches, but over large areas the amount of local relief varies very little (Howell, 1973), and only close to the hill margins does it become more variable. Planation surface features have been identified at 60–75 m O.D. but their form is masked by drift. They do not appear to be very extensive, but if the graben structure was being disturbed at the time of their formation they could be older than some of the more elevated Pennine etchplains. Recent studies by Douglas (1984), and others have shown that the lowland topography, now buried beneath the glacial drifts, could be much older than the upland etchplains, or at least comparable in age. If this hypothesis is correct, then much of the Mesozoic sediment infill of the graben was removed before the Pleistocene Period, and most of the lowland sub-drift surface was reduced to near its present level before the ice incursions occurred.

In Cheshire the nature of the rockhead topography has not been established due to the paucity of borehole data, but in the salt mining areas natural subsidences are known to have occurred and to have affected the sub-drift surface. At Chelford a channel was infilled with sediments of the Chelford formation, and so dates from at least the early Devensian Period. There has also been considerable overdeepening in the Dee estuary by ice. These modifications to the rockhead surface suggest that its history has been a complex one, and that its surface expression is the result of many process conditions operating over very long time periods.

The glacial modification of the lowlands

In the following chapters there is much discussion concerning the effects of glaciation, for without doubt this has been the most important factor shaping the lowland morphology. There are still areas of doubt and controversy remaining and as differing views are expressed in this volume about certain features or events the present review will not attempt more than a very limited synthesis.

In terms of the British Pleistocene chronology, this region has been especially important. It has not only provided a number of type localities, including the internationally recognised Chelford (Interstadial) stage, but it has also given its name, albeit in a classical form, to the Last ('Devensian') Glacial Stage to occur generally throughout Britain. The invasion by the Devensian and also earlier ice sheets caused some erosion to take place in the lowland areas, but it is difficult to assess how much occurred, where it took place and at what time, as the surface over which the ice or its meltwaters moved is now largely buried under glacial sediments. This prevents any detailed assessment of the morphology, but from a study of borehole log data and limited rockhead exposures, it is possible to identify areas which have been more severely abraded by ice from those that were largely modified by sub-glacial meltwater erosion.

In the Manchester area and in south and east Cheshire the surface beneath the drift has been dissected by meltwaters which have eroded deep but narrow channels across its surface. Some of these contain very overdeepened sections within them (see Fig. 13.6), but the surface topography in general appears to have been only slightly modified by ice. Limited erosion took place in the Cheshire salt mining areas, where some initial subsidences may have taken place beneath the ice cover, and close to the proximal flank of the Whitchurch–Ellesmere moraine, where there was extensive excavation along the main flow lines. The mid-Cheshire hills were overridden by ice and meltwater channels were eroded on some of the hill sides, but the relative low strength of the rocks has failed to retain much of the glacial imprint.

The most strongly eroded areas appear to be in west Lancashire, where Irish Sea ice transgressed onto the lowlands from the north-west and in west Cheshire, where the ice invaded the Wirral, and the lower valleys of the Dee and the Weaver from a more northerly direction. In the Chorley region Williams (1978) has shown that the more elevated parts of the rockhead surface was abraded and that selective ice scouring took place along the outcrops of the weaker strata in the locality. He argues that the northwest–southeast flow of the ice imposed new drainage alignments on the rockhead surface, and that these were then utilised by the subglacial drainage as the ice melted. Selective overdeepening by ice is also thought to have occurred in north-west Cheshire where some depressions within the Dee and Weaver valleys reach −75 m O.D. (Gresswell, 1964; Howell, 1973). The preglacial forms here are no longer recognisable but are replaced by tapered interfluves with planar slopes on the softer rocks. The evidence for ice *vis-à-vis* meltwater erosion is difficult to evaluate, but since the channels can be traced westwards, where they are eroded into the Irish Sea rockhead floor, the latter process was probably the more important.

In the past most literature studies for this region have been more concerned with the age relationships of the glacial deposits than with surface forms, and undue emphasis has been given to the establishment of a stratigraphical succession whose sediments were mapped as though parts of a traditional geological (non-glacigenic) sequence. Unfortunately such studies were based on invalid presumptions as they did not take account of the geomorphological conditions which prevailed at the times when the sediments were being laid down.

The oldest known Pleistocene sediments found in this region have been recovered from boreholes, and from the bottom of the silica sand-pits at Chelford. Their age is unknown except that they predate the recognised Chelfordian succession, and therefore date from an as yet unconfirmed Early Devensian glaciation or more probably, from an earlier Pleistocene glacial (pre-Ipswichian) stage (Worsley *et al.*, 1983). One such till – the Oakwood till – is underlain by organic silts and gravels of an unknown age, and its upper surface was strongly denuded by frost-associated processes before being covered by the basal sediments of the Chelford Sands Formation. These too were subjected to frost processes during their deposition, being laid down during a cold arid but not glacial environment. Within this formation there is the intensively studied Farm Wood organic stratotype whose materials include a wide range of north temperature (Boreal/sub-Boreal) faunal and floristic materials which have been radiocarbon dated in excess of 60 000 years, but whose age is now being re-evaluated (Worsley, Chapter 12). Some of the plants appear to have been growing at the time when their sites were invaded by peat, and later this organic soil was itself covered by more sands and gravels of the Chelford Formation, which were formed during a cold, periglacial climatic phase. In these conditions frost wedging and cracking were again prevalent, but the upper part of the succession is not complete, as these sediments were overlain and truncated by later glacial outwash and till deposited by the main Devensian ice sheet.

The main Devensian glacigenic suite of sediments has been termed the Stockport Formation (Worsley, 1967a), and consists of a sequence of alternating glacial and fluvio-glacial sediments. Till and sand members are replaced rapidly as the succession is traced from site to site, and the alternation of strata may be repeated some 4–5 times in one site. To decipher the sedimentary succession in such circumstances has been extremely difficult, but some order is now being established with the recognition that the glacial sediments were formed in a variety of very widely differing environmental situations.

Within an ice sheet, deposition takes place either beneath the ice or at its outer margin, and as it ablates a specific suite of landforms and sediment features are formed. In sub-glacial environments the wet basal tills tend to form a sediment mantle that drapes over the rockhead surface, and presents little distinctive topography, except where drumlins could be moulded under fast flowing ice streams. The tills generally become very compact and impermeable, but sometimes they are overlain by esker and outwash gravels laid down by glacial meltwaters. In contrast, ice ablation at the margin of the ice sheet leaves deposits in supra-glacial positions actually on the ice or covering stagnation ice blocks. Such deposits are quite different in their physical degree of compaction, texture and permeability, and are often water-saturated, being liable to flow across any surface on which they are deposited.

The release of large volumes of meltwater in the supra-glacial environmental conditions permits sediments to be redistributed into stratified deposits, and landforms are constructed which are quite different from those formed at the base of the ice. Masses of sand and gravel overlain by flow tills are accummulated over 'dead' ice as morainic landforms with lacustrine sediments and outwash deposited in hollows between the hummocks. These later become buried by flow tills, but when the ice melts out the relief becomes inverted and the stratification of the sediments greatly disturbed.

These two distinct land systems can be identified in specific areas, and the details of each form the basis for the several chapters that follow; but it is useful here to briefly summarise the main patterns. In south Cheshire, North Staffordshire and Shropshire the relief is dominated by a broad belt of morainic features which extend from the edge of the Pennines through Bar Hill, Woore, Whitchurch and Ellesmere to the Welsh borderlands. There are several factors which have helped in the evolution of this landform.

First, there was, in the western areas, an influx of Welsh ice which overrode and disturbed the drift materials already derived from the Irish Sea ice sheet. The ice from this second source had once extended much further south to Wolverhampton and beyond, but had mostly disappeared before the Welsh ice advance. Its decay resulted in the collapse or flow of large masses of supra-glacial sediment into topographic depressions as the ice decayed from around it. Second, the location of the moraine is close to a major water-divide that separates the Dee and the Mersey catchments on the one hand from the Severn and Trent catchments on the other. This divide is of some considerable antiquity, for it occurs near to the southern edge of the Cheshire Basin structural depression, and along the line of a tectonic boundary at which there is a rise in the level of the rockhead surface. This topographic barrier, although not very high, must have offered some obstruction to the forward movement of the ice across it, and would have encouraged deposition to take place at the base of the glacier and at the ice front where the wasting ice sheet was forced to shear over its basal layers in its attempt to overcome the increased friction offered by the rockhead relief.

The moraine therefore provides a strong contrast in its relief to the lowland areas located to the south and north of it, but there are also contrasting zones to be found within it. On its southern flank considerable volumes of meltwater distributed large quantities of sediment which were laid near to or within an area of stagnating ice and where the water could drain into a number of ephemeral lakes impounded between the ice and the local relief barrier. However, within the moraine itself the terrain is more irregular, the ridges are separated by kettle-hole meres and meltwater channels. Widespread sandur surfaces lie to the north of the moraine, being most extensive in the eastern parts of the plain between Alderley Edge and Audlem and to the east of the Ellesmere ridge. These sandurs have a local slope outwards from the hills but an overall slope northwards, that is, towards the ice front and as would be expected, in certain locations the meltwaters were impounded to form small temporary lakes (Worsley, 1975).

Other morainic features occur elsewhere on the lowlands, and can be traced northwards from the Ellesmere moraines along the foot of the Pennine and Welsh hills. On the Pennine hillslopes, above the moraines, the slopes have been eroded into complex networks of gullies and former meltstream channels which in their lower parts drain into areas of hummocky drift. In most places the supra-glacial

landforms form only a narrow zone but immediately south of Alderley Edge
in the area between Middleton, Oldham and Rochdale they provide an easily defi
element of the lowland topography. At the Welsh borderland similar mora
topographic fatures mark the limit of the plain, and above them many chan
record the one-time presence of either the Welsh or the Irish Sea ice. North of
Wrexham there is an extensive complex outwash terrace whose formation resulted
from the juxtaposition of the two ice sheets (Worsley, Chapter 12). During its construction, sediments were deposited over ice that subsequently melted out to leave
kettle holes and a strong ice-contact face escarpment feature at Marford
(SJ 342557). No similar feature has been recognised anywhere along the Pennine
margin, but small moraines are to be found where ice from the north-west
encroached upon the uplands in east Lancashire and penetrated into the Douglas
valley.

In the Fylde, south-west Lancashire and the Irish Sea area very different glacial
conditions appear to have prevailed from those described for the areas further
south and reasons for this are given in some detail by Thomas (Chapter 8) and
Longworth (Chapter 10). Glacial, glacio-marine and pro-glacial deposits have now
been mapped over extensive areas of the Irish Sea floor and Thomas suggests that
their distribution can be explained by the development of an ice sheet which was
extremely unstable and flowing into an area occupied by an ice shelf or open water.
During the waning of the ice sheet sea level remained 'high' and little isostatic
movement took place within the Irish Sea basin, but on the mainland the melting of
the ice took place extremely rapidly, with large areas of ground becoming ice-free
at the same time. This 'area wasting' of the ice sheet released large volumes of
meltwater and Longworth reports the occurrence of 'stepped' sequences of thick
laminated silts interbedded with clays and coarser sediments which he believes were
laid down in similar conditions to those found by Marcussen in Denmark.

North of the Ribble estuary there is a low undulating series of ridges which can
be traced from Blackpool southeastwards through Kirkham and the northern part
of Preston, to the edge of the Bowland hills. This feature was termed the 'Kirkham
moraine' by Gresswell (1964), and it lies in a location where Irish Sea, Lake District
and Ribble ice must have converged near to a point where the sub-drift surface
forms a saddle ridge separating the Fylde from the south Lancashire lowland.
According to Longworth (Chapter 10), the moraine at its eastern end terminates
in an area where thick laminated clays and silts were deposited by sub-glacial
meltout of ice but at its centre near Kirkham (SD 422342) the sediments were
deposited in a supra-glacial landform environment. Here they were released onto
and buried stagnant ice which later melted out, causing widespread subsidence.
The main direction of ice flow was from the north-west, with the Irish Sea ice
most dominant during the construction of the moraine. It is therefore unlikely
that the Kirkham moraine would represent a stillstand stage as its alignment is
from west to east but it could well have formed as an inter-lobate morainic aggradation where Lake District ice to the north and Irish Sea ice to the south-west
were still in contact with each other, but were separated from the Ribble ice which
had downwasted more rapidly. In the newly exposed pro-glacial area the tills were

overlain or interbedded with glacio-lacustrine sediments, but as yet the field evidence regarding the origin of this feature is not yet sufficient to test this hypothesis and Longworth would argue that the rise of the rockhead surface alone was sufficient to cause the ice during downwasting to release much of its supra-glacial sediment in this area of the southern Fylde.

The lowland till plain of Lancashire, like that in Cheshire, offers little variety in its relief except where the rivers have incised their valleys across its surface. Much of its terrain lies below 50 m O.D. and parts have been invaded by the sea or have been modified as a result of changes which have occurred since the end of the Devensian (see next section). Near Lancaster, however, a major topographic change occurs for the regular even till and outwash surface is here replaced by more undulating terrain and the lodgement tills have been moulded into drumlins. These, together with a number of rock drumlins, form part of a drumlin field which extends across Morecambe Bay to the Furness region of south Cumbria. They are located where local relief helped to impede the movement of ice southwards and according to Boulton *et al.* (1977), this would cause the sliding velocity of the basal ice to decrease to a point to which lodgement tills would be both deposited and moulded into drumlin forms by the ice. Such changes took place during deglaciation and in places where the Cumbrian valley glaciers were coalescing into larger piedmont glacier ice streams.

Late and Post-Glacial environments in Lancastria and Cumbria

The dissipation of the ice sheet occurred under conditions similar to those experienced elsewhere in north-west Europe, with extensive deep permafrost persisting in areas adjacent to the ice margin maintaining ground temperatures at or below freezing point. Ice wedges dating from the early Late-Glacial period were formed on till surfaces at Congleton, Cheshire (Worsley, 1967a), and in overburden materials overlying sandstones and shales at Scout Moor and Whitworth in the Rossendale Forest, Lancashire. But where the permafrost became thinner and more discontinuous, snow meltstreams, solifluction and many forms of rapid mass movement became the dominant agents of debris transportation.

A substantial number of pollen-analytical and related studies by Pennington (1977), Coope (1977) *inter alios* has shown that most of North-west England was ice-free by 14 600 years BP, but that only a sparse tundra vegetation had become established in the severe climate of the time. Birch woodland did not replace the tundra until *c.*13 000 years BP, at which time there was a dramatic climatic amelioration. This change was marked by a tenfold increase in the rate of sedimentation in the Cumbrian lakes and the development of more stable soils under the increasing tree cover on the adjoining slopes. A thousand years later, the soils were more or less stabilised and the sediment rates reduced (see Gale, Chapter 15), but the transition to a more temperate climatic regime was briefly halted by a return to more severe cold climatic conditions. This temperate period is now recognised as the Windermere Interstadial, a relatively brief interlude lasting approximately 2000 years (see Table 1.1).

Table 1.2 Subdivision of Flandrian chronozones (after Mangerud *et al.*, 1974; Godwin, 1977). (See also Table 9.1, p. 163)

Chronozone	Subdivision	Radiocarbon years BP	Pollen zones
	Late	1000	
Sub-Atlantic	Middle	2000	VIII
	to		
	Early	2500	
	Late	3000	
Sub-Boreal	Middle	4000	VIIb
	Early	5000	
	Late	6000	VIIa
Atlantic	Middle	7000	
	Early	8000	
	Late	8500	VI
Boreal			
	Early	9000	V
	Late	9500	IV
Pre-Boreal	Early	10 000	

At *c.*11 000 years BP there was a recrudescence of glaciers in the higher valleys of Cumbria, Wales and Scotland and a major disturbance of slope regolith occurred over much of Britain. The prevalence of cold climates for the next 500 years resulted in the replacement of the light forest cover with vegetation types adapted to arctic and sub-arctic environments. On the flatter slopes of the valley floors and on the plateau benches the soils and sub-soils were convuluted by alternating freeze–thaw conditions and, along with other forms of soil disturbance, these processes gave rise to the deposition of a 500-year succession of varve clays in some of the larger Cumbrian lakes. In addition, much sediment was removed from the smaller corrie lakes and tarns and this was in part due to increased runoff from snow-covered summits and small glaciers in Cumbria. In the higher Cumbrian mountains several fossil rock glacier debris deposits have been dated to this period (Sissons, 1980).

In the Pennines, the climatic regime was less extreme but blockfields and screes, often cemented and stratified, were developed beneath escarpment cliffs or 'crags' and many of the Pennine hill tors were finally exhumed at this time. Landsliding took place often on quite an extensive scale with such movements increasing in scale and frequency as wetter, milder climates prevailed in the succeeding Flandrian period. For at least the next 4000 years the steeper Pennine slopes were characterised by their general instability and susceptibility to this particular hazard (Tallis and Johnson, 1980).

In the Late-Glacial and Early Post-Glacial periods the till and fluvial-glacial out-wash surfaces were exposed to strong, drying winds which removed much of the fine sand, silt and clay fractions from the upper horizons and redeposited them in dune spreads inland from the present-day coastal dune belt of the Lancashire coast. These Shirdley Hill (cover) Sands were formed on both till and terrace surfaces and remained unstable until a vegetation cover developed. At the same time as the aeolian deposits were being formed, the sea level was rising due to the wasting of the terrestial ice sheets and the removal of extensive tracts of sea ice. As a result the coastline receded and is now formed almost entirely from drift materials or modern river sediment, with the exception of local rock exposures on The Wirral, on the northern margin of Morecambe Bay and at St. Bees Head in west Cumbria. The extensive sand beaches and dune belts of Morecambe Bay, the Fylde and Liverpool Bay are derived from outwash deposits from the bed of the Irish Sea which have been moved onshore and, although the beaches are exposed to a considerable tidal range, their beach ridges appear to be semi-permanent, changing in height with the seasons and shifting only slowly along the coast.

Aggradation within the valleys continued intermittently from the end of the Devensian until *c*.7000 BP and during the same time span coastal sand dunes and salt marshes were being formed in spite of frequent periods of marine inundation. Deposition, both in the valleys and along the coast, was not continuous and the major rivers eroded terraces as they cut into and redistributed the glacial outwash that formed their transport load. The return to wetter and warmer climates saw the development and expansion of extensive areas of lowland peat mosslands and the growth of blanket peat on the upland Pennine surfaces. (Table 1.2).[3] The development and changes in these peatlands is described in Chapters 16 and 17 in the light of the increasing effect of man as a geomorphic agent since *c*.7000 years BP. His early forest clearances and his burning off the vegetation increased the instability of some slopes and led to soil erosion on a scale which has only recently been appreciated (Tallis, Chapter 17).

The control by man of his environment is discussed in the final two chapters which examine some of the geomorphological factors affecting urban and industrial developments within the region. Such developments have not always been achieved without detriment to the scenic qualities of the landscape, yet the results of projects such as the control of flooding, the minimisation of salt-mining subsidence, and the improvement of estuarine flows suffering from excessive sediment inputs, and marshland development have enabled widespread environmental stabilisation to be achieved in many parts of the region. Derelict lands and mineral waste tips have also benefited from sound geomorphologically-based reclamation and the water quality of many polluted rivers has been greatly improved though much still remains to be done. Such applied geomorphological projects and an increasing public concern for its environment offer some degree of optimism for the future of the Lancastrian and Cumbrian landscape heritage but this optimism has to be based on a deeper understanding of the intrinsic qualities of the landscape itself. The work presented in this volume will, hopefully stimulate others to further research into the problems outlined and provide a broad base upon which future studies may progress.

Notes

1   This diagram may be interpreted using one of the following hypotheses. Given normal etchplain conditions, any one of the upland stages on this diagram could have developed concurrently with any of those shown as evolving in the graben region, providing that etchplain development was controlled by simple arching of the relief as a result of uplift (cf. Büdel, 1982, Fig. 49). An etchplain escarpment (A–G) would then develop between the two surface levels, and would be analogous to the west Pennine edge as discussed in this chapter. Alternatively, downcutting could have been episodic, with the less resist-ant Mesozoic and Cainozoic sediments of the subsiding graben being removed more readily than the Palaeozoic rocks of the rising tectonic horst. A third alternative would be to attribute the stages to differential uplift between the two structural units, and to argue that the two massifs would have experienced comparable rates of erosion at each stage.

2   Cave-speleothem dating sequences have been determined in a number of Yorkshire and Derbyshire caves, (see Chapter 14), and these are used in Table 1.1. The Pleistocene succession presented is based on data sources currently available, but shows several major inconsistencies in the Middle and Late Pleistocene data. For example, Wymer (1981) cites two estimations of the dates for the commencement of the Anglian and Hoxnian which have an overlap of 80 000 years. A date of 320 000 years BP has been estimated as the start of the Anglian but a date of 400 000 years BP has also been pro-posed for the start of the Hoxnian! The inconsistencies occur because there is little or no sedimentological record in Britain for a period of some 200 000 years and this includes the important oxygen stages 7 and 9 of the ocean floor sediment sequence. Hopefully, the new method of applying amino-acid dating to the Pleistocene deposits of the British Isles will resolve some of these problems.

      Currently work now in progress (Bowen and Sykes, 1984) has already established a more definitive chronology in which many more climatic fluctuations are recognised than were previously considered and in particular evidence is now being presented which shows that the 'Last Glaciation' (Devensian) saw a build up of ice in western Britain on two separate occasions.

3   The interpretation of these and later climate changes are well summarised in detailed tables in Chapters 15 and 16, but for a general reference Table 1.2 is included, although it should be noted that such climatic and floristic changes did not occur synchronously, and cannot be correlated from region to region without very careful evaluation of the evidence.

The geological evolution of North-west England

Introduction

The rocks exposed in North-west England preserve a record of events spanning about 500 million years. During this time there have been periods of deep-seated crustal deformation and melting, times of volcanic activity, intervals of erosion and sedimentation. There has been a continuously changing geomorphology. This chapter outlines the main features of the geological evolution of the area and relates them to current ideas in the concept of plate tectonics.

The oldest rocks of the area

The oldest rocks exposed in North-west England are sedimentary in origin and comprise the poorly-sorted sandstones of the Ingletonian Group in the Ingleton–Settle district (Fig. 2.1, Table 2.1). Although not yet precisely dated these rocks may be of Cambrian or early Ordovician age (Ingham and Rickards, 1974). Yet older rocks almost certainly occur at depth across the entire area but little is known about them. In the Peak District the oldest exposed rocks (Carboniferous Limestone) have been penetrated by two boreholes. In the Woo Dale borehole rocks of possible Devonian or older age were found (Cope, 1979) while in the Eyam borehole marine sediments of Ordovician age were encountered (Dunham, 1973).

Rocks of Ordovician age are extensively exposed in the north of the area, especially in the central part of the Lake District. The oldest of these rocks, the Skiddaw Group, were formed mostly as clastic sediments (muds, silts and sands) of immense total thickness (possibly over 9000 m). Locally they yield graptolites and other fossils of marine origin. A 'flysch' type of sequence (sandstone–shale alternations) is often developed and suggests the periodic operation of turbidity currents from the south to carry coarser-grained, poorly sorted, material into an area normally dominated by mud deposition (Jackson, 1978). The Skiddaw Group of sediments is believed to have accumulated on the floor of the Iapetus Ocean which separated what is now North America, Greenland and north-west Scotland from England, Wales, Scandinavia and the rest of north-west Europe (Fig. 2.2). This ocean occupied roughly the same position with respect to North America and north-west Europe as the present Atlantic, hence the name Iapetus (in Greek mythology Atlantis was the son of Iapetus).

The Skiddaw Group is overlain by the Eycott and Borrowdale Volcanic Groups of early to mid Ordovician age (Table 2.1), comprised mainly of tough volcanic

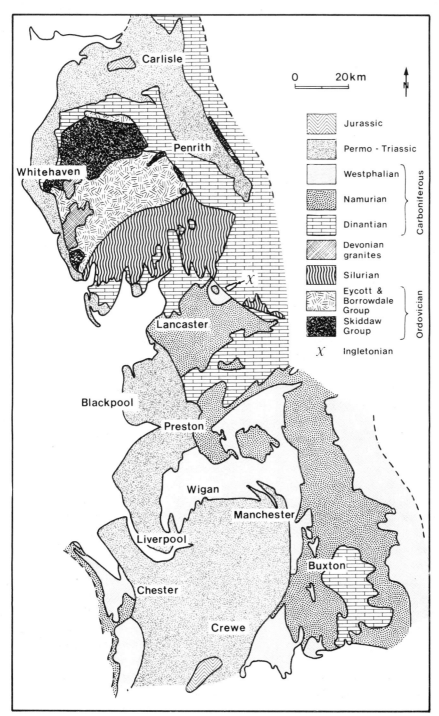

Legend:

- Jurassic
- Permo - Triassic
- Westphalian ⎤
- Namurian ⎥ Carboniferous
- Dinantian ⎦
- Devonian granites
- Silurian
- Eycott & Borrowdale Group ⎤ Ordovician
- Skiddaw Group ⎦
- X Ingletonian

Fig. 2.1 Simplified geological map of North-west England.

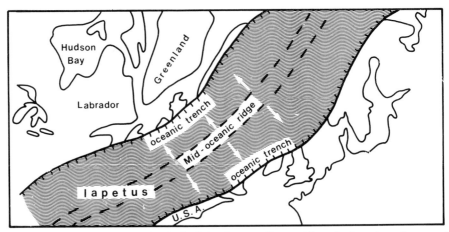

Fig. 2.2   Reconstruction of the Ordovician Iapetus Ocean. Note the location of parts of Britain on opposite shores. Based on Moseley (1978).

rocks which are often highly resistant to weathering and produce the most rugged mountain scenery in the Lake District. Lava flows with a wide range of composition, ranging from basalts to rhyolites, occur and these are associated with extensive pyroclastic deposits. Of particular interest are those rocks which were produced, apparently, by the violent nuée ardente type of eruption, characteristic of island arc systems. Where these pyroclastic deposits accumulated on land a fusion or welding of the components took place, particularly inside the ashfall away from the chilled base and top, to form ignimbrites. The presence of ignimbrites in the Lake District was unsuspected until the 1950s; previously they had been identified as rhyolite lavas. Much pyroclastic material also fell into the sea and shows evidence of reworking by currents to produce structures such as ripple marks. For a general account of the Eycott and Borrowdale volcanic rocks see Millward *et al.* (1978).

It is thought that during the Ordovician the Iapetus Ocean was shrinking by subduction of ocean crust on both sides of the ocean. The Eycott and Borrowdale volcanic rocks were probably produced by the action of the subduction zone on the south side of Iapetus (Fig. 2.3). The Lake District and adjacent areas may then have been part of a volcanic island arc system such as the present-day Indonesian arc. The concept requires the descending ocean slab to have undergone partial melting to generate a basic (high temperature) magma. This ascended towards the surface, yielding up heat to melt other rocks on the way. The result was the wide range of composition now seen in the Ordovician volcanic rocks, ranging from rhyolites to basalts.

The Upper Ordovician and Silurian rocks (Table 2.1), are mostly exposed in the southern part of the Lake District, the Howgill Fells and the Settle area. They indicate a return to marine sedimentation with only minor and dwindling vulcanicity. Evidence from fossils and sediments shows that at times the sea transgressed southwards over a land area and at other times (especially towards the end of the Silurian) withdrew (Ingham, *et al.*, 1978). Being less resistant than the underlying Ordovician volcanics the Upper Ordovician and Silurian rocks today form a more subdued and generally lower topography.

Shrinkage and final closure of the Iapetus Ocean by the end of the Silurian led to several phases of compression in the adjacent continental crust and deposits of the sea floor. The resultant Caledonian structures (folds, thrusts, cleavages)

Table 2.1 Sequence of geological deposits, their absolute age (Harland *et al.*, 1982) and environments of formation, in North-west England.

Quaternary	Recent		
	Pleistocene		Advance and retreat of ice sheets Intervening warm periods
		———— 2 ma ————	
Tertiary	Pliocene/Miocene		Brassington Formation
		———— 65 m ————	
Mesozoic	Cretaceous		If developed now entirely removed by erosion
	Jurassic		Only parts of the Lower Jurassic (Lias) preserved. Marine sediments
		———— 213 ma ————	
	Triassic		New Red Sandstone extensively preserved. Semi-arid conditions.
	Permian		Deserts, flash floods, playa lakes
		———— 286 ma ————	
	Carboniferous	West-phalian	Coal-bearing sequences of fluvial environment dominant
		Namurian	Fluvio-deltaic
		Dinantian	Marine 'block' and 'basin' environments
		———— 360 ma ————	
Palaeozoic	Devonian		Igneous intrusions
		———— 408 ma ————	
	Silurian		Marine sediments
		———— 438 ma ————	
	Ordovician		Upper Ordovician Eycott and Borrowdale Volcanic Groups Skiddaw Group
	?Cambrian ?Ordovician		Ingletonian. Probable marine sediments

trend NE–SW and are recognised not only in North-west England but also in Scotland, Wales, Ireland, Scandinavia and the Appalachians of North America.

The granite bodies of the Lake District (such as those of Eskdale, Shap, Skiddaw) intrude Ordovician and Silurian rocks and so are post-Silurian in age. However, they were already in existence by the Carboniferous, as evidenced by the presence of Shap Granite fragments in Carboniferous deposits. Stratigraphical evidence thus suggests a Devonian age and this has been confirmed by isotopic dating (Firman, 1978). Geophysical (gravity) evidence suggests that the present isolated granite exposures are all part of one extensive underlying granite batholith (Bott, 1978). The generation of granitic magma on such a scale points to the end-Silurian and Devonian as times of deep-seated heat generation, possibly linked to the final elimination of the Iapetus Ocean by the collision of the opposing continents. There are no Devonian sediments known in North-west England, although elsewhere in the United Kingdom they are of major importance (forming the continental Old Red Sandstone, for instance).

Sediments of the Carboniferous System (Table 2.1) are abundantly represented in North-west England. Not only is the area of outcrop large but these rocks have contributed much to economic development in terms of limestone, coal, various metal ores and other materials. The rocks comprise the Lower Carboniferous (Dinantian), where limestones are regionally important, if not dominant, and the Upper Carboniferous (Silesian) dominated by clastic sequences, including coal.

The Lower Carboniferous rocks appear to rest on an irregular, erosional, surface cut across the older rocks. Fossils show that deposition was largely marine and the distribution of sediments indicates that the sea must have progressively flooded the area. Later fluctuations of sea level produced a cyclic pattern of sedimentation recognisable over wide areas (Ramsbottom, 1973, 1979). Sedimentation, in terms of sediment type and thickness, was strongly influenced by the stability of the immediate crustal rocks below. Some areas ('blocks'), probably underlain by relatively low-density granites of Devonian age, remained relatively stable and so accommodated thin deposits, dominated by carbonates. These areas include the

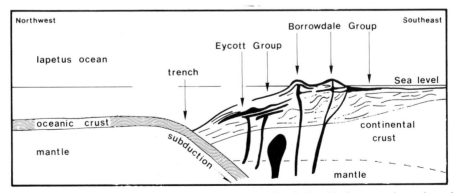

Fig. 2.3 Diagrammatic cross-section to show probable relationship between the early and mid-Ordovician volcanic rocks of North-west England and the Iapetus Ocean subduction zone. Based on Moseley (1978).

Lake District, most of the northern Pennines and a large part of the Peak District. Elsewhere, in the 'basins', subsidence along contemporary faults, such as the Craven Fault system, allowed the accumulation of thick deposits of mud and silt with subordinate limestone. There is, therefore, a marked difference between the deposits of the 'basins' and those of the 'blocks'.

Life was prolific on the 'blocks' and there is an abundant fossil record of invertebrate shell life and fishes together with algae of various types. Although carbonate mounds occur which are often reef-like in form, for instance around Malham and in north Derbyshire, there is little evidence that contemporary organisms could build up wave-resistant structures like present-day reefs. A striking exception to this general situation has recently been described by Adams (1984) from the Furness area, on the south side of the Lake District.

The Lower Carboniferous rocks show abundant evidence of vulcanicity which was no doubt linked to differential movement between 'blocks' and 'basins'. Lavas, ashes and vent rocks are seen particularly well in the Lower Carboniferous of the Peak District. Although deposition of the sediments was generally submarine, many of the lava flows were subaerial in origin and exhibit features such as hexagonal cooling columns which depend on slow cooling (impossible on the sea floor) for their formation. Further evidence that the 'block' areas may have been exposed periodically above sea level is seen in the numerous palaeokarstic surfaces developed beneath contemporary soils (of volcanic origin) in the Carboniferous Limestone of the Derbyshire 'block' (Walkden, 1974; and Fig. 2.4). Adams (1980) has demonstrated the presence of calcrete profiles (caliche) in the Eyam Limestone of Derbyshire.

Marginal areas between 'blocks' and 'basins' in the Lower Carboniferous are well known in the Settle–Malham–Cracoe area of Yorkshire (along the line of the Craven Fault system, active in the Lower Carboniferous and subsequent times) and in the Castleton area of the Peak District. In both these districts cavities within the limestones are often partially filled with sediment and so serve as spirit levels. By reference to these geopetal infills the 30° dips seen in the limestones are recognised to be largely depositional and indicate that the sea floor sloped down from the shallow waters of the 'block' into the deeper regions of the adjacent 'basin'. Around Castleton the virtual coincidence of the slopes of present-day hillsides with bedding planes in the limestone show that the present-day topography is a resurrection of what was once a Lower Carboniferous sea floor. This is a striking example of Carboniferous palaeogeomorphology (Fig. 2.5), which is being progressively exposed as the cover of soft Upper Carboniferous sediments is removed by erosion (Broadhurst and Simpson, 1973). At one stage the Lower Carboniferous sea retreated from the higher parts of the surface and a lava flow (the so-called Cave Dale lava with hexagonal cooling columns) then flowed towards Castleton from the south, reached the steep slopes leading down towards the 'basin' and finally entered the sea. Remnants of the submarine part of the flow are recognised in the form of a fragmented glassy lava, the so-called Speedwell Littoral Cone (Cheshire and Bell, 1977).

The Lower Carboniferous marine environment of North-west England was

eventually invaded by a complex delta system, the chief component of which was built southwards across the area. Growth of the delta can be followed in detail because the changeover from marine to deltaic or fluvial conditions, as indicated by the sediments and fossils, is found to be progressively delayed in a southerly direction. The boundary between Lower and Upper Carboniferous (Silesian) rocks has been arbitrarily taken at the level of a fossil-rich sediment layer (containing the goniatite *Cravenoceras leion*) produced during the course of a marine transgression and regression across the entire area. This and similarly-formed 'marine bands' have proved invaluable for correlation of rocks within the Upper Carboniferous.

Detailed examination of the earliest sediments of the delta complex in the southern Pennines (summarised by Selley, 1978, pp. 111–16) have made it possible to reconstruct the progressive southerly advance of the delta across what is now the Peak District. Before the delta complex arrived the area was part of a sea floor where organic-rich mud was the dominant sediment. The approach of the delta was heralded by deposition of sediments from turbidity currents despatched off the delta to the north (Allen, 1960). These turbidites changed in character with time, due to the progresive approach of the delta complex towards the site of deposition. Finally the sediments indicate a shallowing of the water and the establishment of a delta top with a variety of fluvial features and vegetation. Among the sediments developed on the delta top are extensive sheets of sandstone probably formed in braidplains (see Haszeldine and Anderton, 1980). These sandstones are often coarse-grained and were formerly used for the manufacture of millstones, hence the name of Millstone Grit so often used for the rocks formed in this environment.

Once the deltaic–fluvial environment had become established across northern England the marine influence progressively diminished. Whereas correlation, by means of marine bands, in the earlier part of the Upper Carboniferous (Namurian) is more or less straightforward, it becomes more difficult later on (in the Westphalian). The declining marine influence was accompanied by a greater development of peat so that the Westphalian became the principal repository of coal and forms the coalfields of North-west England, today.

Sediments of fluviatile origin are dominant in the Westphalian. Sandstones occur, some formed in braidplains, others in meander belts with point bars. Between the channel systems of the fluvial environment lay interdistributary bays or lagoons in which thick sequences of mud and silt accumulated. Striking rhythmic bedding is seen in such sediments. One interesting example (Fig. 2.6) has been described (Broadhurst *et al.*, 1980) where the sequence contains escape structures made by bivalves obliged to move upwards in response to high rates of sedimentation. In this case it appears that the rhythmic bedding, comprised of alternate mud-rich and silt-rich bands, resulted from seasonal sedimentation. The coarser-grained (silt-dominated) material was supplied during seasons of high water level and the finer

Fig. 2.4 Palaeokarstic surface on Carboniferous limestone, Derbyshire. Photo by G. M. Walkden.

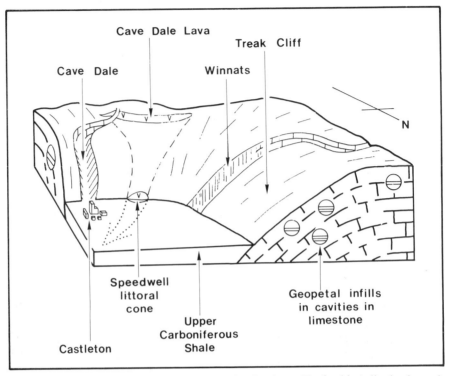

Figure 2.5 Diagrammatic bird's-eye view of the Castleton (Derbyshire) district from the north-east. Erosion of the soft Carboniferous shales is uncovering a sea floor topography of Lower Carboniferous age. See text. Diagram based on Broadhurst and Simpson (1973).

(mud-dominated) sediments during seasons of low water. Such seasonal variations in sediment supply are known in present-day deltas, for instance, that of the Niger (Allen, 1965).

The catchment area for the Carboniferous delta across northern England was immense. Reconstructions of Carboniferous palaeogeography (Smith *et al.*, 1981) place North-west England within a vast continental area including what is now North America, Greenland, northern and central Europe. Iapetus had been destroyed at the end of the Silurian and the North Atlantic was not to open until the Tertiary.

Although movement on faults was of major importance in controlling the development of the 'block' and 'basin' environments in the Lower Carboniferous, the influence of such movements progressively declined through the Upper Carboniferous (Leeder, 1982). Nevertheless some faults were demonstrably active during Westphalian time. An interesting example is the Deerplay Fault of the Burnley area in Lancashire. The trace of this fault closely parallels the line of a split in the Union coal seam for at least 13 km (Wright *et al.*, 1927). This suggests that contemporary movement on the fault was responsible for the formation of the seam split (Fig. 2.7). Of particular interest from the geomorphological point of view are the linear depressions formed in the top of the coal seams close to

Fig. 2.6 Rhythmic sediments, probably of seasonal origin, above the Lower Mountain coal, Ravenhead Quarry, near Wigan. Published by permission of the *Journal of Geology*, University of Chicago Press.

the line of seam split (Fig. 2.7A). These depressions, now infilled with carbonaceous sediment, are known to coal miners as 'rigs' but can be readily matched with the linear pools that form on sloping peat surfaces today (Moore and Bellamy, 1974; Broadhurst and Simpson, 1983).

Towards the close of the Carboniferous the rocks of North-west England were involved in a process of deformation that produced both faulting and folding. This deformation is generally regarded as the consequence of collision between continental plates south of the British Isles to form the complex structural belt across central Europe known as the Hercynides or Variscides. The Hercynian (Variscan) structures in North-west England include most of the structures seen in the coalfields. The dislocation of seams by faulting and the steep inclinations found in some seams are factors which operate against the economic extraction of coal.

Permo-Triassic and Jurassic sediments

Rocks of Permian and Triassic age, often referred to as the New Red Sandstone, are commonly red in colour due to the presence of iron oxide largely developed as coatings around sediment grains. Sedimentary structures suggest that fluvial, lacustrine and aeolian environments were common during the deposition of these sediments and the association of evaporites such as gypsum and rock salt (halite) indicate periods of aridity. Palaeogeographic reconstructions (Smith *et al.*, 1981) place Britain within the trade wind belt as part of a large continental area. The Permo-Trias is developed as a continental type of deposit not only in Britain but over large areas of the world and so was apparently a time of generally lowered sea levels, possibly due to restricted development of the mid-oceanic ridges, which normally serve to displace ocean water and therefore maintain higher sea levels.

In places the Permo-Triassic was accumulated on surfaces of high relief. In and around the Lake District, for instance, the basal deposits are found to be breccias (known locally as Brockram), apparently formed as screes banked against steep slopes. The screes were composed of Lower Carboniferous limestone fragments and these were subsequently infiltrated and bound by fine-grained sediment.

During the Permian much of the crust beneath the British Isles and adjacent areas was subjected to a process of stretching which anticipated the eventual opening of the North Atlantic. In response to this crustal tension normal faulting developed accompanied by subsidence of various basins, sometimes as graben. Faulting continued and maintained the basins into the Triassic despite the accumulation of thick sediment sequences with them. Apart from a brief incursion of the Zechstein (Upper Permian) Sea into North-west England sedimentation remained continental. Conglomerates and pebbly sandstones of the 'Bunter Pebble Beds' in the Sherwood Sandstone Group (Warrington *et al.*, 1980) are interpreted as braided stream deposits (Fitch *et al.*, 1966; Steel and Thompson, 1983) which were transported, in stages, by flash floods from source areas in the Variscan mountains possibly between Brittany and Dorset, through the Worcester Graben to their present sites

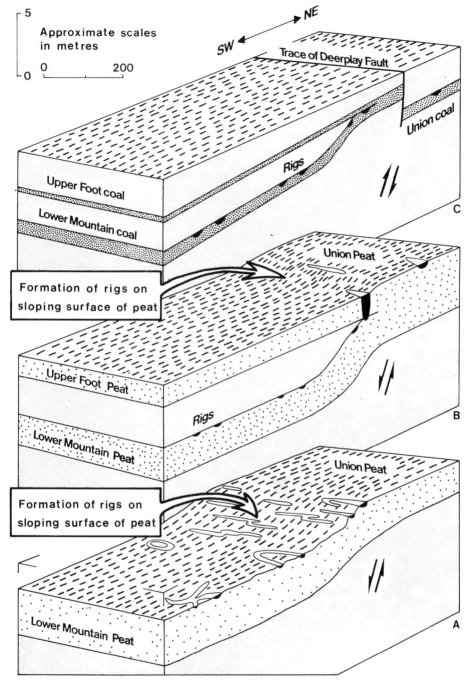

Fig. 2.7 Reconstruction of events which led to the formation of the split in the Union coal seam (Upper Carboniferous) in the Burnley area of Lancashire. By permission of the *Journal of Geology*, University of Chicago Press.

in North-west England (Fitch *et al.*, 1966; Warrington, 1970). The pebble beds fail northwards and pass laterally into sandstones (Colter and Barr, 1975) including the St. Bees Sandstone of Cumbria.

At a later stage in the Trias thick deposits of rock salt and gypsum formed in an area extending from Cheshire, through the Fylde and into the Irish Sea. In Cheshire there are some 6 or 7 beds of rock salt separated by red mudstones. One of the salt beds reaches a thickness of well over 100 m (Evans *et al.*, 1968). These deposits are of considerable economic importance and are generally extracted by solution although one mine extracts rock salt by pillar and stall mining. The evaporites are now generally regarded as being the deposits of playa lakes.

Triassic sediments have played host for the accumulation of economically-important minerals in post-Triassic times. The Helsby Sandstone Formation (formerly the 'Keuper Sandstone') is broken by faults which locally carry ores of copper and other metals, formerly worked in the area (e.g. at Alderley Edge, see Warrington (1965). The same sandstone is also the reservoir rock for the natural gas deposits of the Morecambe Field in the Irish Sea (Colter and Barr, 1975; Colter and Ebbern, 1978).

The continental conditions so prevalent during the Permo-Trias were finally terminated by a major marine transgression during which the sediments of the Rhaetian (the topmost Triassic stage) were formed, some 20–30 m thick. Marine sedimentation continued into the Jurassic, marked by the first appearance of ammonites. Jurassic sediments are clay-rich and the area of outcrop of these rocks has been severely reduced by erosion. They have survived in one small area to the west of Carlisle and in a slightly larger area across the Cheshire/Shropshire border (see Fig. 2.1).

Post-Jurassic sediments

With the exception of the restricted Miocene–Pliocene deposits in the Peak District (Walsh *et al.*, 1972) there is a general absence of sediments belonging to the interval between the Lias and the overlying and generally unconsolidated deposits of the Pleistocene and Recent. It is not clear how much this relates to non-deposition and how much to post-depositional erosion. Sediments of the Pleistocene are well represented and include materials of glacial origin such as tills, sediments formed during times of glacial retreat and sediments deposited during warmer intervals between glacial episodes. It is remarkable that the Pleistocene, with abundant evidence of glacial environments, should rest so frequently on rocks formed under tropical or near-tropical conditions. The Pleistocene is a rich source of geomorphological problems and so figures predominantly elsewhere in this book.

Long-term landform development in North-west England: the application of the planation concept

Introduction

This chapter has three aims. First, to present a review of the literature on the application of the planation concept to the topography of North-west England. Second, to present a brief synthesis of current ideas which may affect the appearance and application of that concept in the future. Third, to outline a modern long-term landform development in North-west England using such ideas.

By way of introduction, some definitions are needed: namely, of long-term landform development, planation and North-west England. Long-term landform development can be defined as the study of the evolution of a region's topography over timespans of millions of years. Crudely, it is large-scale historical geomorphology. It contrasts with the smaller scale study of landforms either historically, as in Quaternary studies, or using a functional approach, as in process work. Regrettably, at present, it is a much less popular field of enquiry in British geomorphology than its smaller scale relatives (Clayton, 1980). However, it is an older area of research, serious investigation dating back to the middle of the nineteenth century. It is also gaining increased attention with the demise of classical concepts and the recognition of the need for cross-fertilisation of new ideas in geomorphology with those in geology (e.g. Summerfield, 1981). Esentially, the aim of a long-term study is to produce a denudational chronology of a region consisting of the major pre-Quaternary erosional events, a case of 'what' was eroded 'where', 'when' and 'how'. The formation of a planation surface has always been regarded as a fundamental stage in large-scale topographical evolution. The recognition of planation surfaces in the landscape has always been the key to establishing an area's denudational chronology. That is why this chapter concentrates on this one concept and the history of its application to the North-West.

Planation, in its simplest sense, means that, over time, the gross surface irregularities of an area's topography will be largely destroyed to produce a relatively flat surface. In the 'long-term' sense, this means the flattening of large areas (tens of kilometres) over millions of years. Evidence of such large-scale planation exists in the stratigraphic column as major unconformities. One notable example is under the Chalk of Southern England (Fig. 3.1a). It suggests widespread erosion of tilted strata to form a plain that was then covered by chalk. It is not inconceivable that similar surfaces existed elsewhere and their influence still persists in the present landscape. Even if they have suffered erosion due to prolonged exposure, they may still govern the outlines of a region's topography. Therefore, their recognition is the key to establishing a region's denudational chronology. Indeed, to the casual

observer, much of the topography of upland Britain displays even skylines due
to flattened summits and hillsides which may represent the partial survival of a
collection of many planation surfaces of different ages at different heights.
Accepting this simple definition of planation, and assuming such surfaces exist,
any researcher trying to construct the denudational chronology of an area has
always had to answer three questions about the surfaces. How were they formed?
How many of them are there and where are they? When were they formed? Simply,
'how', 'where'/'how many' and 'when'. As the rest of this chapter will show, the
generally accepted set of answers to these planation questions has changed considerably over the last 140 years and is likely to change significantly again. Each
change led to the development of a new long-term model in the North-West.

As for the North-West itself, this has not been an unimportant region in British
geomorphology. Its large-scale topography was being analysed and interpreted in

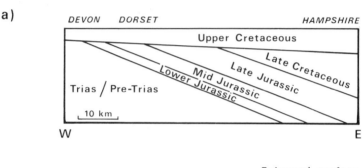

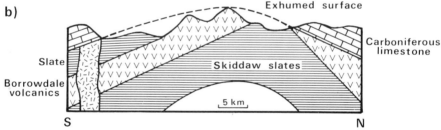

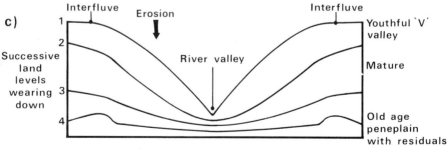

Fig. 3.1   a The Cretaceous unconformity across southern England   *b* The sub-
Carboniferous marine surface in the Lake District (Ward, 1870)   *c* Davisian
peneplanation.

the late nineteenth century, whereas the South-East, which came to be regarded as the most important region, only reached the 'interpretation stage' in the mid twentieth century (Jones, 1981). More than that, interest in the North-West's topography has gone beyond science. In the early nineteenth century, the scenery of part of it, namely the Lake District, had become the focus of the English artistic and literary Romantic movement (Edwards *et al.*, 1972). Briefly, the North-West is an area of varied relief and lithology ranging from 978 m at Scafell Pike in the heavily glaciated Lake District, through the less glaciated dissected plateaux of the Isle of Man and the Pennines, to the undulating till-covered Cheshire Plain. Its geology varies from igneous plutonic and volcanic rocks in Cumbria to the sedimentary sequences of grits, shale and limestone in the Pennines, all covered with a discontinuous layer of glacial drift and alluvium.

Given these definitions, the second section of this chapter traces the application of the planation concept to the North-West. It is subdivided into three parts. Each considers a period when a different model of planation surfaces was constructed in the region in response to a different set of answers to planation questions, that is a different British school of long-term landform development. The subdivision is according to the date of publication of crucial work which initiated the new answers. The third section gives a brief synthesis of current ideas and evidence likely to cause future re-evaluation of the problems. The final section considers how such a synthesis might be used in the construction of a new model for North-western England.

The research history

The three periods are:
1846 to 1895: Marine planation; the influence of A. C. Ramsay.
1895 to 1939: Subaerial planation; the influence of W. M. Davis.
1939 to the present: Attempt at national synthesis; the influence of S. W. Wooldridge and D. L. Linton.

1846 to 1895: Marine planation; the influence of A. C. Ramsay

One of the first planation surfaces to be recognised in the North-West is on the moors of Crin Edge near Buxton in the Peak District (Plant, 1866). Plant recognised an ancient marine beach at about 427 m on gritstone edges above the valley. A litter of rounded pebbles on the 'surface' he regarded as beach shingle; a layer of yellow clay and the valley itself were due to subsequent subaerial weathering and erosion, which was insignificant compared to the marine erosion which had removed great thicknesses of Carboniferous strata to form the surface. The age of the surface was 'post-glacial', having been formed during higher sea levels following glacier melting.

This work was an application of the late nineteenth century British school of long-term landform development. Marine erosion was considered the answer to the 'how' question. Once cut, such surfaces were little affected by subsequent sub-aerial erosion except for dissection. The combination of planation and dissection produced accordant, flattened summits and spurs standing above deep valleys. This approach originated with A. C. Ramsay. In 1846 Ramsay proposed that the isolated but roughly height-accordant summits in South Wales were relics of a widespread plain cut by wave action. This was, in fact, the beginning of the planation concept in Britain.

Ramsay was not the first to recognise that summits could stand at roughly accordant heights in the landscape; Hildreth (1836) had recorded its prevalence in the Appalachian mountains of the eastern United States. Nor was he the first to consider the possibility of planation for Hutton (1788) had suggested that, given time and crustal stability, subaerial denudation 'could produce plains'. But he was the first to synthesise the two concepts into a denudational system. Prior to the publication of his studies the majority of mid nineteenth century geologists had, despite the work of Hutton and Playfair, attributed all erosion to the action of the sea, in what was termed marine dissection; all valleys were due to marine erosion. In this, they followed Charles Lyell (1830) and the earlier schools of geological thought with their insistence of the importance of the Biblical Flood (Buckland, 1824). However, in the late nineteenth century, the 'subaerial approach', championed by Greenwood (1857), received more attention: for example, the work of Geike (1865, 1868), Ramsay (1863) and Jukes (1862). The concept of marine dissection was replaced by one of subaerial dissection. A subaerial origin was proposed for many valleys in the North-West: e.g. Hull (1866) and Sorby (1869) in the Pennines, Tiddeman (1868) in the uplands north of Manchester, and Ward (1870) and Marr (1889) for valleys in the Lake District. Some still adhered to a marine origin for small-scale features, valleys (Mackintosh, 1865a) or tors (Mackintosh, 1865b), but these were in the minority. However, in Ramsay's view the erosive ability of subaerial forces was still considered insignificant when compared to that of the sea: planation surfaces were a product of marine erosion; subaerial forces only dissected and modified but did not destroy.

Following Plant's work, many of the flattened summits and spurs in the upland areas in the North-West were interpreted as being relics of extensive marine planation surfaces at different heights. For example, the flattened summits visible from Helvellyn in Cumbria (De Rance, 1869; Ward, 1870; Ramsay, 1872) or the accordant summits around 549 m in the Forest of Bowland in north Lancashire (Tiddeman, 1872).

Whereas the answer to the 'how' question remained unchanged until the turn of the century, the answers to the 'how many'/'where' and 'when' questions altered significantly over the late nineteenth century. The recognition of widespread ancient marine transgressions which would have cut planation surfaces was primarily due to the progress of geological mapping in the area. Such surfaces still existed in the stratigraphic record as major unconformities and they could easily exist in the landscape if their sedimentary cover was removed by subsequent erosion, i.e. if they were exhumed. Differences of opinion still arose however, when the age of the surfaces was considered. Whereas Plant's (1866) view was of a relatively recent post-glacial surface, Ward (1870) maintained that the summits of the central Lake District, were relics of a much older surface cut by the Carboniferous sea which had also deposited the thick Carboniferous limestone sequence which, he envisaged, had been eroded following doming of the region in Tertiary times. Subaerial forces had not only exhumed the surface on the pre-Carboniferous strata, they had gone on to dissect it to create much of the present large-scale topography. The surface survived on the flattened summits and, of course, buried under Carboniferous rocks on the flanks of the dome (Fig. 3.1b).

Following Ward's Carboniferous incursion, younger marine transgressions were invoked to cut younger surfaces on younger rocks. For example, Goodchild (1889/90) recognised Ward's sub-Carboniferous surface on the pre-Carboniferous strata and then suggested that the post-Triassic (presumably Liassic) and the Upper Cretaceous Chalk Sea marine transgressions had planated surfaces on the Carboniferous rocks of the outer Lake District and in the Pennines. As with Ward (1870), subaerial forces in the Tertiary were invoked to erode any subsequent sedimentary cover of Mesozoic or Tertiary age, to exhume and then to dissect the surface.

This change in the answer to the 'when' question was accompanied by a change in the method of answering the 'where'/'how many' question. Plant sought local marine deposits as a guide to specific surface location, whereas the later authors, such as Ward, relied on the recognition of extensive areas of solid deposits whose surface had been truncated by later, major marine transgressions. Such surfaces *de facto* could only survive on hill summit levels.

As for 'how many', because the later surfaces of Goodchild and Ward were so extensive, two at most could exist in an upland region. This contrasts with the polycyclical step sequences of many surfaces that became so prevalent in the twentieth century as later sections will show.

So, by the end of the nineteenth century, a model of three ancient exhumed and dissected marine surfaces on the peaks of the Lake District and Pennines had developed in response to the evolution of the late nineteenth century British school of long-term landform development with its particular set of answers to the planation questions. However, it was to change as the school changed with the appearance of the general landform development model of W. M. Davis.

1895 to 1939: Subaerial planation; the influence of W. M. Davis

In 1901 Reed interpreted the Pennine summits, including those of the Peak District and Askrigg Block, as relics of one surface of subaerial origin and Early Tertiary age. Tilting in mid-Tertiary times had initiated subaerial dissection to produce much of the present topography. This was a radical departure from the model devised in the same area by Goodchild only eleven years earlier. The change was due to the influence of W. M. Davis. In 1895 Davis introduced into Britain a different set of answers to the planation questions in the form of a general model of landscape evolution. This first attempt at such a model was readily accepted in Britain, initiating the evolution of a new British school. The answer to the 'how' question became 'subaerial' instead of 'marine'. The 'when' answer became 'Tertiary' instead of 'pre-Tertiary'. The general answer to the 'where' question changed to include the recognition of lower levels on spurs and the method of answering the question concerning specific location became a question of morphology rather than lithology, both resulting in the possible recognition of many more surfaces in an area. These answers were based on the work of late nineteenth century explorers, Powell and Gilbert, in the western United States (Chorley *et al.*, 1964).

Briefly, Davis proposed a model for the scarp and vale topography of Eastern and Southern England consisting of two 'cycles of subaerial erosion'. His erosion cycle consisted of rapid uplift which initiated fluvial dissection and interfluve mass wasting. These in combination acting continuously reduced large areas to a plain. But he considered such erosion not to be totally effective in levelling the topography unlike the sea. Hills would still survive on the surface, 'residuals' above it, making it almost a plain, a peneplain (Fig. 3.1c). The level at which the surface would be cut, that is the level at which the fluvial dissection would cease, was, in Davis's view, governed by sea level, the base level of the model. Any changes in this base level due to a rise or fall in sea level, or due to tectonic disturbance, would interrupt the cycle and initiate a new one, that is reset the clock. So such a planation model required stability – sea-level and tectonic stability as Hutton had envisaged.

As for Eastern and Southern England, one of these cycles had produced a peneplain over the Early Tertiary period following retreat of the Chalk Sea – the sea responsible for the last major drowning of the British landscape. But uplift in the Mid-Tertiary, which Davis considered to be due to the compressional forces of the Alpine orogeny further south, had initiated a second cycle which had lasted through the Late Tertiary up to the present, with the Ice Age as a brief climatic interruption. However, erosion in this cycle had not got beyond the early dissection stage. Subaerial forces had attacked the weaker clay lithologies, leaving the first cycle surface preserved above as flattened summits in the 'harder' limestones and sands. Given time and stability, even these would be destroyed by interfluve mass wasting, as the cycle proceeded, to form a new peneplain. It was this model, with its new set of answers, that Reed applied to the Pennine topography.

Obviously, the model was not accepted immediately by everyone. Hughes (1901) interpreted the summits of the Askrigg Block of the central Pennines as relics of a 610 m high 'plain of marine denudation' as in the nineteenth century school. But the general trend over the early twentieth century was for the old ideas to be modified by, and then merged with, the new, to produce a new British approach and a new regional planation model in the North-West. Marr (1906), Marr and Fearnsides (1909), Gibson *et al.* (1925) and Fearnsides (1932) all illustrate the modification of the nineteenth century concepts by the Davisian answers. The work of Hudson (1933) illustrates their merger and the work of Hollingworth (1935) the first example in the North-West of the developed early twentieth century British school.

In 1906 Marr still adhered to the idea of pre-Tertiary age for any summit or envelope surface and suggested a dissected Devonian surface over the summits of the pre-Carboniferous Lake District strata. This surface had been extensively buried by the Carboniferous strata, partly buried by deposits of Permian and Mesozoic age and then, according to Marr, exhumed and dissected during the Tertiary. However, since the Devonian was a period of catastrophic continental erosion following the uplift of the Caledonian Orogeny, this surface must have originated as a subaerial rather than a marine eroded feature.

Marr and Fearnsides (1909) applied the same modified view to the long-term landform development of the Howgill Fells, a small Palaeozoic outlier near the Lake District. They recognised and interpreted a summit surface of Devonian age and subaerial origin, exhumed from its marine cover of Palaeozoic, Mesozoic and Tertiary rocks and dissected during Tertiary times.

Gibson *et al.* (1925) and Fearnsides (1932) applied similar reasoning to the Carboniferous coal measure summits and gritstone edges of the North Staffordshire uplands and Peak District respectively. However, being areas of younger rocks, they invoked a younger, but still pre-Tertiary period of significant subaerial erosion, the Permo-Triassic. Again, any later cover had been removed and the surface dissected by subaerial process in the Tertiary. Supporting lithological evidence for these models came from the occurrence in hollows on the surface of deposits rich in Bunter pebbles which were considered as relics of the former Permo-Triassic cover.

The work of Hudson (1933) re-emphasised the pure Davisian approach in the Pennines. He suggested the existence of a subaerial surface formed in an Early Tertiary cycle of erosion on the flattened summits of the Askrigg Block with peaks such as Ingleborough, at 723 m, as residuals above the peneplain. Like Reed's surface, it had been tilted, initiating a second Late Tertiary cycle which had only succeeded in dissection. But the surface had also been warped, i.e. differentially uplifted. This, in Hudson's view, accounted for the fresh and therefore recently-formed fault scarps in the area, e.g. Giggleswick Scar. However, what really distinguished his work from Reed's, and shows the merger of the new concepts in their pure form was Hudson's recognition on the lower summits of a sub-Triassic subaerial surface near to extensive Triassic rock outcrops. The late nineteenth century concept of the existence of exhumed pre-Tertiary surfaces in the landscape had not been totally brushed aside by the Davisian theory, even though by the thirties their occurrence had been restricted.

With the work in the Lake District of Hollingworth (1935), the new British approach following the modifications and mergers had arrived in the North-West. By this time British researchers were recognising multilevel, polycyclical landscapes of marine and subaerial Tertiary planation surfaces. Recognition of the greater complexity of sea-level change had caused the extension of the simple two-level Davisian model into models consisting of many levels, each the result of an incomplete cycle of erosion. The old answers still persisted. Some surfaces were marine and there was the possibility of pre-Tertiary surfaces exhumed locally. The models were now based on the study of morphology rather than lithology and this was perhaps due in part to the increasing independence of geomorphology from geology in the mid twentieth century.

Using this approach, Hollingworth reinterpreted the topography of the western

Lake District as a staircase sequence of Tertiary erosion surfaces 'younging downwards' all dissected to a greater or lesser extent, by fluvial wash and glacial processes. Which of these surfaces were marine and which subaerial depended on the reconstructed morphology. Those with prominent residual peaks, he identified as being subaerial in the Davisian approach. Those flatter and slightly inclined to sea level, he regarded as marine as in the late nineteenth century school. All were Tertiary in age. As with origin, morphology was the key to distinguishing between them in terms of age. Those Hollingworth considered warped pre-dated the Mid-Tertiary Alpine folding referred to earlier, and were therefore Early Tertiary. Those unwarped post-dated the folding and were therefore Late Tertiary. The result was a step sequence model of surfaces, each surface the result of an incomplete cycle of marine or subaerial erosion in Tertiary times in response to an intermittently falling sea level.

McConnell (1938) and Miller (1938) continued the application of this new approach to the North-West. McConnell, using projected profiles, supported by field evidence, identified a staircase of seven surfaces in the southern Lake District from a summit level at 732–823 m such as on flattened summits at High Street, down to a 122–152 m level found as a prominent step in river valleys such as in Troutbeck near Windermere. Although he did not date or suggest an origin for these levels, his attempt to correlate them with Hollingworth's surfaces suggests he viewed them in the same way.

Miller (1938), again from projected profiles, proposed a staircase sequence of three Tertiary surfaces preserved on the summits, spurs and valley benches of the Lancashire Uplands and on hill summits above the drift-covered Cheshire Plain. These were a 244 m level notable between Rochdale and Oldham, a 152 m level on the Blackburn plateau and Ainsworth plateau, near Bolton, and a 61–76 m level near Runcorn and on Alderley Edge in Cheshire. Based on reconstructed morphology, he considered all of them to be marine.

This, then, was the early twentieth century model of planation surfaces in North-west England: a rather confused one of Tertiary step sequences of marine and subaerial surfaces in the Lake District and in Lancashire and Cheshire, and a mixture of ancient and Tertiary subaerial levels in the Pennines and Staffordshire Uplands. It had arisen out of the theoretical conflagration initiated by W. M. Davis's new set of planation answers. There was a need for modernisation of the models in the Pennines, especially in the Peak District. But, more important, with so many levels replacing the one or two regional surfaces of the nineteenth century, there was a need for surface correlation and synthesis within and outside the region – a Grand Design. This occurred during the next phase of the research history mainly through the work and influence in the North-West of D. L. Linton, but set against the denudational chronology model that he and Wooldridge constructed for the South-East (Wooldridge and Linton, 1939).

1939 to the present: attempt at synthesis, the influence of
S. W. Wooldridge and D. L. Linton

The approach represented by Hollingworth's work continued to be applied to the

North-West into the mid twentieth century. For example, Owen (1947) recognised a Late Tertiary marine surface at 61 m in south Lancashire on summits in the Wirral peninsular. Sweeting (1950) extended the approach into the Pennines by recognising two surfaces on the limestone topography of the Craven highlands in the Askrigg Block, at 457 m and 244–183 m respectively. Clayton (1953) extended it into the Peak District. He reinterpreted the summit surface at around 305 m on the limestone and gritstone north of the River Trent as a Tertiary subaerial surface. His rejection of Fearnsides's (1932) sub-Triassic view was again based on morphology – in this case the discordance between the 'level' 305 m surface and the greatly sloping sub-Triassic surface preserved buried south of the Weaver Hills near Ashbourne (Fig. 3.2a). This discordance suggested a different age. But typifying the merger of old and new, he still regarded the ancient surface to exist in small patches near the Trias outcrop, having been exhumed in the Tertiary. Again following the thirties' approach, he also recognised a step sequence below the 305 m level consisting of a middle surface at around 250 m on the interfluves of the Dove and Derwent rivers and three lower surfaces at 168 m at Darley Moor, at 143 m near Yeaveley and at 98 m at Mickleover Common. Each of these lower surfaces had numerous substages, the result of a complex history of stillstand events in a generally falling base level. But the significant part of his work was the specific dating of the 305 m level. Instead of answering the 'when' question for such high levels as Early Tertiary, as in the earlier work of Reed (1901), Clayton suggested it was Late Tertiary. But this dating was not on the basis of morphology as in the thirties' approach, but by correlating this Pennine level with the Late Tertiary or Mid-Pliocene summit surface of South-East England. This change in the mode of answering the 'when' question was due to the effect on British long-term landform development of Wooldridge and Linton.

In 1939, Wooldridge and Linton, through a merger of the former's work on surfaces and the latter's drainage studies, constructed a denudational chronology model of South-East England which included an already recognised chalk summit surface formed after the folding of the Wealden area in Mid-Tertiary times (Wills, 1929). Wooldridge's concept was that the role of British geomorphology was to explain the geomorphology of South-East England; research in the uplands was secondary (Wooldridge, 1956; Dury, 1983). His pre-eminence in British geography plus his influence through his students meant that his idea became widespread and the South-East model became the linchpin of a Grand Design, which subsequent work like that in the North-West had to be fitted into. As workers like Clayton, following Linton's lead (Linton, 1950), sought correlations with the South-East model, so the surface chronology, the answer to the 'when' planation question changed slightly. This was because, apart from the Late Tertiary surface, Wooldridge and Linton had identified two other surfaces, a sub-Eocene marine surface cut on the Chalk dip slopes exhumed from its Tertiary marine cover, and a later marine surface of Pliocene age cut into, and sometimes destroying, the summit surface at 290 m (Fig. 3.2b). Both had been cut following widespread marine transgressions in the South-East in the Tertiary. The former was not applicable to the uplands north and west outside the Early Tertiary basin that had

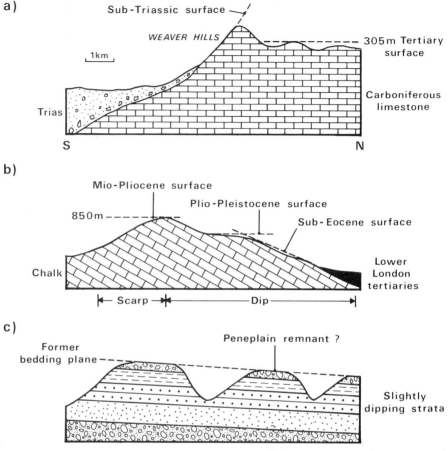

Fig. 3.2 a The two discordant surfaces in the Trent Basin (Clayton, 1953) *b* The three
   Tertiary surfaces in South-East England (Wooldridge and Linton, 1939) *c* An apparent
   peneplain due to structure.

existed over Southern England. But the later surface could be and was recognised
in areas like the North-West. But identification of such a surface at relatively high
altitudes meant that surfaces below, previously recognised as Late Tertiary, given
the younging downward concept, now became Pleistocene in age. For example,
in the work of Gresswell (1953) and Parry (1960).

   Despite this new approach with a change in the answer to the 'when' question
introduced by Wooldridge and Linton, the answers and the mode of answering
the 'how' and 'where'/'how many' questions remained unaltered. However, this
phase did see the further gradual modernisation of the older planation surface
model in the Pennines (Linton, 1956; Johnson and Rice, 1961), the Staffordshire
Uplands (Yates, 1956) and the Forest of Bowland (Moseley, 1961) and the recognition of other ancient exhumed surfaces (Johnson, 1957b). All these combined to
produce a new model of planation levels in the North-West; part of the Grand
Design. This is illustrated by the work of Gresswell (1953) who recognised four

new surfaces in the Rossendale Anticline north of Manchester, at 396 m to 274 m at Anglezarke Moor near Chorley, at 213 m to 152 m around Wheelton, at 76 m to 61 m around Skelmersdale, to the lowest at 31 m to 8 m at Scarisbrook. As in the thirties' approach, he regarded them all as marine, formed during stillstands as the sea fell to its present level. But, following the new approach, he correlated them with the surfaces in the South-East. The highest was the time equivalent of the Chalk summit surface. The second was the Pliocene/Early Pleistocene level. The third and fourth became Pleistocene, the result of later stillstands.

The absence of a sub-Eocene surface from North-west England did not prevent geomorphologists from searching for other pre-Miocene land surfaces that were possibly exhumed in the Tertiary Period. Johnson (1957b) recognised an exhumed Carboniferous (sub-Namurian) in the Wye valley and elsewhere in the Peak District and it was the superimposition of drainage from this surface which provided the pattern of dry valley catchments in that area (Warwick, 1964) but such ancient surfaces were of course limited in their extent. Yates (1956) extended the new approach into the North Staffordshire uplands. Like Clayton, he rejected the earlier sub-Triassic surface on the coal measure plateau around Keele and replaced it with an undulating surface around 152 m which correlated in age and origin with the South-East summit level. In support of this correlation, he noted how the surface cut across post-Triassic faults and an Early Tertiary dyke, suggesting a post Triassic, post Early Tertiary and therefore possibly Late Tertiary age.

Linton (1956) himself continued the modernisation of the planation model in the Peak District by identifying an Early Tertiary subaerial surface on gritstone edges above Clayton's (1953) level between 503 m and 549 m with relics of a pre-Tertiary sub-Cenomian marine surface above that preserved on top of residuals above the Early Tertiary level such as on Kinderscout at 610 m near Glossop. Even this pre-Tertiary marine surface fitted in with the new approach. Wooldridge and Linton (1939) regarded the Chalk marine transgression as having obliterated previous morphologies to form a new surface. This was of course still buried under the Chalk in the South (Fig. 3.1). But in the uplands it had been quickly exhumed, then cut into by successive Tertiary subaerial and marine cycles of erosion. The transgression therefore represented the initial datum plane of British geomorphology. Its surface could still be found on the highest summits as in the Pennines, but it had largely been destroyed. Any older surfaces, such as the Permo-Triassic, had been more recently exhumed and therefore were only important locally at lower levels. However, this concept did mean that at least one of the pre-Tertiary surfaces of the late nineteenth century school (see Goodchild, 1889/90) could affect the overall topography of a region, even if in an indirect way because of the influence of Davisian Tertiary erosion.

Parry (1960) extended the new conceptual framework to the Lake District. He replaced the earlier step sequence with a staircase of ten surfaces from a 457 m level on Red Pike down to a largely drift-covered surface at 88 m. His acceptance of the new approach can be seen in his interpretation of the age of surface number four at 210 m found on valley benches, e.g. in Mitredale. He correlated it with the Plio-Pleistocene marine level in the South-East. Consequently, the six surfaces below it, which earlier would have been regarded as Tertiary (see Hollingworth, 1935), now became Pleistocene in age. The three surfaces above he interpreted as Late Tertiary. These he also tried to correlate with the South-East and regarded them as subaerial rather than marine.

Johnson and Rice (1961) added to the surfaces of the south-west Pennines in an area stretching from Biddulph to the Kinderscout Plateau. They recognised three more levels at 287 m to 274 m, 229 m to 210 m and 201 m to 177 m in the Goyt, Dean and Bollin river valleys. Even though at heights comparable with Pleistocene marine surfaces elsewhere, e.g. the Lake District, they regarded the two lower surfaces as subaerial in origin. Even this was in accordance with the current British approach. Linton (1956) had considered the formation of subaerial surfaces far inland in Pennine valleys likely, being Pleistocene stages in the dissection of the 305 m surface responding to the fluctuating sea level nearer the present coast.

The result of all this work was the construction of a mid twentieth century model of planation surfaces in North-west England. It consisted mainly of Tertiary subaerial surfaces above Pleistocene marine levels, with the possibility of a sub-Chalk surface on the highest peaks and recently exhumed older pre-Tertiary levels at lower altitudes. The differences between this model and the models formed in the early twentieth and late nineteenth century were the result of a different set of answers to the same planation questions.

However, not all the work in the North-West on planation surfaces at this time followed the generally accepted British approach, although they still used Davisian concepts. The work of Moseley (1961), Sissons (1960b) and King (1969) all imply criticism of the existing school, criticism which grew in the seventies generally.

From detailed statistical analysis of slope data, Moseley (1961) produced an alternative to Tiddeman's (1872) model of a marine level at 549 m in the north Lancashire Forest of Bowland. He proposed a step sequence of three levels at 518 m to 366 m, as on Hawthornthwaite Fell, at 335 m to 122 m around the River Raeburn, and at 76 m in Hindburndale. He did not date them by correlating with the South-East and he answered the 'how' question as subaerial, which for the lower surfaces would have been unacceptable to the British school for an area so close to the sea.

At greater variance with the prevailing approach was Sissons's (1960b) planation model of the Pennines. In it, he regarded the surfaces as Late Tertiary in age, but a combination of marine and subaerial in origin. A surface was cut by the sea during a particular stillstand but was then considerably modified by subaerial processes when it was exposed after the sea fell to a lower level to cut the next surface. In doing so, Sissons drew attention to a major inconsistency in the prevailing British approach, that is the survival of marine surfaces or indeed any surface on an interfluve, given the supposedly great efficiency of subaerial downwasting in the Davisian model.

Possibly the work most at variance was that of King (1969). Given the quantitative revolution in British geomorphology at that time (see Dury, 1983), King applied trend surface analysis to summit heights in the Pennines. To her, the results suggested one differentially uplifted summit surface rather than the step sequence proposed by Linton. In a sense this represented renewed interest in the earliest Davisian-influenced approach of Reed as opposed to the staircases devised since the thirties, of which some authors were becoming increasingly sceptical.

During the sixties and seventies, planation in the North-West largely entered a consolidation phase with attempts at a much needed regional synthesis of the individual models (Smailes, 1960; Rodgers, 1962; Johnson, 1965a; King, 1976; Straw and Clayton, 1979). However, such attempts only fired more criticism of the current approach. For example, Johnson (1965a) and King (1976) note first, the lack of evidence for all these higher sea levels and secondly, the general difficulties of altitudinal correlation of surfaces even between areas in close proximity. The latter problem suggested to King that differential uplift over the Tertiary was an important factor unconsidered by the mid-twentieth-century school with their emphasis on widespread changing base level. The actual existence of surfaces on some summits was questioned. Some authors began to record the possibility that some of the flattened summits were due to the level inclination of sedimentary strata (Rodgers, 1962; Sweeting, 1974; Fig. 3.2c). Further, the lack of contemporary deposits, marine or subaerial, on these surfaces made interpretations of origin, the answering of the 'how' question, suspect.

More fundamental, the whole assumption of being able to correlate summits and flattened hillsides in a particular area to construct a planation surface was being questioned – a case of 'did any of these surfaces exist at all' (Smailes, 1960). The underlying themes of all this criticism were that researchers were beginning to state the highly subjective nature of the multicyclical surface approach and the lack of hard evidence in putting so much erosion down to the Late Tertiary and Early Pleistocene, periods for which we have so little complementary stratigraphic record, even in the South-East. Despite these misgivings, such surfaces are still referred to in the literature up to the present, since nothing has replaced them (see Clayton, 1981). However, continued criticism is making many ideas not available or acceptable to the mid-twentieth-century school seem attractive. This, in turn, is leading to another fundamental change in the appearance of the planation concept: a different set of answers to the same planation questions. This will eventually mean a new planation model in the North-West. The third section of this chapter, therefore, attempts a brief summary of what ideas may be important in answering the planation questions afresh.

Current theories

First, the question 'How'. The marine origin of planation surfaces is now in doubt. The concept may simply be the result of a marine obsession in British geomorphology with roots in the Biblical Flood, rather than a scientific theory based on fact. This is indicated in four ways. First, modern process studies have questioned the ability of waves alone to cut a broad plain (King, 1963; Ollier, 1981). Second, Jones (1974) in a study of the drainage pattern of South-east England, while accepting the Plio-Pleistocene marine transgression, rejected its geomorphological significance, i.e. the idea of it cutting a planation surface. This criticism was supported by work on the clay-with-flints distribution on the Chalk Downs (Hodgson *et al.*, 1967; Catt and Hodgson, 1976). Logically such criticism could apply to all recognised wave-cut surfaces. Third, the great lack of marine deposits on any of these levels. With the rise of the glacial theory in Britain in the nineteenth century, (Agassiz, 1840/1) the deposits which had originally been regarded as indicating widespread marine action in landscape formation were reinterpreted as glacial drift. Despite the continued use of the marine theory which such deposits had helped to introduce, very few new marine deposits were discovered to replace them. Fourth, contemporary European schools of long-term landform development, do not, unlike 'maritime Britain', place much emphasis on marine erosion in landform creation.

With the rejection of a marine origin for such widespread inland surfaces, Hutton's 200-year-old emphasis on subaerial denudation has finally occurred. However, the answer to the 'how' question is further complicated by the rejection of the accepted Davisian type of subaerial denudation. Planation in the humid temperate climate by a combination of fluvial dissection and active interfluve destruction is now widely considered unlikely. The ability of humid temperate processes to actively erode uniformly over the whole landscape is in doubt (Crickmay, 1975; Twidale, 1976; Brunsden, 1980; Bradshaw, 1982). The very preservation of relic surfaces on interfluves suggests inactivity at the large scale, as the preservation of glacial striae does at the small scale. One solution to the 'subaerial origin problem' is etchplanation. This can be defined as the removal by wash processes of deeply decomposed rock (weathered in a humid tropical or sub-tropical climate) to expose, to a greater or lesser extent, the weathering front as an etchplain (Fig. 3.3a). Although this is a concept developed to explain humid tropical planation levels (Wayland, 1934; Büdel, 1957; Thomas, 1974), it has been applied to temperate areas. The German school of long-term landform development use it to account for the origin of European surfaces (Büdel, 1979; Bremer, 1980). The use of such a concept to explain the origin of British surfaces is suggested by five interrelated points. First the recognition of the importance of climate and its variation in geomorphology (Schumm and Lichty, 1965). Second, the recognition of the importance of etchplanation in stratigraphy, that is clay deposits being derived from weathered rock (Blatt *et al.*, 1972). Third, the growing evidence that the Tertiary, or at least the Early Tertiary was much warmer and wetter than the present. For example in the Palaeocene, shown by the Antrim Interbasaltic Beds (Montford, 1970) or in the Early Eocene, shown by the London Clay flora (Reid and Chandler, 1933). Fourth, the evidence for Tertiary etchplanation from British deposits (Thomas, 1978), for example in the Palaeocene Reading sands (Buurman, 1980). Finally, the attractiveness of a concept to explain surface origin, which is independent of Davisian base-level, modern data throwing doubt on the concept of stable sea level over long periods (Brunsden, 1980). Even if Britain was never under a humid tropical climate, Britain not having moved south of its present latitude over the last 70 million years and the earth's curvature possibly preventing such extreme conditions penetrating this far north, this does not diminish the

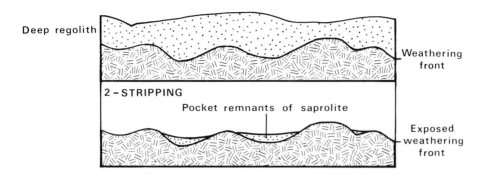

a)

1 – DEEP WEATHERING

Deep regolith

Weathering front

2 – STRIPPING

Pocket remnants of saprolite

Exposed weathering front

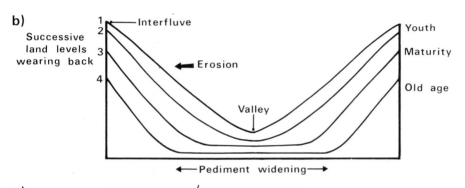

b)

Successive land levels wearing back

1 — Interfluve

Erosion

Valley

Youth

Maturity

Old age

◄— Pediment widening —►

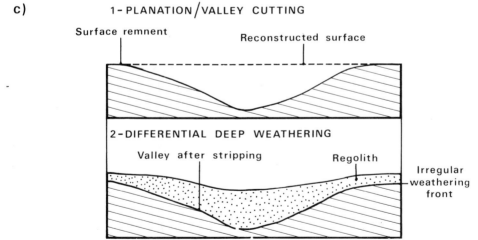

c)

1 – PLANATION/VALLEY CUTTING

Surface remnent

Reconstructed surface

2 – DIFFERENTIAL DEEP WEATHERING

Valley after stripping

Regolith

Irregular weathering front

Fig. 3.3   a Etchplanation: (1) deep weathering, (2) stripping   *b* Pediplanation   *c* Valley/ridge formation: (1) planation/valley cutting, (2) differential deep weathering.

C

concept's applicability. Deep weathering and etchplanation can occur in the humid sub-tropics (Tricart, 1972) and a humid sub-tropical Britain in Tertiary times is possible. The period before the Upper Cretaceous Chalk transgression – the Lower Cretaceous – is considered as having been partly sub-tropical (Allen *et al.*, 1973) with deposits suggesting etchplanation (Allen, 1975). Further, a humid sub-tropical thesis can explain Tertiary floral anomalies, for example, the discordance between humid tropically weathered rock containing warm temperate vegetation as in the Interbasaltic Beds (Montford, 1970) and the high percentage of non-tropical flora in the London Clay deposits (Daley, 1972). Under humid sub-tropical conditions, weathering and landforms development are still of a tropical nature but the tropicality of the vegetation diminishes (Tricart, 1972). Furthermore, etchplanation may not have been restricted to the humid tropical or sub-tropical Early Tertiary. Thomas (1978) suggested the possibility of cooler versions of etchplanation, and Hall (1983) suggested deep weathering and stripping under warm temperate conditions in the Late Tertiary as being important in the landscape development of North-East Scotland

Although etchplanation is a very likely alternative to the Davisian planation concept of subaerial erosion it is not the only one. If erosion in other climates than our own is thought relevant to an understanding of Britain's topography then the semi-arid climate must be considered, since under such conditions erosion is at a maximum (Blatt *et al.*, 1972). Here planation occurs through the parallel retreat of hillslopes through wash action leaving low-angle rock slopes (pediments) which then coalesce to form a large-scale erosion surface, the pediplain (Fig. 3.3b). This pediplanation has already been voiced as an alternative to Davisian peneplanation (King, 1953). But in common with etchplanation and many other concepts of value, such as continental drift, it was either ignored or resisted by the mid-twentieth-century British school with its Davisian-based approach influenced by Wooldridge (Dury, 1983). However, now, with evidence of Tertiary semi-arid phases in Europe generally (Bakker and Levelt, 1964) pediplanation must also be considered. Given that one of the great geomorphological discoveries of the late nineteenth century was the multigenetic origin of landforms (Chorley *et al.*, 1964), a combination of etchplanation and pediplanation should be used to answer the 'how' planation question in Britain in the future. Such a combination has already been used to explain the origin of surfaces on the Western Australian Craton (Fairbridge and Finkel, 1980). It has also been suggested as being possible in Britain (Thomas, 1978).

Secondly, the 'how many/where' question. In general terms, the recent rejection of the threefold step sequence model in the South-East (Hodgson *et al.*, 1967; Catt and Hodgson, 1976; Jones, 1981) throws doubt on the step sequences of Late Tertiary and Pleistocene surfaces recognised in upland regions like the North-West. A multiplicity of surfaces in a region may still be possible, but it is important to note that the increasing and highly subjective, subdivision of former envelope surfaces probably went as far as it could with the British school of the mid twentieth century; compare Parry (1960) with Ward (1870).

As for 'where', this may well be effected by the new answer to the 'how' question in two ways. First, the specific location of surfaces need not be sought just by morphology. Etchplains can be shown by surviving basal remnants of weathering profiles from which they have been cut. Second, and more fundamental, the application of the etchplain concept may change the traditional view, dating back to Ramsay (1846), of planation surfaces only on summits and spurs, the valleys, separating surface relics, having been cut at a later date (Fig. 3.3c(1)). The etchplanation concept means that many valleys may have to be regarded as part of the planation surface, the valley/summit landscape being due to differential weathering, possibly repeated several times. In such a model the deep humid tropical or sub-tropical weathering will be deeper along lines of weakness: fault-rich zones or areas of less resistant lithology. When the saprolite is removed such areas become basins and valleys while the less deeply weathered zones become summits and ridges (Fig. 3.3c(2)). This concept of differential weathering and etchplanation has already been used to explain the origin of the basin and ridge topography in the humid tropics, on areas of varying lithology and on granite (Thorp, 1967; Kroonenberg and Melitz, 1983). Waters (1957) has even suggested it to explain the basin/tor topography on the Dartmoor granite. With the acceptance of etchplanation as an origin of planation surfaces, the idea of large-scale differential weathering will have to be more widely applied. As well as valley formation, deep weathering in the etchplanation context can also be used to account for the origin of flattened summits. Trendall (1962) considered the flattening of summits in the humid tropics to be due to laterisation giving the appearance of isolated relics of planation surfaces, although it may, notably on sedimentary strata, be effected by structure, i.e. level or slightly dipping rocks. The occurrence of such flattened summits at different heights thought by previous schools to represent surfaces of different ages, can also be explained, at least partly, by differential weathering, but differential uplift may also be important.

Despite the importance of differential deep weathering in forming surface height variation, many valleys would still be due to fluvial incision. But in keeping with current theory, that of the restricted lateral effect of fluvial action (Brunsden, 1980), such valleys would be generally narrow. Waters (1957) considered only the wide sections of the river valleys on Dartmoor to be due to differential weathering (as shown by their cover of saprolite). The narrow rock-cut sections linking them were due to fluvial incision.

The 'when' question is perhaps the most intriguing to answer in the light of current concepts. Although stratigraphic evidence, i.e. the Tertiary deposits of South-East England, suggest Tertiary etchplanation/pediplanation has occurred, we should forget the Tertiary emphasis on landscape development introduced by W. M. Davis. Like the pre-Davisian school we should consider more ancient periods as being as important, if not more important in the formation of present topography. After all, Davis originally applied his Tertiary-based model to the landscape of the more recent Mesozoic and Tertiary rocks of southern and eastern Britain, a temporal approach which may only be valid in that area. The Palaeozoic and Pre-Cambrian rocks of upland Britain, as in the North-West, surely suggests Palaeozoic and Mesozoic eras should be more closely considered. The idea that the timescale of planation surface formation in upland Britain should not be constrained by that of lowland Britain is consistent with the rejection of the mid-twentieth-century British school with its emphasis on the South-East. Furthermore, if we reject marine planation, then the importance of the sub-Cretaceous marine transgression in obliterating former relief should also be rejected, making widespread topographical inheritance from earlier periods possible. This plus the current concept of differential spatial subaerial erosion (Crickmay, 1975; Brunsden, 1980) means it is possible that Tertiary etchplanation/pediplanation may have affected some areas while in others, little erosion may have occurred over the same and longer timespan. For example, Early Tertiary volcanic rocks in Australia, suggest some valleys are pre-Tertiary in age and that relatively little erosion has occurred in the Tertiary period (Ollier, 1979). At a larger scale King (1953) considered major scarps in South Africa as dating back to the Mesozoic. The idea that some unburied landforms can be pre-Tertiary in age may be applicable to Britain. In South-West England near Penzance, unburied Early Tertiary gravels exist at low altitude (six to nine metres) (Reid, 1904). These flint-rich gravels can be correlated with the Bullers Hill gravels of the Haldon Hills regarded as Early Eocene in age (Hamblin, 1973). As Reid recognised, the gravels suggest a largely pre-Tertiary origin for the landscape, at least in the immediate vicinity, with relatively little erosion since.

But a pre-Tertiary origin for British landscapes may also be important at the large scale. The modern French concept of long-term landform development considers that peneplains form on individual massifs in immediate post-orogenic times and that any subsequent rapid erosion is mainly of thin layers of unconsolidated sediments deposited on the planated hard core during marine transgressions, the core itself being relatively little affected except by Early Tertiary etchplanation and differential uplift: there is therefore a fundamental topographical development taking place in pre-Tertiary times (Pinchemel, 1969; Beaujeau-Garnier, 1975). Such a concept has already been applied to Britain by Battiau-Queney (1980) who postulated the development of a planation surface in North and Central Wales in Devonian times immediately following the effects of the Caledonian Orogeny. She argued that the stratigraphy and lithology of the Late Palaeozoic and Mesozoic strata surrounding the massif indicates that the surface itself remained little affected by subaerial or marine erosion until it was destroyed by differential uplift in the Late Tertiary. Such a fundamental planation surface could be regarded as an initial form to which all subsequent erosion and deposition would be modifications. Even if the surface was largely destroyed, its recognition would still be of great importance in understanding the long-term development of an area, since its creation means the initiation of form, age and geological structure.

It can be argued that geology provides the spatial and the temporal framework for the initial surface formation and its subsequent modification. The underlying basin/massif pattern establishes a regional large-scale areal differentiation in which the initial surfaces can be recognised, but the geological structure within the massifs and basins also provides an areal differentiation at a smaller scale where, for example, etchplanation, is often effected by varying rock resistance.

The timing of the initial catastrophic erosion and the subsequent modifications would ultimately be related to the timing of the major plate-tectonic events in and around the region. The creation of the initial form on a massif would follow uplift caused by plate collision either on site as in the Lake District, or nearby as in the Pennines. The subsequent erosional and depositional modifications, including the formation of new planation surfaces, become responses to changing tectonic, sea-level and climatic conditions which would all be ultimately related to nearby plate activity. These include prologues to sea-floor spreading (the uplift and collapse of triple junctions) actual sea-floor spreading with its varying rates, and subsequent plate collisions near or far, all noted in the recent geological literature (Burke and Dewey, 1973; Hays and Pitman, 1973; Anderton *et al.*, 1979). Although plate tectonics is recent, the idea of relating landform development to tectonic activity is not – it formed the basis of the work of W. Penck (Penck, 1924). However, his work was either considered secondary to Davisian concepts, as in Britain, or ignored, as in his own country where the climatic geomorphology of Albrecht Penck, which with the later work of Büdel was to include etchplanation, was predominant (Bremer, 1983). However, the above model is not only based on a rejection of Davisian concepts, it also suggests that climatic geomorphology, i.e. etchplanation, is not necessarily incompatible with a scheme of landform development based ultimately on tectonic activity.

The use of plate tectonics in British geomorphological studies is long overdue, and its importance as a major control on long-term landform development has only recently been recognised by British workers (George, 1974; King, 1977; Summerfield, 1981). Forming a major part of the new field of 'evolutionary geomorphology', it offers a new method whereby planation problems may be investigated and as such has already been successfully used in Australian land-form studies (Ollier, 1979). Already some British authors, notably King (1977) and Brunsden (1980) have stressed the role of plate tectonics and tropical climates as elements in Tertiary Britain's landform development, but as yet no regional studies have been made for any part of upland Britain which would demonstrate the value of this type of synthesis. This is remedied in this chapter's final section.

Discussion and conclusion

Given these new conceptual ideas a new model can be envisaged. First there is a need to recognise and use the basin/massif structure to establish the large-scale areal differentiation in the region (see Moseley, 1972; Kent, 1975) and to note the importance of post-orogenic plutons in maintaining this framework by isostatic uplift. Plutons have been recognised under nearly all the massifs in the North-West. The Wensleydale granite underlies the Askrigg Block in the Pennines and a batholith exists under the Lake District, outcropping as the Eskdale and Shap granites and the Ennerdale granophyre. A pluton also underlies the Isle of Man, exposed as the Foxdale granite. These plutons have created crystal mass deficiencies to which periodic uplift has been the response to regain equilibrium (Bott and Masson-Smith, 1957; Bott, 1967, 1974). Correspondingly, the South Lancashire/Cheshire basin and the basins between the coast and the Isle of Man are not underlain by such plutons. They therefore collapsed to form lowlands during periods of crustal tension following pluton intrusion.

Second, there is a need to recognise and date the initial surfaces in each of the subregions. The stratigraphic record suggests an initial surface of Devonian age on the Lower Palaeozoic strata of the central Lake District and the Howgill Fells, and a Permian age for the initial surface on the Carboniferous strata of the Askrigg Block, Peak District, North Lancashire Uplands and outer Lake District. The unmodified survival of these surfaces is unlikely. However, such periods are still of great significance in defining not only the initial form but also the structure which was likely to guide subsequent modifications. It may also have set the overall dimensions of the subregion. Ford and Stanton (1968) in considering the geomorphology of the Mendip Hills, a small massif of Carboniferous strata in the West Country, suggest that although the hills have suffered considerable planation, the surrounding Triassic rocks indicate little change in the overall dimension of the massif in the Mesozoic and Tertiary. In other words, since their uplift and initial surface formation, their areal extent has hardy changed. Furthermore, the destruction, in the North-West, of the initial surfaces does not preclude the possibility of some landforms being of great antiquity, although no deposits comparable with the Penzance gravels have been recorded.

As for the Lancashire/Cheshire basin, the recognition of an initial surface seems at first less certain. Much of it is covered with drift that can be as thick as 90 m (Poole and Whiteman, 1966). The topography of this till plain is largely unrelated to the pre-glacial surface underneath, which occasionally pokes through the drift to form ridges such as at Alderley Edge. Consequently, glacial and post-glacial times seem more important in understanding the topography of this area. However, it is conceivable that the drift, being largely unconsolidated sediment will be removed eventually to reveal the pre-glacial terrain. The drift is analogous to the marine sediments that once covered the planated hard core of the massifs, i.e. it is a transitory depositional modification. Therefore, the surface of the basin should, like the massifs be viewed in terms of the initial surface/subsequent modification model. Crudely the structure of the Cheshire basin is a syncline mainly of Permo-Triassic strata but it also includes a small central patch of conformable early Jurassic (Liassic) beds. Therefore, at some post-Liassic period the area suffered relatively catastrophic erosion to form an initial surface which was then subsequently modified in the Mesozoic and Tertiary. Traditionally the age of such deformation was regarded as Tertiary, being the outer ripples of the Tertiary Alpine Orogeny (Pocock, 1906). However, this section suggests that the initial surface formation could have occurred sometime in the mid-Jurassic, late Jurassic or mid-Cretaceous, all periods of significant plate activity near Britain (see Anderton *et al.*, 1979). But again, although of large-scale significance, the survival of such a surface is unlikely.

Third, there is a need to recognise the occurrence, location, age and structural control of the erosional and depositional disturbances including planation. The stratigraphic record would indicate the occurrence and age of these events, namely periods of higher sea level leading to major deposition on the initial surface, periods of etchplanation and pediplanation and what lithologies were being eroded and consequently which subregions were being effected. The timing of these modifications would in the North-West be ultimately related to the major plate-tectonic activities around Britain in the last 250 million years. These would be the Hercynian Orogeny, the premature spreading of the Rockall trough in the Mesozoic, the opening of the North Atlantic ocean, the possible early Tertiary hot spot activity under western Britain and the repeated effects of the Alpine Orogeny to the south (Brooks, 1973; Pitman and Talwani, 1972; Anderton *et al.*, 1979). The location of these modifications would be indicated primarily by morphology and in the case of etchplanation, by the recognition of relic weathering profiles.

Although much information about these events may be gained from detailed studies of the Mesozoic and Tertiary strata of southern and eastern England, there is already some evidence for such modifications, especially planation, from geomorphology studies in the North-West itself. Corbel (1957) suggested humid tropical and sub-tropical karst landforms exist in the Furness area. The Brassington deposits in Derbyshire, now thought to be relics of a fluvial cover of Pliocene age (Walsh *et al.*, 1972) can be correlated with similar deposits in North-East Wales which have been interpreted as being derived from deeply weathered rock (Walsh and Brown, 1971). Arenaceous saprolite derived from Millstone Grit, as in the Todmorden Valley (Wright *et al.*, 1927) and from the Ennerdale granite in the Lake District (Ward, 1876) may be regarded as basal relics of such deeply weathered rock developed under tropical or sub-tropical conditions. The Millstone Grit and Dolomitic Limestone tors in the Pennines may be resistant sections of the 'weathering front' exposed after the removal of such saprolite (Linton, 1964). All these may point to the occurrence, age and location of etchplains in the North-West. Of course, alternative theories exist to explain these phenomena. The arenaceous weathering could be due to one or a combination of other factors such as rapid leaching in non-tropical conditions or frost-shattering in a periglacial climate. In the case of granite, hydrothermal alteration cannot be ignored (see Palmer and Neilson, 1962) and with sedimentary rocks there is the possibility that clay minerals which suggest intense conditions, such as kaolinite, may have been derived from the parent material. Pennine tor formation could be the result of periglacial, rather than tropical weathering and erosion (Palmer and Radley, 1961), although recent authors have tended to prefer the Lintonian view (Ford, 1963; Cunningham, 1965). Evidence also exists for the structural guidance of these modifications. Johnson (1965a) noted that the major valleys and smaller glacial meltwater channels in the south-west Pennines are consistently in areas of less resistant rock or in fault-shattered belts. The alternating resistant gritstone edges with less resistant clay vales of the Peak District is well known. The subdrift surface of the Cheshire basin also shows structural control since the highest sections of the surface, those that poke through the drift, are composed of the more resistant conglomerates as opposed to sandstones and marls. Structural control is also important in the Lake District. At the large scale, the more resistant volcanic rocks form the highest mountains, whereas the lower hills are composed of slate, but at a smaller scale, Marr (1906) recognised the importance of shatter belts in influencing the location of major valleys and gullies.

Finally, one can regard these modifications, although not etchplanation and pediplanation, as continuing up to and beyond the present, albeit at a smaller scale and less intensely. Contemporary weathering and soil formation, fluvial and coastal erosion and deposition all with the effects of man can be viewed as the most recent additions to the long-term story.

In conclusion, this chapter has shown that through the merger and modification of new and old ideas, the form of the planation concept and of the long-term geomorphological model in the North-West derived from its application, has changed considerably over the last 140 years and is likely to do so again. At present, the literature suggests that planation should be regarded as a combination of etch-planation and pediplanation in short phases throughout the last 200 million years and as one of a series of erosional and depositional modifications to the topography of areas whose form, age, structure and dimensions were initiated in periods of catastrophic crustal and erosional activity. The recognition of planation surfaces will still be of great importance in attaining an understanding of a region's long-term landform development. This is because of the great areal effect etchplanation has compared to other modifications. But in the future, the key to forming a denudative chronology of an area should perhaps rely more heavily on the induction of events from stratigraphy and the interrelating of form with saprolites and deposits rather than what is now considered the controversial recognition of planation surfaces from morphology alone.

Glacial and post-glacial climatic conditions in North-west England

Introduction

Climate is not static but is constantly changing on a variety of timescales. In recent years there have been anomalous seasons in North-west England including the severe winter of 1962–3 and the summer drought of 1976, but these can be considered as fluctuations about the long-period average climatic conditions and are part of the natural variability of the climatic system which is to be expected. However over longer periods of time there is unequivocal evidence that the climate has undergone significant changes, not least of which have been those producing the glacial conditions of the last Ice Age over much of Britain, which ceased some 10 000 years ago.

Knowledge of these climatic changes and fluctuations has increased considerably in the last thirty years, and has accrued from an integrated approach to the subject involving many disciplines. The knowledge is well synthesised in the classic work on climatic change by Lamb (1977a).

Figure 4.1 shows some of the different influences which operate on the complex earth–atmosphere–ocean system responsible for climatic changes and fluctuations. The forcing mechanisms shown may be classified according to the timescales over which they operate and include: fluctuations of solar output; changes in the earth's geography; changes in the earth's orbital parameters; changes in surface snow and ice cover; changes in the circulation and heat stored in the oceans; variations in volcanic dust, smoke from forest fires and pollution; and seasonal/daily changes in the radiation budget and the responses induced by these in the atmospheric circulation. These mechanisms do not act in isolation however, and there are complex feedback mechanisms operating between them which induce non-linear responses to any one change in the system.

The aim of this chapter is to evaluate the available direct and indirect (proxy) evidence of changes in the climate of the British Isles, and of North-west England in particular (where the evidence allows this). From this available evidence a review of the climatic history of the area together with an assessment of the changes in the atmospheric circulation will be presented.

The evidence for climatic change

It is not possible to provide a detailed critique of the quality of all the evidence used in reconstructing the climatic past because of constraints of space; for further

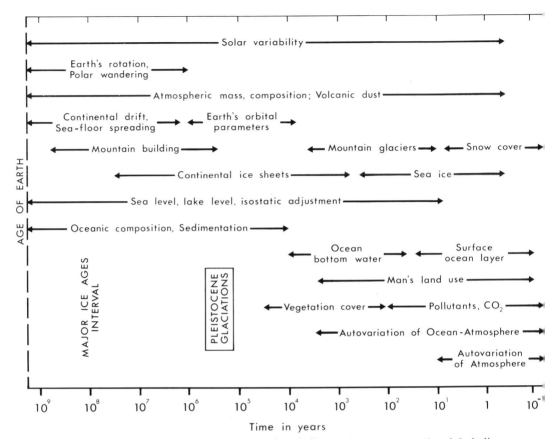

Fig. 4.1   The time-scales over which various influences operate upon the global climate system (adapted from Lamb, 1977a).

detail the interested reader is referred to texts such as those by Lamb (1977a), Frenzel (1973) or Goudie (1977). The variety of indirect (often termed proxy) and direct evidence used includes the following.

(i) *Geological deposits*   Much information can be inferred about the environmental conditions in which sediments have been deposited if one assumes the principle of uniformitarianism that 'the present is the key to the past'. Glacial depositional and erosional features provide important evidence about the areas occupied by former glaciers and their movement, while temporary readvances can be dated by radiocarbon testing of the remains of the vegetation that was overwhelmed.

(ii) *Fossils*   These are useful if the climatic tolerance of the particular fossilised plant or animal species is known and is unambiguous. However, climatic requirements of species may change over time and extinct species may have adapted to a climate different to that in which their modern counterparts now live.

(iii) *Changing sea levels* may be deduced from raised or submerged beaches and evidence of former marine transgressions resulting from changes in the global water balance. Changes in sea level may not be synchronous with changes in climate, however, for isostatic readjustment may continue long after the climatic change has taken place.

(iv) *Archaeological evidence* of former settlement and land-use may allow useful inferences to be made of climatic fluctuations in the past. However, it is not always easy to isolate climatic factors responsible for change from social, economic and other factors.

(v) *Pollen analysis* and analysis of macroscopic vegetational remains allow deductions to be made of the former extensions of species whose climatic tolerance (in terms of summer or winter temperatures, rainfall, or other conditions) is known. But care is needed in the interpretation of the evidence, for while it may be safe to argue from the known existence or abundance of certain species, it may be unsafe to argue from the apparent absence of species, for the traces of some species perish sooner than others. Different botanical species may have different migration rates, so that some may fail to exploit situations which are climatically favourable for them. Furthermore the development of a climax vegetation for given soil and climate conditions is a gradual process and the emergence of a dominant species is not an instantanous response even where human interference is absent (Prentice, 1983). Figure 4.2 compares the faster response time of vegetation with the slower accumulative response of glaciers to changes in climate from glacial to non-glacial conditions. From the evidence of prevailing plant communities and particularly of successive developments in the natural forests and peat bogs, workers such as Godwin (1975) have produced a consistent picture of the climatic sequence of the period since the last Ice Age, and the parallelism of botanical trends has led to the definition of the sequence of pollen zones describing the succession of vegetational communities.

(vi) *Tree ring analysis*   The varying widths of tree rings undoubtedly reflect the climatic character of the years in which they were formed, but dendrochronology only yields unambiguous climatic information where tree growth is affected by just one factor, such as rainfall at the desert fringe or temperature at the latitudinal limit of a species. Such information is of restricted value and covers only a limited time period in this country compared to, say, south-west USA.

(vii) *Diaries and archives*   Diaries (such as the weather diary kept by the Rev. W. Merle at Driby, Lincolnshire in the years 1337–44) may provide useful evidence of the occurrence of exceptional weather conditions, but the comments may lack objectivity and yield only qualitative information. Records of past grain, fruit or vine harvest yields may provide indirect evidence of former climates.

(viii) *Oceanographic data*   Ships' logs and other records contain important information on sea temperatures and the former extent of drift ice.

(ix) *Isotope analysis*   Ice cores from Greenland and Antarctica reveal a stratigraphy of annual ice accumulations in layers. Dansgaard *et al.* (1971) have shown the value of measurements of the oxygen-18: oxygen-16 ratio in the water substance of the ice. These measurements reveal a clear measure of the temperature at the time of deposition (the concentration of the ^{18}0 being directly related to temperature). They obtained a useful chronology of temperatures (and snow accumulation) extending back for thousands of years. Such cores as these also record major volcanic events of the past in their layers.

(x) *Historical climatic data*   Manley (1974) has constructed a record of monthly mean temperatures for central England (using combinations of records from different meteorological stations) extending back to 1659, which is the longest such record in the world. A record of temperature in Manchester exists from 1765. Monthly rainfall data for England and Wales are available from 1727, but in general there are few instrumental records extending back before the eighteenth century

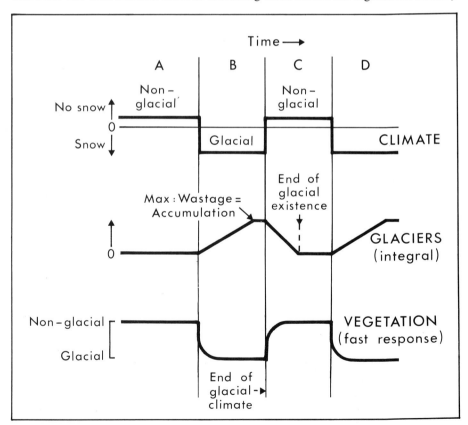

Fig. 4.2   The different response curves of climate, glaciers and vegetation over time (after Bryson and Wendland, 1967).

(and most of these are European, there being little from the rest of the world).

Instrumental records of prevailing climatic conditions exist only for some 200 years at best (for certain Europan sites), and at worst for less than 100 years for many parts of the world. Because of this fundamental data deficiency, the climatologist has to draw on information culled from a wide variety of sources to make inferences about past climatic conditions. In some areas the information is not unambiguous, in other areas it may not be easy to date the evidence (where errors arise due to the difficulties of experimental measurement and uncertainties in the basic assumptions about the constancy of the geophysical environment), but hopefully the suite of available evidence converges on a reasonably coherent picture of past conditions. It must be stressed that certainty and precision are not always possible, and that numerical values of past climatic conditions are estimates which may have wide margins of error; in general the further back one attempts to trace the climatic record the greater these problems become.

The evolution of glacial and post-glacial climates

From the variety of evidence described above a reasonably clear picture has emerged of the climatic history of the British Isles over the last 100 000 years or so.

It is difficult to define the onset of the last (Devensian) ice age with any precision. By 82 000 BP (before present) the British Isles were relatively clear of ice. During the period 75–70 000 BP there was a rapid fall of temperature to near the lowest levels of the last Ice Age. Between 70 000 and 60 000 BP evidence from botanical indicators and oxygen isotope measurements strongly suggests that ice sheets were quickly established over north-west Europe. This relatively rapid build-up of ice may have been assisted by the process of instantaneous glacierisation whereby snow and ice fields coalesced over wide areas assisted by persistent blizzard conditions, but this is contentious (mainly because such an event has never occurred during the period of instrumental record).

It is thought that the prevailing atmospheric circulation during the onset of the ice age was highly meridional (Lamb and Woodroffe, 1970) rather than zonal, with a prevalence of northerly and southerly winds over Britain, associated with the upper cold trough extending further south than 'normal'. The main moisture supply came from the warmer sea waters to the south.

The last Devensian glaciation had two maxima: a secondary maximum around 55 000 BP followed by glacial retreat, and the main maximum around 18 000 BP. Between the climaxes of low temperature and ice cover in the early and late glacial there were thousands of years of variable, rather temperate, conditions, though generally it was colder than at present.

Two main interstadials are recognised in Britain, the Chelford interstadial (60 000 BP) and the Upton Warren interstadial (43 000 BP). The Chelford interstadial was a period when trees appeared on the landscape and summer temperatures are estimated to have been around 12 °C or even higher in July in Cheshire (similar to those of today), but winter temperatures remained low with January mean temperatures of -10 to -15 °C in Cheshire (West *et al.*, 1974). Coope (1977a) has suggested that mean summer temperatures of up to 14 °C occurred in lowland central Britain. The climate was of a continental character in this interstadial, the extensive ice sheet was not far away, and there was little circulation from the Atlantic over Europe. Investigations at the type site at Chelford, Cheshire (Simpson and West, 1958) showed that the vegetation was dominated by pine (*Pinus*) with some birch (*Betula*) and a little spruce (*Picea*); the spruce was making its last appearance as a native British tree. Following this interstadial, in fact, was a period with no forest flora in England for some 50 000 years, until the return of the predominantly birch forest at the time of the Allerød oscillation.

A period of relatively cold climatic regimes followed the Chelford interstadial until the Upton Warren interstadial (43 000 BP) when mean summer temperatures were perhaps even higher than those observed today. Coope has suggested that the mean July temperatures were 17–18 °C in lowland central Britain, while mean winter temperatures were still around 0 °C. The mean annual temperature was about 1–2 degrees C below those of the present; the climate was continental in character and similar to that occurring in central Europe today. The landscape was treeless, however, for conditions did not ameliorate long enough for forest vegetation to spread from southern Europe. The Coleoptera remains from the period show a total lack of species dependent on trees either as a source of food or as a habitat. Lowland precipitation at the time was probably in the range 450–650 mm per year (Lockwood, 1979).

The glacial maximum (18 000 BP)

There followed a cold period which culminated between 15 000 and 20 000 BP (the maximum of the Devensian Ice Age was reached in 18 000 BP) when the lowest temperatures of all occurred and the greatest extent of the ice sheets was attained. These dates have been obtained from radiocarbon dating of ocean bottom sediments (Emiliani, 1961) and analysis of vegetational remains near the limits of the advancing ice sheets (West, 1960). From the evidence of radiocarbon dating of deposits in kettle holes, North Wales at least was ice-free again by 14 500 BP.

The deterioration in climate following the thermal maximum of the Upton Warren interstadial was more gradual than the amelioration before, but none the less was a sudden event (Coope, 1977a). Shotton (1977a) supports the contention that the climate deteriorated very rapidly after the interstadial to become arctic and eventually to pass to polar desert conditions before the glacial maximum. It is evident that this later cold period was both drier and more continental than the earlier cold period.

Estimates of the prevailing temperatures in Britain at the glacial maximum vary. Lamb (1977) considers that the mean surface air temperatures in the unglaciated parts of Britain were probably 10–12 degrees C below those of today (only a few degrees less in summer but more in winter). Goudie (1977), however, suggests on the basis of snowline evidence that temperatures were only lowered by about 6 degrees C. Manley (1964) considers that mean July temperatures were about 8.5 °C in the south and south-west of Britain, with January mean temperatures of − 5 °C on the extreme south-west coast and − 12 °C inland. The general consensus is that average July temperatures in the unglaciated parts of Britain were about 8 °C at the glacial maximum, and that winter temperatures were very cold, with mean February temperatures below − 20 °C. There was a very rapid decrease of mean winter temperatures eastwards towards the continent. On the uplands and ice sheets of North-west England summer temperatures were at or below 0 °C, allowing an annual accumulation of snow.

It is clear that the unglaciated part of Britain provided a raw, cold and stormy sub-arctic margin facing the North Atlantic. But Lamb points out that occasional warm days would have occurred in summer when light southeasterly winds and strong sun would have brought temperatures up to 30 °C, as they do in Finnish Lapland today. He suggests that the standard deviation of the mean temperature for the same month or season in different years in Ice Age Britain was probably twice as great as it is today.

According to Lockwood (1979) the strength of the hydrological cycle generally decreases with falling temperatures (because of the relationship between the water vapour content of the air and temperature) and he suggests that it was reduced by 14 per cent during the maximum of the last Ice Age. Precipitation is generally more difficult to estimate than temperature, but the presence of ice wedges in Britain suggests a low winter snowfall when they were growing. He suggests that during the cold periods precipitation was between 260 and 370 mm per year, with winter precipitation much lower than present at 80–120 mm. The summer maximum of precipitation is caused by the development of instability and showers in cold arctic air as it moves southwards over the warmer land surface. Precipitation was at a minimum in April/May when the circulation was probably weakest.

Table 4.1a illustrates the components of the water balance at a site in lowland Britain at this time for conditions of both low and high summer precipitation (the two regimes are shown because of the uncertainty of summer precipitation – in analogous tundra environments today there is a large variation in summer precipitation) which can be compared with the water balance at a similar site today in Table 4.1b. Winter evaporation values are small because of the low temperatures and the high albedo of the snow. Most of the snow would melt in May, providing snow-free ground by June and during the summer the only snow remaining would be on the uplands and ice sheets. At the height of the glacial maximum North-west England was almost certainly ice-covered and conditions were much more severe.

Computer simulations by Manabe and Hahn (1979) of conditions at the glacial maximum endorse the conclusions regarding the summer maximum of precipitation. Analysis by Williams (1975) concludes that total precipitation was 50 per cent of present values in eastern England and one-third of present values in the Midlands. However, on the ice sheets of North-west England mean annual precipitation may not have been greatly different from that at present (Manley, 1975) while in the upland accumulation areas totals may have been 150 per cent of those at present.

At 18 000 BP there was greatest precipitation just to the west of the British Isles (Williams *et al.*, 1974), and a marked decrease eastwards towards the continent; this rainfall maximum coincided with a region of strong cyclonic activity in the region, but few depressions actually crossed into the continent.

Table 4.1 a. The water balance of a site in lowland Britain during the last glaciation for conditions of high and low summer precipitation (from Lockwood, 1979). (In rainfall equivalents in mm.)

Month	Precipitation	Snow storage	Direct evaporation of snow	Potential evaporation	Runoff	Soil moisture	Temperature (°C)
J	21	70	3			0	frozen	− 17 to − 23
F	14	81	3			0	frozen	− 17 to − 22
M	16	94	3			0	frozen	− 14 to − 20
A	13	102	5			0	frozen	− 7 to − 13
M	9–10			50	10–57	57–100	0 to − 1
J	13–40			65	0	5–75	5 to 7
J	39–60			70	0	0–65	8 to 12
A	46–70			55	0	0–80	8 to 10
S	26–35			20	0	6–95	4 to 6
O	27	22	5			0	frozen	− 2 to − 5
N	17	36	3			0	frozen	− 9 to − 14
D	19	52	3			0	frozen	− 15 to − 21
Total	260–367		25	210	10–57	Average	− 4 to − 7

Table 4.1 b. The water balance of a site in lowland Britain today (from Lockwood, 1979; and Smith, 1976). (In rainfall equivalents in mm.)

Month	Precipitation	Potential evaporation	Runoff	Soil moisture	Temperature (°C)
J	60	1	59	100	3
F	47	10	37	100	3
M	46	30	16	100	6
A	45	58	0	87	8
M	55	81	0	61	11
J	50	91	0	20	15
J	58	93	0	0	16
A	68	76	0	0	16
S	57	47	0	10	14
O	56	20	0	46	11
N	68	4	10	100	6
D	59	0	59	100	4
Total	669	511	181	Average	9

Consequent upon the reduced precipitation eastwards there was an eastwards rise in the snowline. The snowline was at 150–200 m in the Hebrides but rose to 500–550 m in the Cairngorms. The snowline in the southern Pennines was relatively high at 450 m during the glacial maximum; Kinderscout in Derbyshire at 635 m shows no sign of ice accumulations.

At the glacial maximum, sea level was reduced by some 85 m (Gates, 1976) because of the transfer of water from the oceans to the ice sheets; some authors have suggested lowerings of up to 130 m. Much of the present North Sea was therefore a land surface. Because of the reduced sea level, surface pressure was generally some 12.7 mb higher. Excluding the regions covered by sea ice, North Atlantic sea surface temperatures were reduced by 3.8 degrees C in August, but there was a marked steepening of the thermal gradients towards the ice margins, with a particularly steep gradient near 42 °N where the polar surface waters converged with the warmer circulations from the south.

The atmospheric circulation at the time of the glacial maximum is shown in Fig. 4.3. In the vicinity of the British Isles the mean circulation was much more meridional than at present with a polar anticyclone extending south towards the mid-Atlantic blocking the progression of most of the cyclonic systems from the southwest, many of which moved into the Mediterranean region. Over Britain there was a general easterly surface wind regime in winter, with lighter and more variable winds in summer (there was much less seasonal change in the vigour of the circulation than today). At both seasons northerly flow over the sea area between Greenland and northern and western Europe is indicated. There was also a strong component of windflow off the ice sheets at their margins.

Late glacial and post-glacial climatic change

Geological evidence shows that most of England and Wales was free of ice by 15 000 BP, and the last major ice sheet disappeared by 10 000 BP. Lockwood (1982) has argued that an increase of only 2–3 degrees C would be sufficient to cause melting of the ice sheets; but since the glaciers had already abandoned much of lowland Britain before the rise in temperature at the start of the next interstadial, it seems likely that precipitation deficiency also played a significant role in the waning of the ice sheets. Mean temperatures rose rapidly at the end of the last glaciation (15 000–8000 BP) but there were large-amplitude temperature fluctuations in the post-glacial period, each lasting for several centuries. These post-glacial climatic changes all lie within the range of behaviour shown by extreme years and extreme seasons today; the problem for the climatologist lies in explaining the persistence of the circulation type responsible.

Knowledge of post-glacial climatic change came first from pollen analysis (showing shifts in vegetational assemblages); this has since been amplified by foraminiferal analysis of cores from the ocean bed, oxygen isotope work on these cores, and other studies. The names of the various climatic periods so defined and their pollen zone numbers are listed in Table 4.2. The main temperature changes and shifts in the locations of the subpolar low-pressure belt and the subtropical

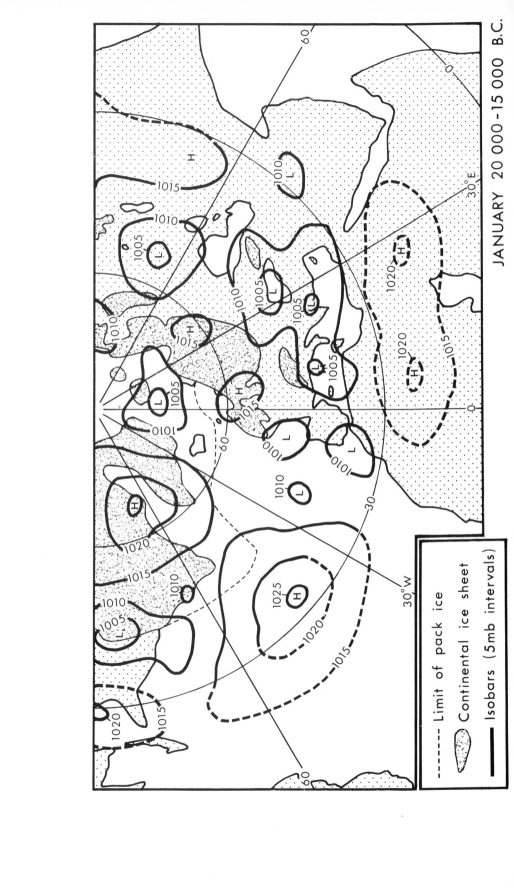

JANUARY 20 000-15 000 B.C.

----- Limit of pack ice

Continental ice sheet

—— Isobars (5mb intervals)

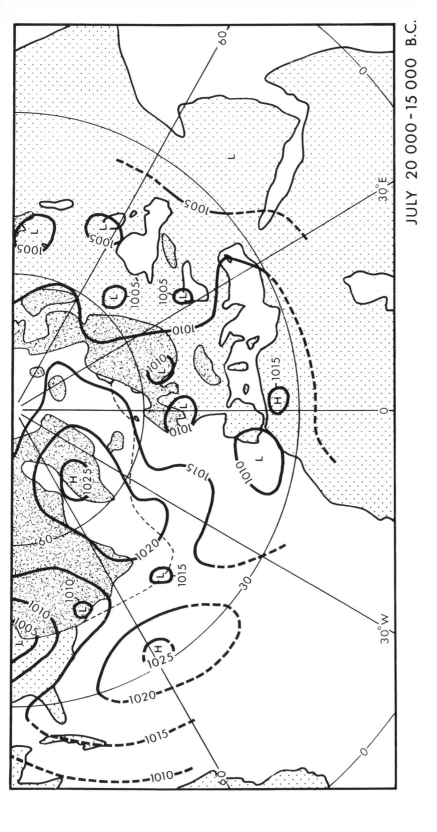

Fig. 4.3 Probable mean surface pressure distribution and implied prevailing surface winds at the glacial maximum (20 000–15 000 BC) in January and July (from Lamb and Woodroffe, 1970).

JULY 20 000–15 000 B.C.

high-pressure belt over the last 20 000 years are shown in Figs. 4.4 and 4.5.

In the earliest post-glacial times the atmospheric circulation exhibited great vigour. This was a consequence of the temperature gradients that developed between the ice-covered and ice-free areas. At all seasons there was a very marked steering of depressions from the south-west to the north-east which partially resulted from the extension of the anticyclonic area from the Azores–Biscay region to join with the continental anticyclone over north-east Europe.

Table 4.2   Climatic period and pollen zones in the late-glacial and post-glacial in the British Isles (from Lamb, 1977a; and Godwin, 1977).

	Period	Climatic Period	Zone Number	Dominant species	Radiocarbon years BP
F		Sub-Atlantic	VII	alder, oak, elm, birch, beech	post 2450
L	Post-				
A	Glacial	Sub-Boreal	VIIb	elm decline	2450–4450
N		Atlantic	VIIa	oak, elm, linden, alder	4450–7450
D					
R		Late Boreal	VI	pine, hazel	7450–8450
I		Early Boreal	V	pine	8450–9450
A		Pre-Boreal	IV	birch, pine	9450–10250
N					
D		Younger Dryas	III	tundra (local birch)	10250–11350
E	Late				
V	Glacial	Allerød/Windermere	II	birch	11350–12150
E		Lower Dryas	Ic	tundra	12150–12350
N		Bølling?	Ib	local birch	12350–12750
S					
I					
A	Full	Oldest Dryas	Ia		
N	Glacial				

The Oldest Dryas period marked the early stages of ice retreat. Landscapes were dominated by plants of the arctic tundra, with grasses on higher ground and sedges in wetter places. Temperatures continued to rise into the Bølling period and a park landscape developed with a few birch trees. Tree birch pollen frequencies reached their maximum frequency in the Windermere basin between 13 000 and 12 000 BP (Tooley, 1977).

During the Allerød/Windermere warm phase, the interstadial carried close birch woodland over a large part of Britain and tree birch grew as far north as central Scotland. The climate was warm with temperatures in England of 14–17 °C in summer. Mountain glaciers remained in the Scottish Highlands but they disappeared in the Lake District before re-establishing themselves later.

Lamb (1970) considers that the duration of the Allerød period of warmth may not have given sufficient time for the spread of botanic species to reach their geographical limits which the prevailing temperatures would have allowed. The warmth may therefore be underestimated, but nevertheless the epoch was the first to bring abundant woodland from the North European Plain to Britain. Evidence from Coleoptera remains confirms this (Coope, 1975), indicating the greater facility and speed with which insects respond to climatic fluctuations. At the time sea levels were still low and Britain was still connected to the continent.

The warming of the Allerød ended abruptly and between 10 700 and 10 100 BP there was a return to colder climates with prevailing summer temperatures in central England at least 4–5 degrees C below those of the previous millennium. Lamb considers that this was a climatic disaster of unparalleled abruptness accomplished within 50–80 years. Northern forests died and glaciers returned to the Lake District valleys, although they did not grow into major ice sheets. There was a lowering of the snowline here: at 485 m in the central Lakes where rainfall is highest, but it rose northwards and eastwards to 750 m in regions where rainfall is less today. The cooling led to a halt in the retreat of the Scandinavian ice sheet and the glaciers in Scotland (where the period is known as the Loch Lomond readvance stage (Sissons, 1974b)). Tundra again dominated the landscape.

The cause of this recession was a weakening of the upper circulation over the Atlantic and an increase in meridional flow. The prevailing surface winds and the

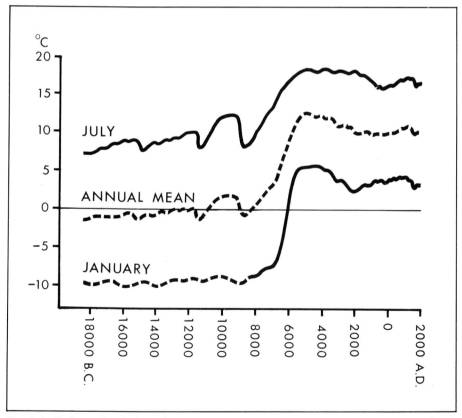

Fig. 4.4   Estimated variation of the mean January, July and annual temperatures in central England during the past 20 000 years (from Manley, 1964).

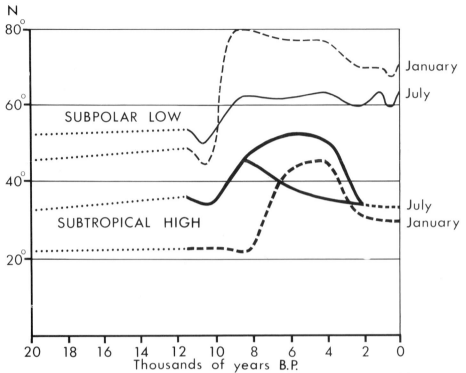

Fig. 4.5 Mean latitudes of the subpolar low-pressure belt and the subtropical high-pressure
   belt in longitudes 0–30 °E (British Isles and Europe) over the last 20 000 years (Lamb,
   1977b).

European anticylone were displaced westwards, which first introduced a phase of
prevailing southerly winds during the warm period, followed later by northerly
winds. There was a notable increase in pack ice south and west of Iceland during
the period of cooling.

   The end of the Allerød is interpreted in the Windermere basin as a period in
which the ecological balance was disturbed, caused by the fall in temperature and
increased soil erosion, presaging the climatic deterioration of the Younger Dryas.
During this final cold period involutions formed in Dalton-in-Furness (Johnson,
1975) and coversands accumulated in south-west Lancashire.

   A renewed rise in temperature set in at the start of the Pre-Boreal period leading
to the sustained warmer climates of post-glacial times. The birch became dominant,
with pine abundant in south-east Britain. Solifluction largely ceased at this time
and the grassy landscapes disappeared. The temperature rise continued during the
Early Boreal. At this time the winters became generally milder, although there
were still some dry and frosty periods. The summers are considered to be generally
warmer than those of today, and the climate tended to be more continental than
at present. The Late Boreal period was the last time at which the pine grew generally
in England on soils of all types; subsequent to this it only grew locally on poor soils.

   During the Atlantic period (7450–4450 BP) the elm (*Ulmus*) and oak (*Quercus*) spread further north, and warm wet conditions prevailed in Britain. The moister conditions were conducive to the growth of wet, peat-forming vegetation communities on flatter surfaces above 360 m, while grassland habitats were rare except above 900 m. Trees grew to about 300 m above their present level in Scotland.

The Atlantic period was the warmest period in Britain since the Ice Age. Annual temperatures were some 2–3 degrees C higher than they have been since, summer temperatures were generally 2 degrees C higher than present and winters 1 degree C warmer (see Fig. 4.4). Lamb (1977) estimates July temperatures at 6500 BP to be 17–18 °C, with mean winter temperatures around 5–6 °C. However this was a wet period in which lowland Britain received some 10 per cent more rainfall than at present, and upland Britain some 20 per cent more. At this time mountain glaciers had disappeared, the arctic sea ice had almost disappeared, and sea levels rose as the ice melted. The North Sea and Baltic Sea filled up, the Straits of Dover were formed and Britain became an island.

The atmospheric circulation of the Atlantic period was generally less vigorous than at present, with the westerlies further north over Europe than at present, and the subtropical anticyclone also further north, as shown in Fig. 4.5.

The Sub-Boreal period was generally cooler (by 0.5 degree C) and drier in the winter months, but warmer (by 0.5–1 degree C) and drier in the summer. During this period the atmospheric circulation had reached its weakest post-glacial phase. Sea ice reached its minimum development between 3950 and 3450 BP. The alder (*Alnus*), birch and pine became widespread in the drier areas.

The Sub-Atlantic period however marked the beginning of a decline from the post-glacial climatic optimum from which there has never been a full recovery. During the period 700–400 BC in particular there were frequent storms and floods associated with a predominantly cyclonic, zonal circulation regime. A maritime climate developed, characterised by abnormal wetness, mild winters and cool, damp and overcast summers. By 500 BC temperatures were generally 1 degree C less than those of today, and there was renewed peat bog growth especially in upland Britain. The birch increased in the lowlands, with oak and elm on the wetter soils. The general treeline fell by some 300 m from its level at the climatic optimum. There is archaeological evidence of a marked impoverishment and decline of the upland Pennine population, as opposed to the situation in the Sub-Boreal (Bronze Age) when upland settlement was widespread. Pack ice re-formed in the Arctic Ocean.

The Secondary Climatic Optimum 400–1200 AD

This period of warmth in Europe and the British Isles reached a maximum between about 1000 and 1200 AD (see Fig. 4.6a). It was a time of dry, warm and relatively stormfree conditions in the eastern Atlantic and North Sea areas. In general, summers were warm, dry (some 10 per cent drier than today as shown in Fig. 4.6b) and anticyclonic, while the winters may have been cold (because of the anticyclonic tendency), but from historical accounts were seldom severe.

The warming was associated with a poleward shift in the mid-latitude westerlies and associated depressions by some 3–5 °N in summer relative to the long-term

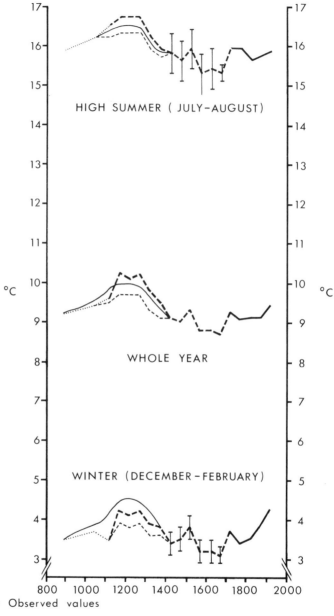

HIGH SUMMER (JULY–AUGUST)

WHOLE YEAR

WINTER (DECEMBER–FEBRUARY)

——— Observed values

——— Unadjusted values based on purely meteorological evidence

- - - Preferred values including temperatures adjusted to fit
botanical indications

··········· Connects points corresponding to 100-200 year means indicated
by sparse data

——— Analyst's provisional recommendation

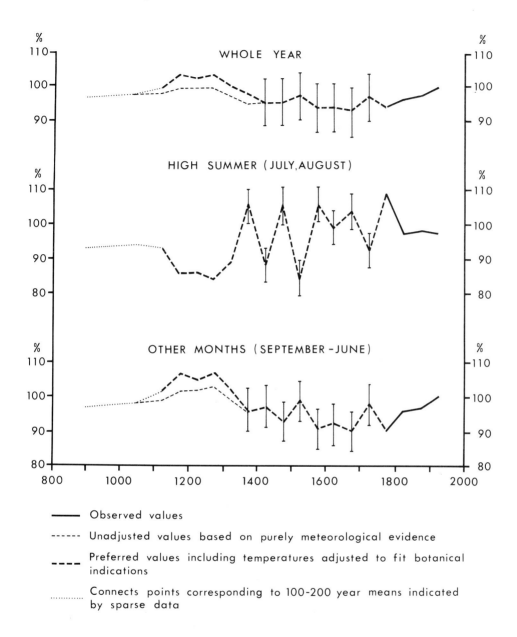

%
110
100
90

WHOLE YEAR

%
110
100
90

%
110
100
90
80

HIGH SUMMER (JULY, AUGUST)

%
110
100
90
80

%
110
100
90
80

OTHER MONTHS (SEPTEMBER – JUNE)

%
110
100
90
80

800   1000   1200   1400   1600   1800   2000

———— Observed values

------ Unadjusted values based on purely meteorological evidence

- - - Preferred values including temperatures adjusted to fit botanical indications

.... Connects points corresponding to 100-200 year means indicated by sparse data

Fig. 4.6a (opposite)   Central England temperatures since 900 AD (50-year averages since 1100 AD and longer-term averages before that).

Fig. 4.6b (below)   Mean rainfall over England and Wales (expressed as a percentage of the 1916–50 mean) (from Lamb, 1965).

mean; in winter the northward shift of the main depression track may have taken it into the Barents Sea region, allowing more anticyclonic winters in temperate Europe. The arctic pack ice melted so far back in this period that drift ice in the waters near Iceland and Greenland south of 70 °N was rare in the period 800–900 AD and unknown between 1020 and 1194 AD.

On land, the period from 750 to 1200 AD marked glacial retreat, which was slightly more marked than that of the first half of this century. Trees grew on sites where today they have either not had the time or the necessary conditions to grow again. In the Domesday Book of 1086 some 38 vineyards were recorded in England; vines were grown as far north as Herefordshire and Gloucestershire and there was one vineyard in the vicinity of York (although none was recorded in North-west England). Decline set in around 1310 (Lamb, 1977), but at their peak the vines were being grown some 500 km further north than the modern limit of commercial vineyards on the continent. This implies that: summer temperatures were some 1–2 degrees C higher than today; there were no late spring frosts; there was sufficient summer and autumn sunshine and warmth to raise the sugar content; and there was not too much rain. During this period there is also evidence of mediaeval cultivation in Britain at altitudes far higher than would be reasonable now (or even during wartime). The thirteenth century limit of tillage in Northumberland was around 300–350 m above sea level, which is 120–150 m higher than today's limit (Parry, 1975).

Climatic decline 1200–1400 AD

Worldwide cooling set in from 1200 AD onwards accompanied by a particularly stormy period in western Europe which brought floods and changes around the low-lying coasts of Britain. World sea level was at its highest this time, following the earlier climatic optimum. Generally this was a time of great climatic instability: floods and droughts occurred and there were both severe and mild winters.

The storms and floods were particularly frequent after 1250 AD, and there was an extraordinary sequence of wet summers between 1310 and 1319. Trevelyan (1928) believes that English rivers were deeper and more navigable in the fourteenth and fifteenth centuries than they are now – perhaps as the result of this increased summer rainfall?

The cause of this climatic deterioration was the migration of the main Atlantic depression track southwards at a period when there was still much heat stored in the ocean; the strong thermal gradients over the oceans and the more intense polar anticyclones of the time invigorated the westerly flow and the associated depressions.

The Little Ice Age 1430–1850 AD

Historians differ about the precise dating of the so-called Little Ice Age because the long cold period was punctuated with warmer decades and the effects differed from place to place. In general this was a period of widespread cooling in which

European glaciers reached their most advanced positions since the glacial maximum. Arctic pack ice underwent a great expansion, especially affecting Iceland and Greenland. In England in the 1430s there was a run of very severe winters in which most of the major rivers froze over, as happened again between 1565 and 1709. The frost fairs on the Thames were begun in the sixteenth century. Snowbeds became common on north-facing slopes in Scotland, and in 1769 Thomas Pennant noted that the highest Cairngorm summits had perpetual snow.

In general, conditions were harsher than those existing this century, but there were also periods of more equable climate (especialy in the 1630s, 1730s, 1770s and 1840s), and Lamb (1977b) suggests that there was greater variability of temperatures from year to year than today. Overall, mean annual temperatures were about 1 degree C below those of this century, and the ocean temperatures were lowest between 1600 and 1800.

The wet summers and very cold winters caused severe problems for the upland farmers of the Pennines and Scotland. Heavy losses of livestock occurred in the winter snows and cereals failed to ripen in the cool, cloudy summers; the 1590s, 1690s and 1780s were particularly hard times. In a study of the cultivation of oats in the Lammermuir Hills of south-east Scotland, Parry (1975) has suggested that the deterioration of climate since the early Middle Ages caused much of the high-lying cultivation to become very sub-marginal in the seventeenth century. Around 1150–1250 AD the frequency of failure of the oat crop was less than 1 year in 20, by 1350 this had increased to 1 year in 5, by 1450 to 1 year in 3, and by 1700 to more than 1 year in 2. Consecutive harvest failures proved the ultimate stimulus to farm abandonment and some 4890 ha of cultivated land were eventually abandoned at levels up to 400 m above sea level.

Lamb and Johnson (1959) have analysed the atmospheric circulation regimes during the Little Ice Age and suggest that there was a highly abnormal frequency of blocking anticyclones in the period from the late 1600s to about 1712 AD; these produced more than the normal frequency of northerly winds over Britain. The circulation was much more meridional than at present, and the available evidence suggests that either the summer circulation was weakened by about 30 per cent or it shifted south by 5 ° latitude, with the main depression track at 57–60 °N. In winter the flow was weaker by some 5–10 per cent at the height of the Little Ice Age, with a southward shift of some 3–5 ° latitude; the main depression track occurred around 64 °N, with frequent blocking, and many more depressions than now passed into the Mediterranean region.

Lamb suggests that there was exceptional volcanic activity in the period 1550–1831 (tailing off thereafter until 1912), which maintained frequent dust veils in the stratosphere, contributing to a reduction of 1–2 per cent in the available insolation, and may well have been a possible cause of the circulation change and reduced surface temperatures.

The period of instrumental record

The barometer was invented in 1643 and the thermometer at about the same time.

A rain gauge was set up (with a roof-top exposure) in Burnley, Lancashire as early as 1677, and the record was maintained until 1704 (Lamb, 1977a). However, problems exist in the use of early data from non-standard instruments, sites and exposures when they are to be compared with modern observations. By the eighteenth century the instruments were generally sufficiently reliable and reasonably exposed for monthly values to be useful.

Using a composite series of observations from a number of sites in central England (which includes sites in Lancashire), Manley (1974) has produced a valuable series of monthly mean temperatures for the period since 1659. Annual figures from this series, together with earlier values estimated by Lamb (1965) for the period back to 900 AD have been included in Fig. 4.6a. Craddock (1976) has produced a rainfall series for England back to 1725 using a composite record from a number of sites, and this is shown together with estimates of earlier figures by Lamb back to 9000 AD in Fig. 4.6b.

Figure 4.7 shows annual rainfall in Manchester since 1765 based on data from Manley (1971). It shows the values for individual years, together with decadal means. Barrett (1964) has analysed local variations in rainfall trends in the Manchester region using data for the period 1890–1959, but visual inspection of Fig. 4.7 for a longer period reveals no obvious long-term trend.

Few major climatic changes have occurred during the period of instrumental records although there have been short-term fluctuations. It can be seen from Fig. 4.6a that the last 100 years has been a period of relative warmth, particularly in the winter months. Over the Northern Hemisphere the mean air temperature rose 0.6 degree C between 1890 and 1940. Glaciers receded in Europe and other parts of the world, while snowbeds tended to disappear from the Scottish Highlands. The first half of the twentieth century was marked by a dominance of westerly winds and a mild oceanic climate with increased rainfall (Fig. 4.6b). It was suggested that increased emissions of carbon dioxide from the burning of fossil fuels by domestic and industrial consumers may have had a role in this through the 'greenhouse effect', but such influences have been masked by other influences producing a slight decrease in temperature since 1950.

There has been a return to more variable atmospheric circulation patterns since 1950, marked by much more prominent meridional surface winds. This has produced a renewed increase in arctic ice in recent years and much greater variability in the weather occurring from one year (or run of years) to another. There has been a remarkable incidence in the last twenty years or so of months and seasons of extreme cold or warmth, or of extreme wetness or dryness, but apart from the 1975–6 drought, such variability is not unusual (Ratcliffe, 1981).

Perhaps the most significant change in the climate of North-west England and other parts of the country in the last thirty years has been the improvement in the quality of the air, following the implementation of Clean Air Act legislation. In Manchester there has been an overall 91 per cent reduction in smoke concentrations, and a 77 per cent reduction in sulphur dioxide levels, since measurements began. As a consequence there has been a sharp reduction in the incidence of fog. The number of days with fog at Manchester Airport was reduced by 54.7 per cent

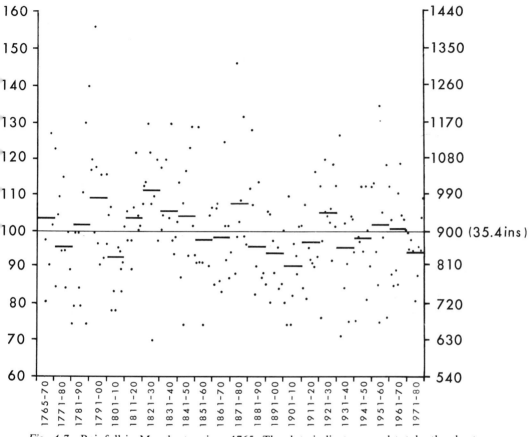

Fig. 4.7 Rainfall in Manchester since 1765. The dots indicate annual totals, the short horizontal bars indicate decadal means, and the horizontal line indicates the overall mean (100% on the vertical scale).

between the 1950s and the 1970s, thick fog was reduced by 65.5 per cent and dense fog reduced by 100 per cent, i.e. a dense fog has not been recorded since 1970. Over North-west England as a whole the number of days with fog has been reduced by 49.7 per cent between the 1950s and the 1970s.

5   *B. S. Kear*

Soil development and soil patterns in North-west England

Introduction

The region is dominated by fine and medium textured soils derived from a wide range of Pleistocene drift deposits (Fig. 5.1). The effects of the Pleistocene glaciations and intense Devensian periglacial activity virtually obliterated any Tertiary or Inter-glacial soil profiles.

With the return of more equable climatic conditions in the early post-glacial soil formation recommenced on surfaces being stabilised by newly colonising vegetation. By the end of Boreal times a deciduous forest covered the region save on the highest and steepest rocky faces up to almost 800 m. Into this forested landscape, especially the less dense woodlands on the upland plateau (350–500 m), came Mesolithic man. His burning of forests on the plateau and the increasing oceanicity in Atlantic times paved the way for the widespread development of peat on flat and gently sloping surfaces above 300 m. Changes in soil hydrological conditions as forest clearances extended down the adjacent footslopes and on to the lowlands have determined the evolution of soil patterns evident in the modern landscape.

The classification of the major soils in North-west England

Soil sequences in which the intensity and extent of gleying in the soil profile are determined by topographic position and site drainage make up the range of soils found in any *one* soil association. Some soil associations, however, have no gleyed soils in them, others are dominantly gleyed soils. Each soil association, however, is related to a particular parent material and the dominant or most commonly occurring soil (*soil series*) in the association lends its geographical name to the associations shown on the recently published (1:250 000) National Soil Maps.[1]

In view of the wide distribution of drift deposits throughout the region and their importance as soil-forming materials it is appropriate to consider their properties and the soil associations developed on them at the outset of this paper.

The surficial deposits

The Northern Drift

An extensive till deposit blanketing lowland areas of North-west England from the west coast reaches inland up to 200 m in places on the Pennines. This reddish-

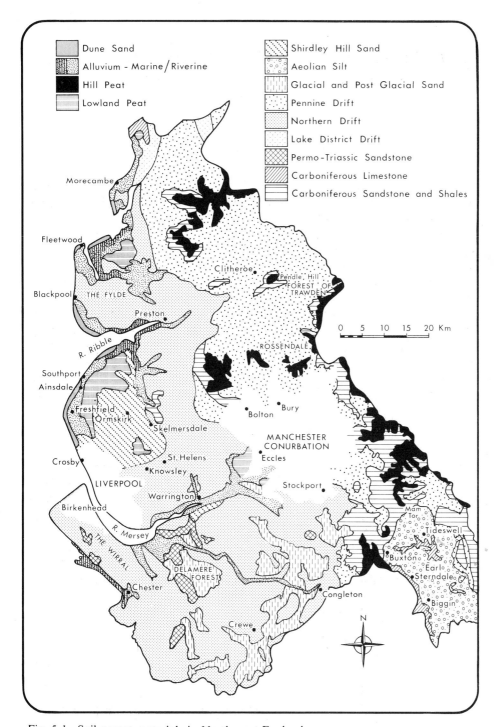

Fig. 5.1 Soil parent materials in North-west England.

brown, slightly stony and slightly calcareous Northern Drift, deposited by Irish Sea Ice, is said to derive its reddish colour from Triassic materials which floored the Irish Sea Basin. Shell fragments and secondary carbonate deposition occur within the till. The slightly stony soils developed on this till, named the *Salop association*, have dark greyish-brown clay loam topsoils over reddish-brown clay subsoils. They are slowly permeable, with seasonal waterlogging reflected in the distinct large grey mottles in the subsoil, and mottling patterns throughout the profile. Occurring mainly on flat land or in localised depressions in the till surface, they receive surface runoff from surrounding areas (Fig. 5.3). Ideally suited to grass, their widespread distribution on the Cheshire Plain has facilitated the development of the dairy farming industry (Furness, 1978). Where drainage is adequate in the relatively drier parts of the region, these naturally fertile soils support a wide range of crops. This soil association extends into Staffordshire, and northwards onto the Lancashire coastal plain and the Fylde. In Lancashire, where fine loamy topsoils over clayey subsoils have developed on gentle to moderately sloping land, seasonal waterlogging is less pronounced and the *Flint association* is recognised.

Three other soil associations are recognised in the areas which are covered by Northern drifts (Fig. 5.1). They are the *Clifton, Salwick* and *Crewe* associations and all show some degree of gleying in the soil profiles. In Cheshire, the loamy Clifton soils occur in areas underlain by, or adjacent to red-Triassic sandstone outcrops, as in the Wirral peninsula and along the mid-Cheshire ridge. They are surface-water gleys, having rusty, mottled, sandy-clay loam topsoils, dark, greyish-brown in colour passing down into a strongly mottled, reddish-brown, sandy-clay loam. Where a combination of underlying sand or sandstone occurs with gentle or moderate slopes then seasonal waterlogging is less pronounced and brown soils become dominant. In such conditions the Salwick soils replace those in the Clifton association. These Salwick soils have dark brown, slightly stony, sandy-clay loam topsoils which merge to slightly mottled, brown sub-surface horizons over mottled reddish-brown, sandy-clay loam subsoils.

The intensely gleyed soils of the Crewe association are extremely localised and appear to have formed only where pro-glacial lake sediments were deposited (Furness, 1978). The soils are formed on stone-free, finely laminated, glacio-lacustrine clays which form a number of very flat terrains near Crewe. The soils have a high clay content, a uniform texture and are highly impermeable.

The Pennine Drift

This is made up of medium and fine textured till and head materials derived from the local Millstone Grit Series and Coal Measures. Dominated by grits, sandstones and shales, the drift is found mainly on gentle to moderate hill slopes ($< 10°$) in the Pennine foothills and the uplands of the Forest of Bowland, Pendle, Trawden and Rossendale. (Fig. 5.1). The Pennine Drift usually succeeds the Northern Drift above altitudes of 60 to 90 m.

The *Brickfield association*, recognised up to about 230 m, is dominated by soils

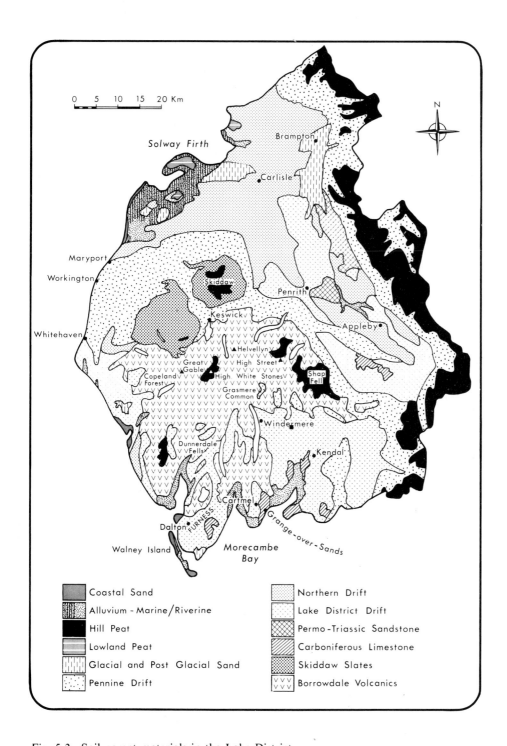

Fig. 5.2 Soil parent materials in the Lake District.

The following labels appear on the map:

Solway Firth

Brampton

Carlisle

Maryport

Workington

Penrith

Whitehaven

Keswick

Appleby

Helvellyn

Great Gable

High Street

Copeland Forest

High White Stones

Shap Fell

Grasmere Common

Windermere

Dunnerdale Fells

Kendal

Cartmel

Dalton

FURNESS

Grange-over-Sands

Walney Island

Morecambe Bay

0   5   10   15   20 Km

N

Legend:

- Coastal Sand
- Alluvium – Marine/Riverine
- Hill Peat
- Lowland Peat
- Glacial and Post Glacial Sand
- Pennine Drift
- Northern Drift
- Lake District Drift
- Permo-Triassic Sandstone
- Carboniferous Limestone
- Skiddaw Slates
- Borrowdale Volcanics

which are slowly permeable, seasonally waterlogged fine loamy textures. These surface-water gleys commonly have greyish-brown to dark-grey, mottled sandy-clay loam topsoils, merging to grey, prominently mottled sandy-clay loam sub-soils. Moderate rainfall (1000–1015 mm) is augmented by runoff from adjacent slopes and intensifies the surface wetness problem. Tile drainage, subsoiling and liming are all essential if good pasture and high crop yields are to be achieved on these strongly leached and gleyed profiles.

Above about 230 m cooler temperatures and higher rainfall (1020–1520 mm) lead to the formation of acid root mats or peat accumulation, particularly on flattish and gently sloping surfaces (Fig. 5.4). The upland soils formed under these conditions, termed the *Wilcocks association*, are slowly permeable, seasonally waterlogged peaty surface-water gleys with very dark grey clay loam topsoils over grey, distinctly mottled clay loam or clayey subsoils. These wet and acid peaty soils are colonised by moorland plants including mat grass (*Nardus stricta*), purple moor grass (*Molinia caerulea*) and cotton sedge (*Eriophorum vaginatum*). Peat becomes extensive with increasing elevation and organic soils of the *Winter Hill association* occur. Depending on local site conditions, aspect and rainfall, the change takes place between 300 and 550 m.

The Lake District Drift

This is made up of medium to coarse textured, very stony till and head derived from Lake District sources, particularly slaty mudstone of the Silurian System and

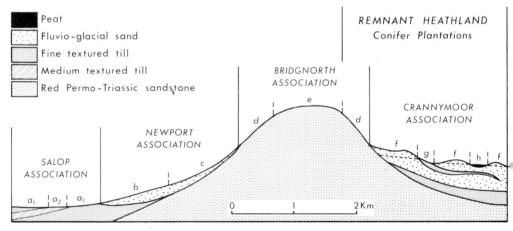

a₁a₂ Argillic stagnogley (SALOP and CLIFTON)

b. Typical sandy gley (BLACKWOOD)

c, e. Typical brown sand and Typical brown earths (NEWPORT and BRIDGNORTH)

Names in brackets represent soil series

d, f. Humo-ferric podzols (DELAMERE and CRANNYMOOR)

g. Typical gley podzols and Typical sandy gleys (SOLLOM and BLACKWOOD)

h. Oligo-fibrous earthy peat soils (TURBARY MOOR)

Source : Adapted from Furness, 1978

Fig. 5.3   Soil associations on Triassic Sandstone and fluvio-glacial materials in Cheshire.

from the Borrowdale Volcanic Series. It occurs widely in local valleys and on the Furness and Cartmel peninsulas, extending southwards into the Fylde where it becomes mixed with the Northern Drift. A strongly undulating relief with north–south trending drumlins or hillocks is fairly typical, with enclosed hollows filled with alluvium and peat. The *Denbigh association* dominated by freely drained brown earths with fine loamy and fine silty soils is recognised. They occur on the tops and sides of drumlins and have dark brown, stony, silty clay loam topsoils over very stony, strong brown, clay loam subsoils. Gleys and peaty gleys, affected by ground water tables occur in the hollows.

In a limited area around Dalton-in-Furness a distinctive medium to coarse textured, very stony reddish drift occurs. Of local derivation it includes haematite-enriched material derived from the underlying limestone. Free draining brown earths with distinctive red coloration throughout the profile dominate this, the *Newbiggin association*. Limestone underlies the solum at about one metre though shallower stony soils or bare limestone outcrops do occur.

Fluvio-glacial sands and gravel

These deposits, although less extensive than the various tills provide a strong contrast in the conditions for soil formation. Coarse textured and highly siliceous, they generally occur as undulating hummocky relief, but they also occur on the sides of valleys such as the Douglas, Irwell and Ribble. Widely distributed between Brampton in the extreme north to the Staffordshire border in the south they provide free draining profiles in which leaching processes are moderately intense. Two relatively extensive soil associations have been recognised in this region. First, the *Newport association* dominated by deep free-draining brown sand soils with either sandy or coarse loamy textures. Dark brown sandy loam topsoils usually overlie reddish-brown weakly structured subsoils merging downwards into a coarse structureless sand. Locally, groundwater influences create sandy gleys.

In areas with a previous history of heathland vegetation, particularly heather (*Calluna vulgaris*) and bilberry (*Vaccinium myrtillus*), humo-ferric podzols which dominate the *Crannymoor association* are recognised. They are most extensive in Delamere Forest with limited areas to the north-west of Congleton. These very acid well-drained sandy soils have bleached sub-surface horizons underlain by humus and iron-enriched zones. Hydrological sequences including humus podzols and gley podzols occur on the lower slopes of the hummocky relief in Delamere Forest (Fig. 5.4). Enclosed hollows filled with deep acid peats have posed severe problems for the Forestry Commission in parts of the Forest.

The Shirdley Hill Sands

This deposit, which blankets more than 200 km^2 of the south-west Lancashire coastal plain to a depth of one to three metres has been the subject of considerable controversy (Gresswell, 1958b; Kear, 1968). Godwin (1959) was the first to demonstrate a Late-Glacial age for the sand from pollen studies on peat layers

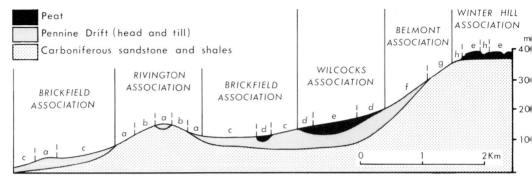

a. Stagnogleyic brown earths

b. Typical brown earths (*RIVINGTON*)

c. Typical stagnogley soils (*BRICKFIELD*)

d. Stagnohumic gley soils (*WILCOCKS*)

e. Raw oligo-fibrous peat soils (*WINTER HILL*)

f. Ironpan stagnopodzols (*BELMONT*)

g. Podzols, bare rock and humic rankers (*ANGLEZARK* and *REVIDGE*)

h. Truncated ironpan stagnopodzols

Names in brackets represent soil series

Source: Hall and Folland, 1970.

Fig. 5.4   Soil associations on Carboniferous materials and derived drift in the Pennines.

below and within it. This was subsequently confirmed by pollen and carbon-14 dates (Tooley and Kear, 1977). In more recent sedimentological and mineralogical studies Wilson *et al.* (1981) have established the provenance of the sand from fluvio-glacial deposits by wind and water sorting. Considerable post-depositional changes in the heavy mineral content have resulted from weathering of the sand.

The sand overlies the fine textured reddish-brown till of the Northern Drift on flat to gently undulating surfaces. Despite its permeability, the underlying till impedes the downward movement of drainage water so that gley podzols dominate the *Sollom association* developed on the sands. Regionally perched water tables above the till resulted in the widespread development of surface peat in Sub-Atlantic times. Artificial drainage, peat wastage and cultivation have incorporated the shallow peats with the sand. These inherently infertile, very acid soils were reclaimed in the nineteenth century by nightsoiling, marling, liming and drainage so that they have been transformed into highly productive arable soils. The thick bleached eluviated and gleyed horizons in the gley podzols also provided sand for the development of the glass industry based on St. Helens. Topsoils were scraped away and stored in temporary mounds until the bleached sand had been removed. A network of tile drains were then installed above the till prior to the restoration of topsoil to the site. The fields were then returned to arable use within a very short time.

Alluvium and lowland peats

The sequence of recent deposits along the north-west coast reflects changes in sea level at the end of the Last Glaciation. Raised beaches, bordering the estuaries of the Leven valley, the Duddon estuary, the Fylde coast and south of the Ribble estuary are composed of Downholland silts, and this calcareous marine alluvium (*Wisbech association*) is distinguished from the alluvial silts of floodplains, valley floors and infilled lakes throughout the region, (*Enborne association*). The latter includes ground-water gleys which are subject to periodic flooding and water table fluctuations as well as the more freely draining brown earths on terraces along the Ribble and Lune valleys.

The raised beaches were once colonised by deciduous forest which was subsequently replaced by varying thicknesses of basin peat. The peat overlapped onto the adjoining till plains of the Lancashire lowlands and the ensuing development of raised peat bogs created localised 'mosses'. Two distinct stages have been recognised in their development (Hall and Folland, 1970) influencing the nature of the organic soils produced. The initial coastal and low moor stage was entirely dependent on the presence of high water tables and favoured the growth of reed-swamp (*Phragmites communis*). Natural succession often replaced this by fen and fen carr peat as the swamps dried out. Their location, mainly near the Lancashire coast, and marginally above sea level, made pumping essential in their reclamation. These eutrophic peats have been mapped as the *Altcar association* and it overlies a variety of mineral deposits, both in lowland Lancashire and around the Lakes in Furness.

The second stage in the development of deep peats on the plain is the raised moss stage and is climatically controlled. Dependent solely on plentiful rainfall, these oligotrophic acid peats are made up mainly of bog moss (*Sphagnum*), cotton sedge (*Eriophorum*) and heather (*Calluna vulgaris*). In some areas they overlie the fen carr peats, but they also occur widely on the till plains and are mapped as the *Turbary Moor association*. Large areas of mossland have been reclaimed and now support highly productive arable farming. The best examples are between Ormskirk, Upholland and Knowsley and the well known Chat Moss between Warrington and Eccles. In this area they overlie coarse textured fluvio-glacial sand and gravel of the Mersey terraces.

Recent blown sand

To complete the pattern of recent deposits are the youthful profiles developing on sand dunes along the coast. These are mapped in the *Sandwich association*. They occur north of the Esk estuary, on Walney Island, east of Fleetwood, and most extensively between Southport and Crosby. At Ainsdale, a complex of stable and unstable dunes form a soil chronosequence culminating inland in podzols under sandy heath at Freshfield. High water tables affect the development of soils in the intervening slacks. Where the dunes have been afforested with conifers, incipient podzolisation is clearly evident.

The solid rock as soil-forming material in North-west England

The course of soil formation in upland areas has been largely determined by the impact of altitude on climate, both past and present (Fig. 5.5). Summers were much

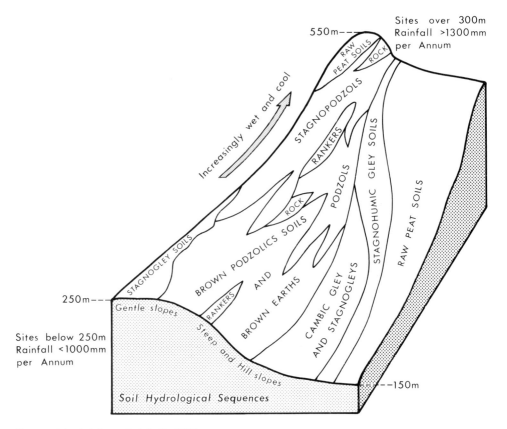

Source : Adapted from Rudeforth, 1970

Fig. 5.5   Soil patterns as related to climate and relief in upland areas of North-west England.

warmer and winters were drier in Boreal and Sub-Boreal times so that the treeline was higher than today. The deterioration in climate in Sub-Atlantic times resulted in lower temperatures with increased incidence of frost and much wetter conditions then hitherto. This intensified leaching of free draining soils and caused water-logging or peat development on slowly permeable materials, especially on flat or gently undulating ground. The continuation of oceanic conditions up to the present day has been accompanied by the progressive clearance of forest which has modified the hydrological cycle and sustained these processes. Freeze–thaw cycles are intensified on denuded soils, whilst in summer these same soils are vulnerable to desiccation and wind erosion, especially on exposed slopes.

The parent materials, derived from hard rocks in upland regions of the north-west, belong mainly to the resistant varieties of the Palaeozoic and Mesozoic. A combination of moderately coarse texture materials developing on steep slopes and sustained and heavy rainfall for much of the year has produced intensely leached soils of low base status. Two main areas concern us in this region: the Lake District and the Pennines (Figs. 5.1 and 5.2).

The Lake District

(i) *Skiddaw Slates* This upstanding area has the oldest rocks in the region, including Skiddaw Slates and Borrowdale Volcanics of the Ordovician and the massive limestone of the Carboniferous. The grits, sandstone flags, shales and mudstones of the former give rise to steep and very steep slopes as typified by the exposed bare landscape of Skiddaw Forest. Soils are very shallow with very acid, often peaty topsoils, resting directly on solid rock. These have been mapped as the *Skiddaw association*. Where the peat thickens to 40 cm organic soils of the *Winter Hill association* are recognised. They tend to blanket all but the steepest surfaces above 460 m. The steep and moderately steep slopes below 300 m are mantled with deeper free-draining fine loamy or fine silty brown podzolic soils of the *Manod association*. This mantle is broken by patches of shallower soils with local outcrops of solid rock (Fig. 5.2).

(ii) *Borrowdale Volcanics* The soils developed on igneous rocks, mainly andesites and rhyolites of the Borrowdale Volcanics, cover a much more extensive area, south and south-west of Skiddaw Forest. A bleak landscape of bare, long steep slopes, and craggy fells extends between High Street and Helvellyn in the east through Grasmere Common, Kirkfell and Copeland Forest in the west, to Dunnerdale Fells in the south. These resistant hard rocks, have yielded shallow mineral soils usually covered with a very acid peaty horizon. They are mapped as the *Bangor association* and give way to limited areas of acid peats on gentler slopes above 300 m as in Copeland Forest and on steep slopes above 460 m around High White Stones. The largest areas of peat are found further east on Shap Fells. Deeper freely draining, very stony loamy soils of the *Malvern association* occupy the steep bouldery slopes below 300 m. These brown podzolic soils are developed on the footslopes of the glaciated valleys where they are interspersed with local rock outcrops and screes.

(iii) *Silurian Shales* The soils of the southern Lake District uplands are developed either on softer shales, mudstones, siltstones and flags of the Silurian System or on glacial drift derived from it. A distinction is made between the *Denbigh association* on the former with predominantly free-draining fine loamy and fine silty brown earths on hillslopes, and the poorly drained peaty, surface water gleys of the *Wilcocks association* over derived drift.

(iv) *Carboniferous Limestone* The massive limestone facies of the Carboniferous Series is limited to a few outcrops west of Kendal on Whitbarrow and around Grange-over-Sands and Urswick. Shallow, freely drained loamy soils over hard limestone dominate the *Crwbin association* which has been mapped on a landscape of moderate and steep slopes. Rendzinas, with dark coloured topsoils, occur intermittently with rock outcrops or exposures of limestone pavement, as above Whitbarrow Scar. They occur commonly with shallow humic rankers where well humified organic matter has accumulated on the hard limestone under heath or represent shallow peat remnants. Deeper free-draining brown soils with slightly acid topsoils and distinct yellowish-brown subsoils form part of this association.

The Pennines

(i) *Carboniferous Limestone*   The Carboniferous Limestone outcrops more extensively in this upland region. Very shallow loamy humose soils over limestones, sometimes calcareous, have been mapped as the *Wetton association*. They outcrop adjacent to the Craven Fault between Ingleton and Grassington, but are found also in the southern Pennines in Derbyshire. Here, they are located mainly on the steep sides of the Manifold, Dove and Wye valleys and the network of dry valleys incised into the limestone plateau. Bare rock and screes also feature along the steep dale sides.

The southern Pennines were little affected by the glaciers and ice sheets of the last glaciation but there was intense periglacial activity in the Late-Glacial which initiated the build-up of solifluction deposits in many dry valleys. More importantly, there was a considerable accumulation of sub-aerial deposits on the plateau surface, varying in thickness from one to two metres. This cover of aeolian silt and the absence of glacial scour limits the exposure of limestone pavement to one or two isolated areas, notably Biggin Dale and Earl Sterndale near Buxton in Derbyshire. Pigott (1962) and Sweeting (1966) agree that the lowering of the limestone plateau surface since the Late-Glacial could have produced little more than 0.9 cm of residual clay. The abrupt textural change from silt to clay close to the limestone surface, accompanied by a change in the heavy mineral suite, supports an extraneous origin for the silt. Free-draining soils of the *Malham association* predominate on the plateau, and are described in the detailed soil survey of the Tideswell area of Derbyshire (Johnson, 1971). This non-calcareous, stone-free silt has brown earths developed in it, though some of the higher areas which are very cherty have not been reclaimed and so have thin ironpan stagnopodzols persisting under limestone heathland. These are relicts of a once extensive heathland that covered large tracts of the limestone plateau from late Sub-Boreal times until the major phase of Parliamentary Enclosure between 1760 and 1820 (Anderson and Shimwell, 1981). The sides of the plateau have complexes of rendzinas, non-calcareous humic rankers, deep brown soils, and screes.

(ii) *The Millstone Grit and Coal Measures*   The Millstone Grit Series of the Carboniferous Formation is made up of mudstones, shales and the more resistant sandstones and grits. The latter produce steep craggy outcrops or 'edges' as well as upstanding 'tors' which surmount some plateau tops. They are covered with bleak moorland fells, extending almost unbroken from the Scottish border in a SSE direction to the Dark Peak area of Derbyshire. Substantial areas above 460 m are blanketed by variable depths of oligofibrous peat. This restricts the area where soils are developed directly on the solid rock, to a high-level discontinuous fringing belt, usually above 300 m, along the western margin of the Pennines. These high-level sites are occupied by free or excessively draining soils; their coarse, loamy textured profiles with bleached sub-surface horizons overlie a humus-enriched zone which passes down into a dark reddish-brown iron-enriched horizon. These humo-ferric podzols dominate the *Anglezarke association*, but because erosion is severe in this fringing belt, the underlying bleached horizon and the more resistant iron-enriched horizon are frequently exposed at the ground surface. Other soils in this association include peaty or humic rankers. These drier moorland habitats provide poor grazing with wavy hair grass (*Deschampsia flexuosa*) predominating, and invasions of bilberry (*Vaccinium myrtillus*) occur where grazing is minimal.

Downslope, as the solum depth increases, limited areas of free-draining soils of the *Withnell association* occur. They can sometimes be distinguished by their bright sesquioxide-enriched subsoils. Below 300 m these brown podzolic soils grade into free-draining brown earths which dominate the *Rivington association*. Both associations occur on moderate or steep slopes in the Pennines where loamy soils overlie sandstone or gritstone. They are fairly extensive in the Forests of Rossendale, Pendle and Trawden in upland Lancashire, and in the Macclesfield Forest area of Cheshire. Brown earths also occur on isolated areas of Carboniferous sandstone protruding through the drift cover on the Lancashire plain. The best example is on the Parbold–Billinge ridge north of Skelmersdale. These free-draining brown earths have dark brown stony topsoils passsing into yellowish-brown weathered B horizons. Fragments of sandstone usually increase with depth so that between 40 and 60 cm they dominate the solum.

An interesting variant of these soils can be seen in the Wildboarclough valley east of Macclesfield Forest where reddish staining of the Millstone Grit has produced red (Munsell Colour Code 2.5YR 4/6) subsoils.

In the Forest of Bowland, around Todmorden and southwards to Mam Tor, the mineral soils fringing the peat on steep scarp slopes are dominated by thin ironpan stagnopodzols of the *Belmont association* (Fig. 5.4). They occur along the valley sides in Longdendale, Alport and Derwent valleys which penetrate deeply into the 'Dark Peak'. These steep scarp slopes have all been subject to solifluction and strongly bedded, alternating bands of sandstone and shale have been thoroughly mixed to produce medium textured soils. Many slopes have terracettes and are dissected by gullies with rock outcrops or screes on the steepest slopes. Mineral soils have a peat or raw acid humus mat resting on a thin gleyed bleached horizon. This, in turn, is underlain by a very thin (< 5 mm), cemented ironpan, often undulating, sometimes contorted. It impedes water movement through the profile and dead roots accumulate on its upper surface following seasonal water-logging above the pan. Below the pan, drainage is normally free and reddish or strong brown subsoil colours prevail. The underlying grey to yellowish-brown parent material usually contains abundant shale and sandstone fragments. The formation of these thin ironpan soils has been described by Crompton (1956). Basically, there must be an excess of rainfall over losses due to evapotranspiration to maintain waterlogged conditions in the peaty surface horizon. This encourages reducing conditions in the top of the mineral soil so that iron passes into solution and is leached downwards. This process is complemented by the acid conditions prevailing in the peaty or humic layer, and by the existence of free-draining conditions in the subsoil. Iron is precipitated to form a pan at the interface between gley conditions in the surface mineral topsoil and the oxidising conditions in the subsoil. Many areas of thin ironpan soils occur in upland areas where heath has replaced deciduous forest. This intensifies leaching and encourages the formation of an acid peaty layer. The resulting wet moorland conditions are dominated either by mat grass (*Nardus strictus*) or by purple moor grass (*Molinia caerulea*) communities because waterlogging is never persistent enough to produce permanent anaerobic conditions. Lateral water movements maintain oxygen levels sufficiently to favour the *Molinia* communities (Anderson and Shimwell, 1981).

(iii) *The upland peat* Upland peat forms almost continuous cover on flat to undulating upland areas above 300 m, particularly on the Pennines, where it is most extensive. This oligofibrous peat is therefore the parent material for organic soil development and is represented on the soil maps by the *Winter Hill association*. These peats are usually laminated with fibrous and semi-fibrous layers of cotton sedge (*Eriophorum*) and *Sphagnum*, remnants passing downwards into a black amorphous peat at 3 m. Near the base, occasional woody material, chiefly birch and pine, is preserved within it. The surface horizon is usually reddish-black, very thin, humified peat with abundant fibrous roots. Upland peats are extremely acid throughout with pH values of less than 3.5.

The combined effects of high-rainfall low-temperature very acid conditions has reduced biological activity so that partially decomposed organic material has accumulated on the surface. This process, set in motion in Atlantic times about 5500 BC and operating with renewed intensity in Sub-Atlantic times (about 600 BC) created the blanket bogs on the upland plateau. Averaging between 3 and 4 m, they thin out towards the plateau edge, where they seldom exceed 1 m. The dominant peat-forming species on this plateau has been *Sphagnum* with cotton sedge and heather (*Calluna vulgaris*) contributing to the build-up until recent times. It is generally recognised that peat formation requires at least an annual rainfall of 1000 mm, together with cool summers which reduce evapotranspiration rates. Only the tops of the moorland plateau receive more than this (1250 mm to 1620 mm) in the southern Pennines, so that it is questionable whether peat formation is still occurring over wide areas (Anderson and Shimwell, 1981; Tallis, Chapter 17). The northern Pennines, with a common history of peat development, has been less subject to pollution, so that the peat-forming species are still evident on today's blanket peat surface. Although peat erosion may be locally severe, it is not as widespread as in the Dark Peak of Derbyshire.

The Cheshire–Lancashire lowlands

Permo-Triassic Sandstone outcrops in lowland regions Where these sandstones protrude through the glacial drift cover of the Cheshire and Lancashire plains they become the parent material for soil development. The Mid-Cheshire Ridge is perhaps the best example. Freely draining brown sand soils developed on these outcrops dominate the *Bridgnorth association*. They have dark reddish-brown loamy sand topsoils merging to yellowish-red sandy subsoils on the weathered sandstone. There are a range of other profiles developed on these coarse siliceous materials of which the humo-ferric podzols are the most distinctive. These soils represent areas with a history of heathland vegetation (Furness, 1978). On the footslopes profiles are sometimes affected by regional water tables so that sandy gleys or gley podzols have developed in these siliceous materials.

Conclusion

The wide range of parent materials described and a diversity of soil-forming conditions has produced a very complex pattern of soils in this north-west region. In lowland areas, the overwhelming influence of heterogeneous drift deposits dominates the pattern of soil associations. Soil sequences may have developed on these desposits that reflect local hydrological conditions so that some associations are dominated by poorly draining soils and others by free-draining soils.

In upland areas, on the one hand there is the extensive cover of peat or aeolian material of the limestone plateau that gives a relative uniformity of soils, while on the other there is the complexity of soils developed on drift under diverse hydrological conditions, related to altitude and slope.

Details of these complex soil relationships and patterns have been steadily unfolding through the progressive publication of soil maps by the Soil Survey of England and Wales. Their latest publications, particularly the National Soil Maps (1:250 000) and supporting Bulletins (Jarvis *et al.*, 1984; Ragg *et al.*, 1984), provide a comprehensive assessment of the nation's most valuable resource, its soils.

Notes

1 There are six National Soil Maps which provide coverage for the whole of England and Wales. They are supported by Bulletins which describe the properties, limitations and uses of each soil association. Soil association names used in this paper relate to the 1:250 000 soil maps and do not always correspond with soil association names used by Hall and Folland (1970).

Sea-level changes and coastal morphology in North-west England

Introduction

The coast of North-west England from the Solway Firth to the Dee estuary (Fig. 6.1) comprises a range of morphological features that are characteristic of mid-latitude humid coasts. The majority of the coast is formed of unconsolidated sediments that are readily eroded and form distinctive features, such as the sand dunes of the south-west Lancashire and Fylde coasts, the salt marshes of the estuaries and around Morecambe Bay, the shingle banks of the Fylde coast, Walney Island and the rest of the Cumbrian coast, and the till cliffs of Blackpool, Gutterby and St. Bees.

The coast of North-west England lies along the eastern margin of an epicontinental sea, and is protected from waves generated in the Atlantic Ocean by Ireland and the Isle of Man: the maximum wave fetch at Formby is about 200 km.

Wave heights of 0.6 to 1.0 m are characteristic off the coast of south-west Lancashire (Parker, 1975), and in Morecambe Bay the most frequently occurring waves are only 0.3 m high (Darbyshire, 1958, in King, 1962). Under extreme conditions, waves up to 9.4 m in height have been recorded off south-west Lancashire and 8.5 m in Morecambe Bay. The consequence of these generally low-amplitude, short-period waves and macrotidal conditions is the development of a characteristic beach profile, known as a ridge and runnel beach, and a wide intertidal zone, of which many examples have been described from the coast of North-west England (Gresswell, 1953; King, 1959; Parker, 1975).

Water depths in the east Irish Sea are shallow: depths greater than 60 m are rarely exceeded (Fig. 6.1). The bathymetry of the Irish Sea is quite different from that of the North Sea. The west Irish Sea is characterised by a broad central region running north–south of deep water in excess of 100 m with localised shallower peaks and a linear central deep to the north–west in the North Channel with maximum depths of some 260 m (Caston, 1976), and the east Irish Sea is conspicuous for its shallow water and gentle floor gradient. The North Sea is characterised by a deep-water zone north of a line joining Flamborough Head and North Jutland and shallow water to the south.

Fig. 6.1   The bathymetry of the east Irish Sea based on a map by Pantin (1977) and reproduced by permission. More detailed bathymetric contours at 5 m intervals are shown on the three British Geological Survey 1:250 000 series maps of sea bed sediments and Quaternary geology that cover this area (Liverpool Bay, Sheet 53 °N – 04 °W, 1983; Lake District, Sheet 54 °N – 04 °W, 1983; Isle of Man and part of Ulster, Sheet 54 °N – 06 °W, 54 °W – 08 °W, in press).

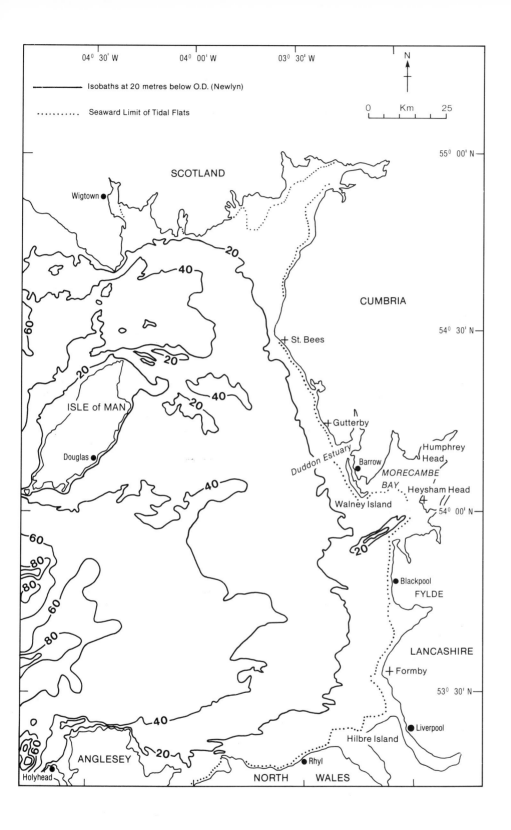

04° 30' W   04° 00' W   03° 30' W

N

0   Km   25

55° 00' N

SCOTLAND

Wigtown

20

40

CUMBRIA

54° 30' N

60

20

St. Bees

40

ISLE of MAN

20

Gutterby

Duddon Estuary

Humphrey Head

Douglas

Barrow

MORECAMBE BAY

Heysham Head

40

Walney Island

54° 00' N

60

20

80

80

Blackpool

60

FYLDE

80

LANCASHIRE

Formby

53° 30' N

40

Liverpool

Hilbre Island

20

ANGLESEY

20

Rhyl

Holyhead

60

NORTH   WALES

The coast of North-west England falls within the category of coasts defined by Davies (1964) as macrotidal and characterised by a storm wave environment. Whilst the enclosed nature of the east Irish Sea moderates the storm wave environment to some extent, the large tidal amplitude is a conspicuous feature of this coast. Bowden (1955) has published a map of co-range lines showing the average range of tides in the Irish Sea: to the east the value of the lines ranges from *c*.3 m to *c*.6 m from the open sea to the coast and the range of spring tides in Morecambe Bay exceeds 8 m at present. At Liverpool, the maximum tidal range of 8.3 m in the east Irish Sea is recorded and Kidson and Heyworth (1979) have demonstrated the dramatic increase in tidal height and range as estuaries, such as the Solway Firth and Morecambe Bay, narrow landward. The consequences of this range are the extensive tidal flats in the Solway, Duddon, Ribble and Dee estuaries and within Morecambe Bay.

The coast of North-west England is separated into two distinctive parts by Morecambe Bay – a large embayment fed by six estuaries. North of the southern shore of Morecambe Bay there are several rocky headlands such as Heysham Head, Humphrey Head and St. Bees Head, whereas to the south there are no solid rock outcrops on the coast until Hilbre Island at the mouth of the River Dee is reached.

Morecambe Bay also separates that part of the coast to the north where there are modestly elevated raised beaches of Flandrian Age, and that part of the coast to the south where all the beaches are buried and lie below the altitude of the Mean High Water Mark of Spring Tides.

The whole coast and the continental shelf sea to the west have been affected by repetitive glaciations during the Quaternary Period; the record of the Late Devensian glaciation is well preserved in the sediments of the Irish Sea and its coastal margins. In Fig. 6.2 the distribution of sediments on the floor of the east Irish Sea is shown: there are two outcrops of solid rock and thirteen outcrops of boulder clay which has been overlapped largely by unconsolidated marine and brackish water sediments of Late Devensian and Flandrian Age.

The Irish Sea and the coast of North-west England have also been affected by repetitive sea-level changes – a consequence of the build-up and decay of high-altitude and high- and mid-latitude glaciers and ice-caps. Whilst the record of pre-Devensian and Devensian sea-level changes is limited and fragmentary (Kidson and Tooley, 1977; Huddart, 1981) there is a rich record of Flandrian sea-level

---

Fig. 6.2   The distribution of sediments on the floor of part of the Irish Sea and the inter-tidal zone, based on a map published in Pantin (1978) and reproduced by permission of the Director of the British Geological Survey. The sediment distribution in Morecambe Bay is based on a map published in Anderson (1972). The symbols are those used in the scheme of Troels-Smith (1955):

1. *As* Clay: particles < 0.002 mm; 2. *Ag* Silt: particles 0.06–0.002 mm; 3. *Ga* fine and medium sand: particles 0.06–0.6 mm; 4. *Gs* coarse sand: particles 0.6–2.0 mm; 5. Gg (min) fine gravel: particles 2–6 mm; 6. Till; 7. Solid.

The size classes and sediment distributions differ slightly from those shown on the three British Geological Survey 1:250 000 series maps of sea bed sediments and Quaternary geology that cover this area: for example, on the main maps the silt and clay size fractions are not discriminated.

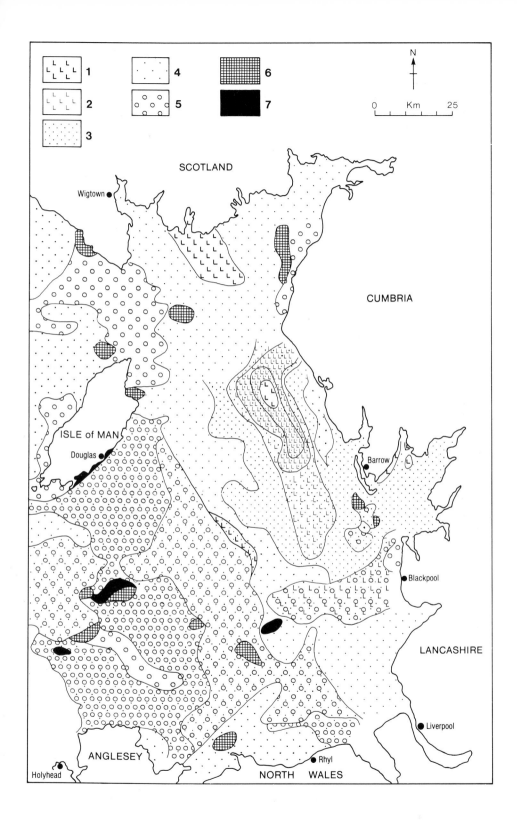

changes, particularly from the coasts of south-west Lancashire, the Fylde, Over Wyre, Morecambe Bay and the Solway lowlands.

The coastal morphology of North-west England is explained by a variety of processes operating at different scales and at different time periods. In part the morphology is explained by contemporary coastal processes related to wave energy, wind and tidal currents, tidal range and sediment supply; in part by an increase in the intensity of these phenomena and in the incidence of storm surges associated with secular changes in climate (Tooley, 1985); in part by changes in sediment supply caused by vegetational changes in the catchments feeding into the Irish Sea, especially after 5000 BP when forest clearance was widespread and the sediment budget of rivers would be at a maximum for the Flandrian Age. Bowden (1955) has calculated that the size of the Irish Sea catchment at 93 080 km^2 (35 800 square miles) is almost as great as the area of the sea itself at 100 100 km^2 (38 500 square miles). The coastal morphology will also be affected by changes in sea level and changes in the tidal range consequential upon a change in the geometry of estuaries and tidal inlets. Finally, the coast will have been affected by recent and longer-period vertical movements of the land occasioned not only by the mass transfer of load during a glacial/interglacial cycle but also by long-period subsidence within and adjacent to a sedimentary basin.

The repetitive reoccupation of coasts during successive interglacial ages has been explored and elaborated by Kidson (1966, 1968), and examples of inheritance can be found on the coasts of North-west England. The south-west coast of Cumbria from St. Bees Head to Walney Island runs along the north-east margin of the Manx–Furness or east Irish Sea Sedimentary Basin of post-Lower Palaeozoic Age (Bott, 1964), and is filled with sediments of Permo-Triassic age (Bott and Young, 1971; Dobson, 1977). The orientation of the present coast is parallel to a major fault line and this coast may be an example of inheritance during each interglacial age. At Strandhall and elsewhere on the Isle of Man, a wave-cut rock platform at +2.4 to +3.5 m O.D. (I.O.M.) is being exhumed by contemporary wave and tidal action from beneath a beach, locally cemented, the surface of which ranges from +6.0 to +9.0 m O.D. (I.O.M.) (Phillips, 1970a,b).

In this chapter the evidence for pre-Devensian, late Devensian and Flandrian sea levels is evaluated, and the effects of sea-level changes on near-coast, terrestrial and limnic sedimentation are described.

Pre-Devensian and Devensian sea-level changes

There is some erosional and depositional evidence of pre-Flandrian sea-level changes from the coast of North-west England and the floor of the Irish Sea.

Around Morecambe Bay caves, water-eroded notches and stacks in Carboniferous Limestone have been described (Ashmead, 1974; Tooley in press) and attributed to marine conditions. At Whitbarrow there are notches at +30 m O.D. and the floor of Kirkhead Cavern lies at an altitude of c. +30 m O.D. (Fig. 6.3). Below the cavern on Kirkhead Hill and on Warton Crag there are notches at +15 and +30 m O.D. Most of the benches and notches are fragmentary, and the benches below Kirkhead Cavern dip towards Humphrey Head and appear to be structurally controlled. However, Edgar's Arch on the west side of Humphrey Head is unequivocally a sea cave with a floor at *c.* +35 m O.D. and a blowhole on the western slope of Humphrey Head at an altitude of *c.* +46 m O.D.

The age of these features is open to speculation. At Kirkhead Cavern, the lower cave earth yielded nine flint artifacts that Mellars (1970) assigned to the late Upper Palaeolithic, about 14 000 years BP, which is consistent with an ice-free Lake District, certainly in the valleys, by 14 300 years BP (Pennington, 1975a, 1977). However, Gale *et al.* (1984) have re-assigned the deposition of the flints to *c.* 10 500 BP. Beneath the lower cave earth are gravels to which Ashmead (1970, 1974) assigns an age greater than 28 000 years BP and hence they belong to the Upton Warren Interstadial Complex. If the cavern is a product of marine erosion, then this episode must pre-date the formation both of the cave earth and the gravels and may be of Ipswichian or Hoxnian age or older. West (1972, 1980b) concluded that the minimum elevation of sea level during the Ipswichian was probably +7.5 m O.D. and the maximum in late Hoxnian times was at least +23 m O.D., and the erosional and solutional marine features around Morecambe Bay are more likely to belong to the Hoxnian interglacial age, as is the platform in Shippersea Bay on the coast of Co. Durham cut in Magnesian Limestone at an altitude of *c.* +27 m O.D. and overlaid by a poorly cemented coarse sand and fine gravel with marine shells and shell fragments (Tooley, 1984).

An Ipswichian or intra-Devensian Age low sea level may be recorded from marine beds reported from six boreholes between Luce Bay and Ramsey Bay. Marine-silts with a maximum proved thickness of some 6 m underlie *c.*20 m of boulder clay and themselves overlie probably non-marine sands and gravels or proglacial deposits over an older boulder clay or rock. The micro- and macrofauna has low boreal affinities and indicate water depths of 20–50 m during their deposition. The marine beds occur at altitudes of *c.* −50 to −70 m O.D., and a sea level of *c.* −30 m O.D. during this intra-Devensian or Ipswichian marine episode. (Personal communication: R. T. R. Wingfield *et al.*).

Synge (1977a) has described two marine limits on the Isle of Man: one at +21 to 25 m (I.O.M. datum) along the east coast of Man and into the highest out-wash terraces of the Bride Moraine: and a second at Peel on the west coast at an altitude of +14 m (I.O.M. datum), which suggest that ice occupied the west coast while the east coast was ice-free and affected by a cold-water marine trans-gression. Synge (1977a) argues for an east coast transgression at *c.*18 900 BP and for a west coast transgression before 14 500 BP

Some support is given to this from sedimentary evidence in the east Irish Sea, from the Isle of Man and from the Wirral, Crosby and Blackpool (Fig. 6.3).

To the south-east of the Isle of Man, Pantin (1978) has described an area, named Vannin Sound (Fig. 6.3), in which marine sediments overlie proglacial lagoonal sediments. The latter sediments are fine-grained with laminations of silt and sand. One sample of proglacial lagoonal sediment yielded fresh-looking foraminifers, of which *Elphidium clavatum*, *E. bartletti* and *Protelphidiun orbiculare* were

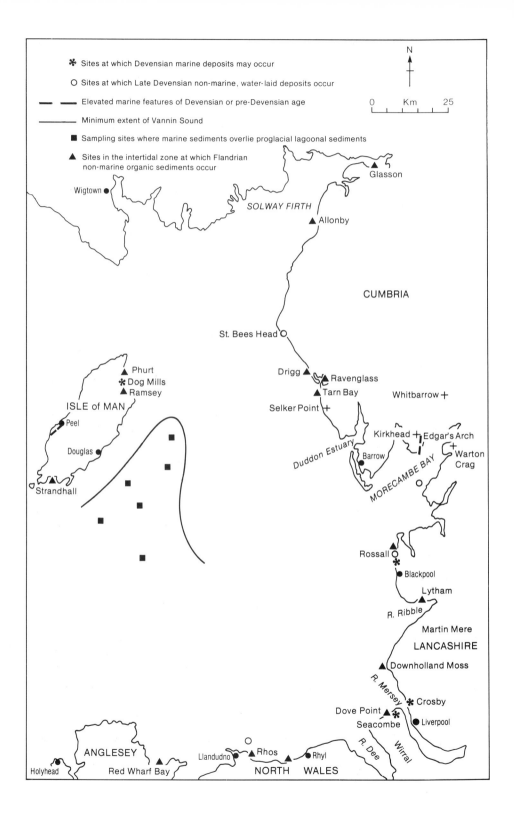

* Sites at which Devensian marine deposits may occur
O Sites at which Late Devensian non-marine, water-laid deposits occur
— — Elevated marine features of Devensian or pre-Devensian age
—— Minimum extent of Vannin Sound
■ Sampling sites where marine sediments overlie proglacial lagoonal sediments
▲ Sites in the intertidal zone at which Flandrian non-marine organic sediments occur

N

0   Km   25

▲ Glasson

Wigtown ●

SOLWAY FIRTH

▲ Allonby

CUMBRIA

St. Bees Head O

Drigg ▲   ▲ Ravenglass
▲ Tarn Bay   Whitbarrow +
Selker Point +

▲ Phurt
* Dog Mills
▲ Ramsey

ISLE of MAN

● Peel

Douglas ●

Kirkhead + + Edgar's Arch
Duddon Estuary   + Warton Crag
Barrow ●

MORECAMBE BAY

○
▲
Strandhall

O

Rossall O
*
● Blackpool

▲ Lytham
R. Ribble

Martin Mere

LANCASHIRE

▲ Downholland Moss

R. Mersey   * Crosby

Dove Point ▲ *
Seacombe   ● Liverpool

R. Dee   Wirral

ANGLESEY

○
Llandudno ▲ Rhos ▲ ● Rhyl

Holyhead ●   ▲
Red Wharf Bay   NORTH   WALES

dominant and interpreted as an assemblage from a subtidal environment colder than the present Irish Sea (less than 10 °C summer) and with a water depth of 10–20 m (Hughes, 1978).

To the north-west of Vannin Sound, Thomas (1977) has described a series of clay, silts and laminated sands, lying between the Shellag Till and thin flow tills at the Dog Mills on the north shore of Ramsey Bay, Isle of Man. The altitude of the Dog Mills series is *c*. + 14 m (Thomas, 1977), and this is the same as the altitude of the marine limit recorded by Synge (1977a) at Peel on the Isle of Man. From these intercalated deposits, the foraminiferal assemblage indicates a cold-water estuarine to intertidal environment. The site may have been close to the ice front and have provided the marsh and intertidal habitats marginal to Vannin Sound that were lacking from the sound.

On the coast of North-west England, there are three sites at which cold-water marine conditions can be inferred. At Seacombe, on the Wirral, Reade (1894) has described a clean sand with some gravel and many shell fagments at an altitude of *c*. + 18 m O.D. overlaid by boulder clay. At Crosby and Blackpool, Reade (1895) recorded a foraminifer-rich boulder clay at altitudes of *c*. + 12 m O.D. in which cold-water estuarine and subtidal forms have been recorded. It is not clear whether the samples from Crosby and Blackpool came from a sediment that was unequivocally till. If so, the presence of foraminifers would derive from pre-Devensian Age Irish Sea sediments.

There has been considerable debate about the age and provenance of the Shirdley Hill Sand formation in south-west Lancashire (Huddart *et al.*, 1977; Tooley, 1977). Whilst there is no doubt that it is not a coastal facies of mid-Flandrian Age (Tooley, 1976) and that the stratigraphy of the formation indicates periods of stability and instability during the Flandrian Age, associated with climatic changes and the effects of prehistoric man, there is some evidence to support the argument that it may be in part a coastal facies of Late Devensian Age, associated with the high sea level recorded at the Dog Mills, Isle of Man, at Seacombe on the Wirral and at Crosby. At Moss Lake, Liverpool, Godwin (1959) recorded a grass pollen grain comparable to that of the sand couch grass *Agropyron junceiforme* from sediments of Allerod age, and speculated that it might have grown in the Late Devensian on inland dunes comparable to the cover sand dunes of Holland. However, it can also be posited that the dunes in this area were along the edge of a cold-water sea, and as the present distribution of *A.junceiforme* reaches 63 °N it would be possible for this plant to have grown here in the Late Devensian.

Elsewhere outcropping on the floor of the Irish Sea and in the intertidal zone at depth or as outcrops, non-marine, water-laid deposits of Late Devensian age have been described (see Fig. 6.3 for distribution). Off Heysham in Morecambe

Fig. 6.3   The distribution of pre-Devensian, Devensian and Flandrian Age sediments on the floor of part of the Irish Sea, adjacent intertidal zones and along the coast. Sub-seabed sediment distributions for sections of the Irish Sea are shown on the three British Geological Survey 1:250 000 series maps of seabed sediments and Quaternary geology that cover this area (see caption to Figure 6.1), and show an increased number of sites falling within each category.

Bay, some 16 m of varved clays overlaid by 2–47 m of clays, silts and sands have been described and interpreted as proglacial lake sedimentation (Knight, 1977), succeeded by sedimentation of a large complex sandur supplied by wasting ice sheets to the north and west (Vincent and Lee, 1981). The varved clays have not been examined for their microfauna, and there is a possibility that like the sediments of Vannin Sound they were laid down in a brackish-water environment. However, the absence of dropstones in the varved clays of Morecambe Bay militates against this explanation.

North of Llandudno (Fig. 6.3), a peat has been reported from 3.5 to 4.0 m below the sea-bed and underlaid by a clay with shell fragments and rare small pebbles, interpreted as possibly a proglacial lake deposit (Wingfield, personal communication) and at 7.5 m below the sea-bed by a red–brown, stiff sandy boulder clay about 16 m thick. The surface of the peat lay at an altitude of about −20.25 m O.D. (Wingfield, personal communication). The microfossils from the peat were dominated by colonies of *Pediastrum* and other taxa associated with fresh-water conditions were also recorded: *Typha angustifolia*, *Caltha*, *Myriophyllum spicatum*, *Potamogeton* and *Isoëtes*. Pollen taxa were dominated by non-arboreal pollen, and of the tree pollen only *Betula* and *Pinus* were present. Organic sedimentation probably occurred during the first millennium of the Flandrian Age in a depression in the boulder clay, at a time when sea level was not directly affecting the site.

Submerged forests as indices of sea-level changes

The distribution and characteristics of submerged forests of England and Wales were described by Reid in 1913, and of the British Isles by Wright in 1914. In the intertidal zone of the east Irish Sea there are many records of submerged forests (reviewed in Huddart and Tooley (1972), Huddart *et al.* (1977), and Tooley (1978b, 1979)) and organic lenses exposed in cliff sections at the head of contemporary beaches, some of which have been used as indices of sea-level changes (Andrews *et al.*, 1973; Gresswell, 1967b; Heyworth, 1978), and their distribution is shown in Fig. 6.3.

In a general way, both Reid (1913) and Wright (1914) have related the submerged forests of the British Isles to a change in sea level, and Dunham (1972) argued for a relative rise in sea level consequent upon subsidence to explain the distribution of submerged forests in the Britsh Isles. Heyworth (1978) concluded that the altitude of trees from submerged forests is closely related to their age and that the tree fossils now found in submerged forests were killed by a rising water table

---

Fig. 6.4 The age and variations in pollen frequencies from 'submerged forests' in the intertidal zone of part of the east Irish Sea. The mean altitudes of the peat surface or the radiocarbon-dated levels are as follows:

Drigg −1, +7.0 m O.D.   Rhyl Beach, +2.4 m O.D.
Drigg −2, c. +7.0 m O.D.   Alt Mouth, +3.1 m O.D.
Annas Mouth, +6.6 m O.D.   Formby Foreshore, +5.0 m O.D.
Tarn Bay, +0.9 m O.D.   Rossall Beach, −0.4 m O.D.

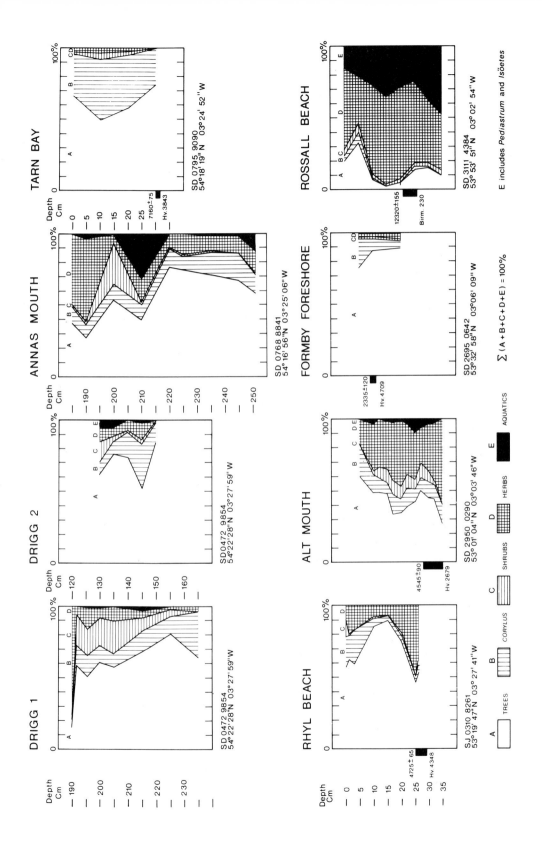

DRIGG 1

Depth Cm
— 190
— 200
— 210
— 220
— 230

A B C D

SD0472 9854
54°22′28″N 03°27′59″W

DRIGG 2

Depth Cm
0
— 120
— 130
— 140
— 150
— 160

A B C D E

SD0472 9854
54°22′28″N 03°27′59″W

ANNAS MOUTH

Depth Cm
— 190
— 200
— 210
— 220
— 230
— 240
— 250

A B C D E

SD 0768 8841
54°16′56″N 03°25′06″W

TARN BAY

Depth Cm
0
— 5
— 10
— 15
— 20
— 25
7160±75
Hv.3843

A B CD

SD 0795 9090
54°18′19″N 03°24′52″W

RHYL BEACH

Depth Cm
0
— 5
— 10
— 15
— 20
— 25   4725±65
— 30   Hv.4348
— 35

A B C D

SJ 0310 8261
53°19′47″N 03°27′41″W

ALT MOUTH

A B C D E

4545±90
Hv.2679

SD 2950 0290
53°01′04″N 03°03′46″W

FORMBY FORESHORE

2335±120
Hv.4709

A B CD

SD 2695 0642
53°32′58″N 03°06′09″W

ROSSALL BEACH

12320±155
Birm. 230

A B C D E

SD 3111 4384
53°53′51″N 03°02′54″W

E includes *Pediastrum* and *Isöetes*

∑ (A+B+C+D+E) = 100%

A TREES
B CORYLUS
C SHRUBS
D HERBS
E AQUATICS

consequent upon a rise of sea level.

These conclusions can be tested by examining the data that have been assembled from submerged forests in the intertidal zone of the east Irish Sea.

In Fig. 6.4 the variations in pollen frequencies of five different life-form classes are shown from submerged forests and cliff exposures of the east Irish Sea coasts. Some pollen assemblages are dominated by the pollen of tree taxa (Rhyl Beach, Formby Foreshore and Tarn Bay) whereas others are dominated by the pollen of non-tree taxa (Rossall Beach). The pollen assemblages and the radiocarbon dates indicate that the deposits making up the submerged forests and organic lenses were laid down at different times and under markedly different palaeoenvironmental conditions. These comprise kettle-hole deposits (Rossall Beach, Ravenglass Harbour, Drigg Beach, St. Bees Beach, Ramsey Harbour and Glasson Shore), tidal flat and lagoonal zone deposits (Rhos Beach, Rhyl Beach, Dove Point, Altmouth), perimarine deposits (Tarn Bay, Selker Point), fossil dune slacks (Formby Foreshore), limnic and terrestrial deposits (Drigg Cliff) and moor logs (Allonby Beach).

There is no consistent relationship between the age and altitude of submerged forest deposits, many different palaeoenvironments are represented and the presence of submerged forests in the intertidal zones or exposed in cliff sections is a consequence of recent processes exposing the sites that have been recorded.

Godwin (1943, 1945, 1956) reviewed the evidence for submerged forests around the North Sea Basin, and his conclusions are supported by work from North-west England. 'It is even possible, in some circumstances, for a submerged peat bed to appear upon our coast line between tide marks, without any relative shift of sea- and land-level: this may occur where shore erosion exposes the freshwater muds and peats of lake or river systems at low level behind the coast' (Godwin, 1956). 'Submerged peats may be shown by pollen analysis to have formed at all stages of the late glacial or post-glacial period, and although perhaps more frequently at some times than others, it is clear that not the slightest justification remains for using the concept of a "submerged forest period" . . .' (Godwin, 1943).

Sea-level changes during the Flandrian Age

Three distinctive coastal landscapes have been identified in Holland (Hageman, 1969; Jelgersma *et al.*, 1979) and can also be identified in North-west England. These landscapes are the coastal barriers and coastal dunes, the tidal flat and lagoonal zones, and the perimarine zones. The differences between these two areas are the extent and distribution of the three landscapes: in Holland most of the coast and the coastal hinterland can be assigned to these three landscapes, whereas in North-west England, their distribution is restricted north of Morecambe Bay by solid outcrops and exposures of till and to the south by outcrops of till. Only in the south-west Fylde and in south-west Lancashire are the landscapes comparable to those in Holland, but on a much reduced scale.

In Holland, the coastal barriers consist of fine to medium grained sands with shells (van Straaten, 1965), overlaid by blown sand. This sand is interrupted by palaeosols of mor soil or dune slack peats that allows a chronostratigraphic sub-division into older and younger dunes (Jelgersma *et al.*, 1970). In south-west Lancashire, beneath the coastal dunes, 7 to 12 m of sands and sandy silts with occasionally clay laminae and shelly sands have been described (Tooley, 1974) and these may be equivalent to the coastal barriers in Holland. North of the River Ribble and Morecambe Bay, where till outcrops occur along the coast, marine erosion ensured a supply of larger grade size material and shingle barriers are characteristic along this coast and at depth. The barriers served as a locus for the accumulation of blown sand, which appears to have begun about 4000 BP and been interrupted by two periods of dune stability when dune slack peat accumulated about 2300 BP and about 800 BP (Tooley, 1978b). Palaeosols in sections in the dunes of the coast of North-west England bear witness to many additional periods of dune stability, but probably of short duration.

The coastal dune sediments overlap tidal flat and lagoonal deposits and, in Holland, the lithostratigraphic units comprise low and high tidal flat deposits, marine, brackish water and freshwater clays, intrcalating saltmarsh, telmatic and terrestrial peats and lake muds. In Holland the marine deposits comprise the Calais and Dunkerque Members of the Westland Formation (Jelgersma *et al.*, 1979), laid down during marine transgressive intervals and the Holland Peat Member comprises organic material of different compositions laid down during marine regressions. The equivalent formation of its constituent members in North-west England lack detailed field mapping and dating, but similar facies have been recorded (Tooley, 1978a,b, in press) especially in Downholland Moss in south-west Lancashire and Nancy's Bay and Lytham Common in the south-west Fylde.

In the perimarine zone, the main lithostratigraphic units are sand, clay and peat, laid down under fluvial or limnic conditions at altitudes controlled by effective sea levels to seaward in the tidal flat and lagoonal areas. The clastic layers correlate with the marine transgressions and the organic layers with marine regressions. In Holland the Gorkum Member of clay and sand in the perimarine zone correlates with the Calais Member of marine sediments in the tidal flat and lagoonal zone, whereas the Tiel Member of clay and sand correlates with the Dunkerque Member. In North-west England, these facies are found where rivers and estuaries occur, but away from their influence changes in water depth and water quality are preserved in changes in organic sediment types, as shown by the example from Martin Mere described later (see also Tooley, 1985).

Of these three distinctive landscapes, only the tidal flat and lagoonal landscapes provide direct evidence for sea-level changes in the alternating layers of organic and inorganic sediments.

The utility of sedimentary changes and inferred palaeoenvironmental changes in the other two landscapes – the barriers and dunes landscape and the perimarine landscape – to sea-level studies can be seen when the sea-level tendency concept is applied.

Streif (1979–80) was the first to use the term 'tendencies of sea-level fluctuations', but Shennan (1982a, 1983) has defined and applied the term. A positive tendency of sea-level movement is an apparent increase in the marine influence whereas a negative tendency is an apparent decrease in the marine influence (Shennan, 1983). At the scale of a single site in the tidal flat and lagoonal zone, transgressive and regressive overlaps can be demonstrated where organic sedimentation gives way to clastic sedimentation and vice versa. On a regional scale, not only can this type of sea-level index point be used but also additional data from the perimarine and dune landscapes from a natural or archaeological context can be used to reinforce the index points showing the tendency of sea-level movement.

This methodology has been applied to the Fenland (Shennan, 1982b) and northern England (Tooley, 1982), and has allowed correlation between uplifted and subsiding areas (Shennan *et al.*, 1983).

In Fig. 6.5, time–altitude graphs using sea-level index points from Liverpool Bay to the catchment of the Ribble estuary (A) and from Morecambe Bay to the Solway Lowlands and including the Isle of Man (B) have been constructed using this methodology. Twelve periods of transgressive and regressive overlap tendencies have been identified using a variety of sea-level index points, the altitudes and radio-carbon ages of which points have been determined. No attempt has been made to fit a curve to the variates plotted on the time–altitude graph. Superficially, both graphs (A and B) show a similar pattern: a rapid rise of sea level from *c.*9000 to *c.*6000 BP and thereafter the sea-level index points fall within an altitudinal band of no more than 4 m. North of the Fylde watershed, from *c.*9000 to *c.*6000 BP the rise of sea level appears to be more rapid.

This superficial pattern ignores the consistent palaeoenvironmental changes that can be inferred from the transgressive and regressive tendency chronologies shown for each sea-level index point on Fig. 6.5. The data cluster together within the limits set by the overlap tendencies and periods of transgressive overlap tendencies are followed by periods of regressive overlap tendency. This clustering of data from a variety of coastal landscapes points to an overriding process such as positive and negative sea-level movements. These sea-level movements override the range of vertical land movements along the coast of North-west England.

The effects of these movements on sedimentation in the perimarine zone of south-west Lancashire will be the subject of the succeeding section.

Perimarine zone sedimentation in south-west Lancashire

In 1869 the Geological Survey of England and Wales published a superficial geology map of south-west Lancashire (Sheet 90 SE Southport 1″:1 mile) on which C.E. de Rance had plotted the probable eastern limit of grey clays. Although the limit requires slight adjustment, it effectively marks the limit of the tidal flat and lagoonal zone to the west and the perimarine zone to the east, where perimarine sediments overlap Shirdley Hill Sands and till.

In 1957 Gresswell mapped the limit of the 'Hillhouse coastline' well beyond and to the east of the limit of grey clays, but as has been shown elsewhere (Tooley, 1976) the 'Hilhouse coastline' is not related to a marine event. In the Martin Mere basin for example, there is a marked concave break of slope around part of the mere, which was mapped erroneously as a marine limit by Gresswell (1957): the feature, in fact, was formed during a period of elevated lake level.

The area of perimarine sedimentation in south-west Lancashire can be identified fairly accurately and the nature of the sedimentation and the connection to sea-level tendencies can be demonstrated by reference to Martin Mere.

Martin Mere was an extensive freshwater lake before its drainage at the end of the seventeenth century (Fig. 6.6). It measured approximately 6 km by 3 km (Brodrick, 1903), occupied an area of about 687 ha (1700 acres) and the altitude of the water surface varied from +2.7 to +3.4 m O.D. (Liverpool: on the Lancashire coast, differences between Liverpool and Newlyn datum are +0.1 to −0.1 m), which gave a maximum water depth of about 6 m (Brodrick, 1902). The natural drainage of the mere appears to have been both south-east into the catchment of the River Douglas and north-west via a meandering tidal inlet to the estuary of the River Ribble (Fig. 6.6). The place-name evidence corroborates the existence of a tidal inlet in that the name 'Wyke', incorporated in the place names Blowick, Wyke House Farm, Wyke Hey Farm and Wyke Lane, is Old Norse for bay and Blowick has been rendered as the Old Norse 'blá-vík' or 'the dark bay' (Cheetham, 1923), bearing witness to the peat-stained waters discharging from the mere and thence to the estuary of the River Ribble at Crossens along the course shown in Fig. 6.6. The limit of the mere is based on a description by Brodrick (1902) and refers to the period immediately before the first attempt was made to drain the mere by Thomas Fleetwood in 1692. The limit transgresses the boundaries of land that lay between +2.4 and +3.0 m O.D. (Newlyn) in 1954 (Prus-Chacinski and Harris, 1963), and although sediment consolidation following drainage and dewatering will have affected the ground altitudes during the past thirty years it is probable that, at its maximum extent, Martin Mere occupied the ground within the +2.4 m O.D. contour with seasonal or longer period dilations to the +3.0 m O.D. contour. In this dilated area fen and mire vegetation would have been characteristic.

Thirty-six borings have been put down across the mere and in adjacent areas (Fig. 6.7). The borings lie in part within the historical mere as defined by Brodrick (1902), in part within the Wyke or bay and in part on the marginal mosses such as Churchtown Moss and Scarisbrick Moss.

The sediments (Fig. 6.8) in and adjacent to the mere can be subdivided into those types characteristic of perimarine zone sedimentation and those types characteristic of tidal flat and lagoonal zone sedimentation. The sampling sites from the Wyke or bay (MM-18 to 22 on Fig. 6.7 and 6.8) show quite a different series of lithologic units from the units recorded from the balance of the sampling sites. The characteristic sediments from the Wyke are grey clay silts and sands with some iron staining near the surface, marine shells at depth and thin organic strata. To the south-east and east of the Wyke beyond Mere Hall the basal sediments are dense clays, occasionally stony and bleached white sands: these are either tills, weathered tills or the Shirdley Hill Sand Formation. The undulations of these deposits have an amplitude of about 1.5 m and within the depressions, limnic

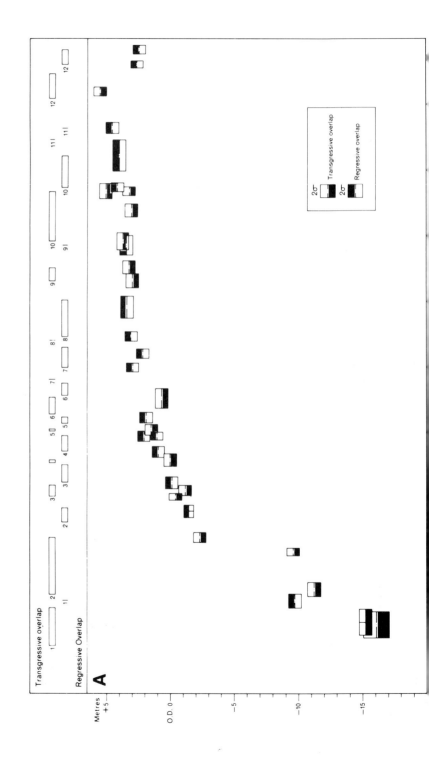

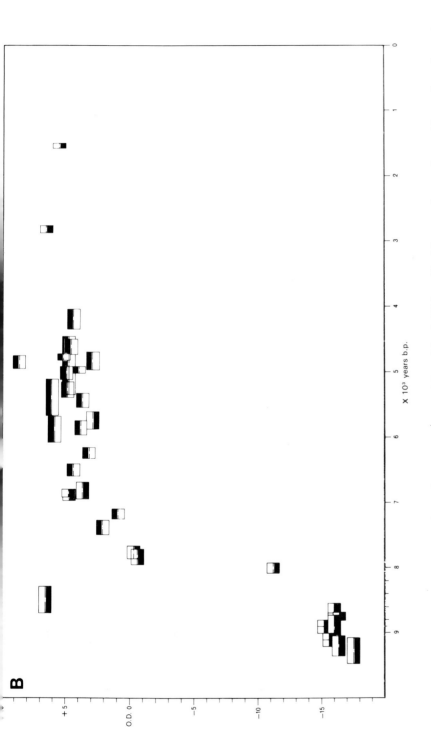

Fig. 6.5 Time/altitude graphs from (A) sites adjacent to Liverpool Bay (North Wales, Cheshire, south-west Lancashire and south-west Fylde) and (B) sites in and adjacent to Morecambe Bay, the rest of the Cumbrian coast and the Isle of Man. Each sea-level variate is a radiocarbon-dated sample, the altitude of which has been related to Ordnance Datum (Newlyn). The radiocarbon dates are shown with two standard errors. (Reproduced by permission of the editor. *Proceedings of the Geologists' Association*. See Tooley, 1982)

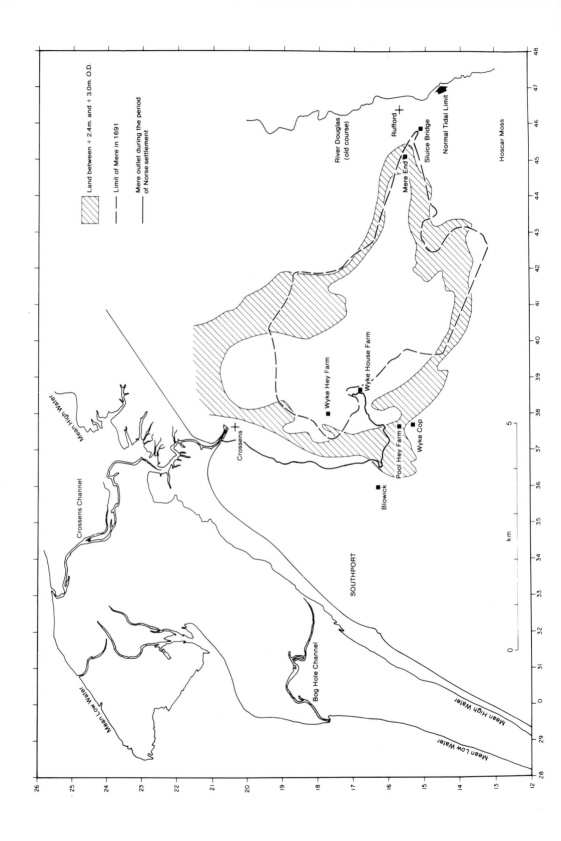

Land between + 2.4m. and + 3.0m. O.D.

Limit of Mere in 1691

Mere outlet during the period of Norse settlement

Mean High Water

Crossens Channel

River Douglas (old course)

Rufford

Mere End

Sluice Bridge

Normal Tidal Limit

Hoscar Moss

Wyke Hey Farm

Wyke House Farm

Crossens

Pool Hey Farm

Wyke Cop

Blowick

SOUTHPORT

Mean Low Water

km

Bog Hole Channel

Mean High Water

Mean Low Water

26
25
24
23
22
21
20
19
18
17
16
15
14
13
12

28   29   0   31   32   33   34   35   36   37   38   39   40   41   42   43   44   45   46   47   48

sediments accumulated. Changes in water depth and quality are reflected in the record of *Cladium mariscus, Phragmites, Equisetum, Menyanthes*, woody detritus, charcoal and *Sphagnum*. The maximum thickness of organic sediments from the mere has been recorded as 320 cm from MM-1. Brodrick (1903) reported 579 cm, Hall (1954, and in Gresswell, 1957) 457 cm in the vicinity of Dohyles and Hibbert (personal communication) has recorded 750 cm also in this area.

A core from MM-1 near the south-western limit of the mere was examined for its pollen content (Fig. 6.9). The presence of the pollen of all the thermophilous tree taxa characteristic of the Flandrian Stage and the absence of a level that can be referred to the *Ulmus* decline indicates that organic sedimentation was confined to Flandrian Chronozone II (Hibbert *et al.*, 1971). The basal organic sediments of stratum 2 are no older than 7000 BP and the near-surface sediments of stratum 13 are no younger than 5000 BP. Organic sedimentation was initiated by a rise of the freshwater table, to which the increase in the frequency of the pollen of reedswamp and obligate aquatic taxa such as *Typha angustifolia, T. latifolia* and *Hydrocotyle, Lemna, Nymphaea* and *Myriophyllum* bears witness. The increase begins at − 1.42 m O.D., reaches a peak at − 1.12 m O.D. and declines at − 0.72 m O.D. The rise of the freshwater table was consequent upon marine transgressive and regressive overlaps which have been dated on Downholland Moss to 6890±55 and 6790±95 at altitudes of − 0.87 to − 0.36 m O.D. at DM-11, and in Martin Mere basin (Fig. 6.8, MM-18 to − 22): the second transgressive overlap recorded has a mean altitude of − 1.23 m O.D. (Tooley, 1974, 1978b). This and subsequent tendencies of sea-level movement can be inferred from the changing ratio of the pollen of *Quercus* to that of *Alnus*. Shennan (1981) has drawn attention to the fact that Godwin (Godwin and Clifford, 1938; Godwin, 1978) used this ratio as an indicator of water-level movements in the Fenland, and this relationship has also been demonstrated by Iversen (1960). Although the record of the pollen of reedswamp and obligate aquatic taxa declines during the sedimentation of strata 7 to 13 (Figure 6.9) periods of wetness are indicated by higher frequencies of the pollen of *Alnus*, and the intervening dry periods that are more strongly marked by isolated peaks of *Quercus*, poor pollen preservation and sparse pollen, to which the column showing the sum of the slide traverses bears witness in the large number of traverses at 60 cm (20), 80 cm (61), 90 cm (15), 110 cm (14), 180 and 190 cm (13), 230 cm (21), 250 cm (11), and 290 cm (18).

Elsewhere in the catchment of the mere basin, organic sedimentation began much earlier. For example, at Sugar Stubbs and Mere Sands Wood, organic sediments of Late Devensian and early Flandrian Age have been recorded. In the Shirdley Hill Sands beneath the organic material at Mere Sands Wood, frost cracks and cryoturbated features bear witness to extreme low temperatures during the Late

Fig. 6.6 A map of Martin Mere to show the extent of the mere in 1691 according to Brodrick (1902): the south-west limit of the mere is shown following the railway line from Ormskirk to Southport, as Brodrick described in 1902. The area of ground between + 2.4 and + 3.0 m O.D. is based on a contour map of the Martin Mere area by Prus-Chacinski and Harris (1963). The outlet of the mere at the time of Norse settlement is speculative and based on Cheetham (1923).

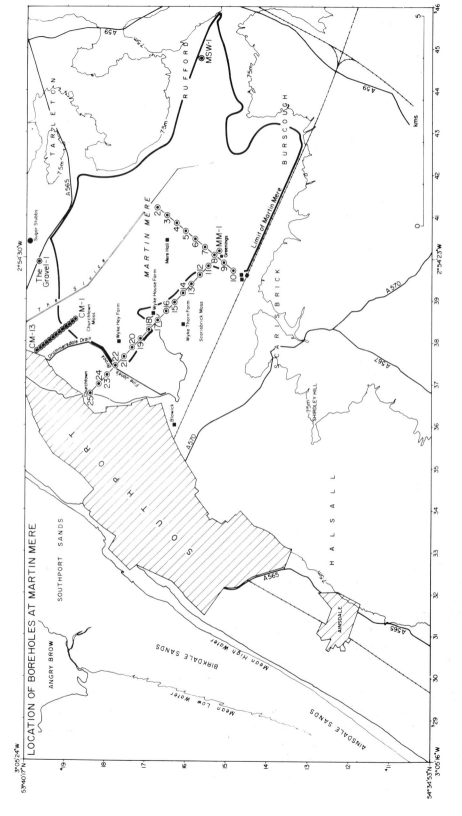

Fig. 6.7 A map of the Martin Mere area showing the location of borings and their relationship to the limit of the mere in 1691 (see Figure 6.6).

Devensian Age. The record of early Flandrian Age changes in the mere basin is shown in the pollen diagram from Mere Sands Wood (Fig. 6.10). The pollen assemblages from strata 2, 6, 7 and 8 show an early Flandrian Stage succession, but strata 3, 4 and 5 show disturbance and reworking with a younger pollen assemblage containing *Ulmus, Quercus, Alnus* and *Tilia* insinuating an older one. It is posited that strata 3–5 were laid down when stratum 6 was floated up as Martin Mere expanded as a freshwater lake. It is significant that stratum 3 lies at an altitude of +2.5 m O.D. which is close to the altitude of the limit of the mere described above (see also Fig. 6.6). At this altitude also are sheets of rounded pea-size gravel along the eastern margin of Mere Sands Wood, and this may be evidence of accelerated discharge from the mere in an easterly direction to the River Douglas catchment when the tidal inlet at the Wyke was blocked by silts during a period of positive marine tendencies.

Whilst the fine details of the palaeohydrology of Martin Mere and the longer record of palaeoenvironment changes require detailed investigation, this mere provides evidence of the nature of sedimentation in the perimarine zone during Flandrian Chronozone II, and its relationship to tendencies of sea-level movement.

Other sites have been described from the perimarine zone of south-west Lancashire (Tooley, 1978b), and within the catchment of Morecambe Bay, there are further sites, such as Rusland Lake (Pearsall, 1918; Dickinson, 1973) and Helton Tarn (Smith, 1958). Further north there are few opportunities for perimarine sediments to accumulate, but there is some limited development in the catchment of Kirby Pool (Huddart *et al.*, 1977) and in the Solway Lowlands where Oulton Moss and Martin Tarn (Walker, 1966) are located in a similar position to marine sedimentation as Martin Mere is to the marine sediments in the Wyke.

Conclusions

The coastal morphology of North-west England is an expression of a range of processes operating at different scales and over different time periods. Coastlines and shore platforms inherited from earlier stages of the Quaternary Period may characterise parts of the Cumbrian and Isle of Man coasts and elevated erosional and depositional marine landforms are found sporadically around the coasts of the east Irish Sea. Their age has not been established, but their elevation and microfauna indicates a pre-Flandrian Age.

Submerged and buried marine landforms also occur. The existence of the Late Devensian Vannin Sound south-east of the Isle of Man, where proglacial lagoonal sediments were laid down in a subtidal environment, permits a closer integration of the depositional and erosional evidence for the marine accompaniment to deglaciation in the East Irish Sea. In this context, it is unfortunate that no samples of the proglacial varved clays from Morecambe Bay and the proglacial silts and clays shown in the sections on the British Geological Survey Sea Bed Sediments and Quaternary Geology maps (Lake District, Liverpool Bay and Isle of Man) have been preserved and their microflora and microfauna analysed.

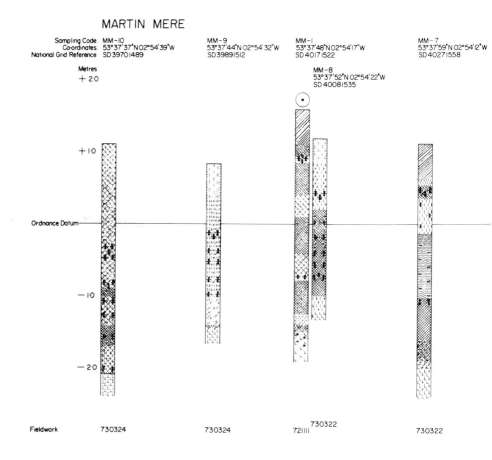

Fig. 6.8   The stratigraphy across Martin Mere (*above*) from MM-1 to MM-10, and (*overleaf*) from MM-2 to MM-10. Symbols are according to Troels-Smith (1955). The stratigraphy was recorded in the field and samples were taken with a gouge or Russian-type peat sampler.

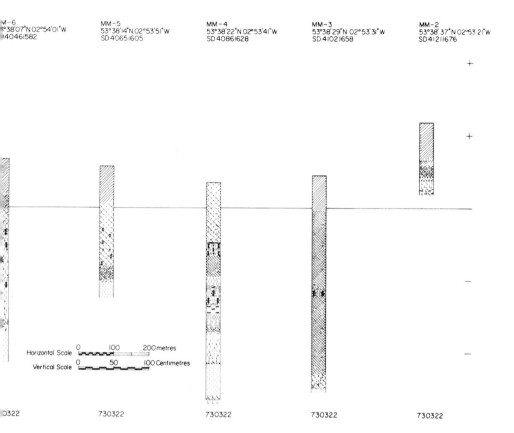

M-6
9°38'07"N 02°54'01'W
40461582

MM-5
53°38'14"N 02°53'51"W
SD 40651605

MM-4
53°38'22"N 02°53'41"W
SD 40861628

MM-3
53°38'29"N 02°53'31'W
SD 41021658

MM-2
53°38'37"N 02°53 21'W
SD 41211676

Horizontal Scale 0   100   200 metres
Vertical Scale   0   50   100 Centimetres

0322   730322   730322   730322   730322

E

MARTIN MERE

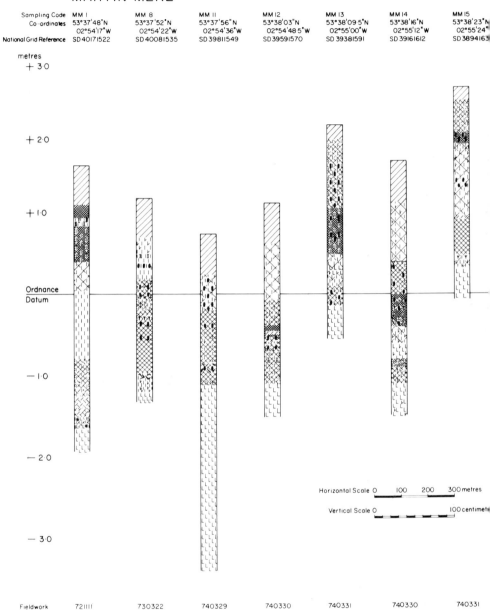

Sampling Code	MM 1	MM 8	MM 11	MM 12	MM 13	MM 14	MM 15
Co-ordinates	53°37'48"N	53°37'52"N	53°37'56"N	53°38'03"N	53°38'09.5"N	53°38'16"N	53°38'23"N
	02°54'17"W	02°54'22"W	02°54'36"W	02°54'48.5"W	02°55'00"W	02°55'12"W	02°55'24"W
National Grid Reference	SD 40171522	SD 40081535	SD 39811549	SD 39591570	SD 39381591	SD 39161612	SD 3894163

| Fieldwork | 721111 | 730322 | 740329 | 740330 | 740331 | 740330 | 740331 |

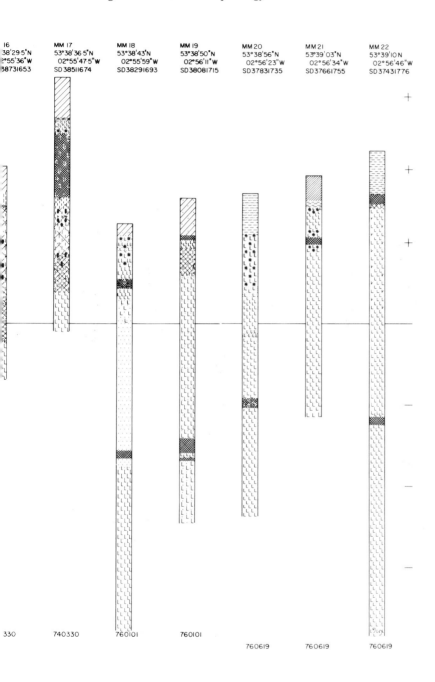

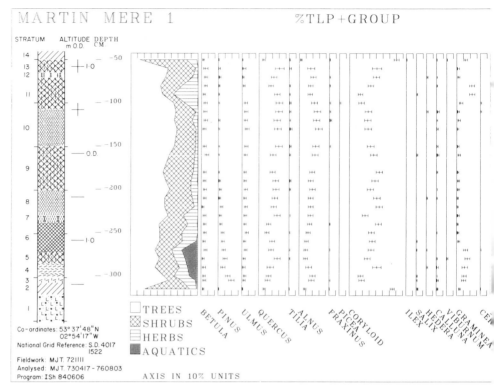

Fig. 6.9 A percentage pollen diagram from Martin Mere-1, the location of which is shown on Figure 6.8. The percentage frequencies are calculated with a denominator of total land pollen (TLP) and group. There are 5 life-form or taxonomic groups: trees, shrubs, herbs, aquatics, and spores. 95% confidence limits are shown for each percentage frequency value (Shennan 1981). A full pollen diagram is shown in Tooley (1980).

The submerged forests of the intertidal zone fall into several different palaeo-environmental categories and only some can be used to establish tendencies of sea-level movement. Others are unrelated to sea-level movement, and occur in the intertidal zone as a consequence of contemporary coastal erosion.

The three genetic coastal landscapes identified in Holland can be identified also in North-west England, although their distribution here is restricted. All three land-scapes provide data that show tendencies of sea-level movement, but the litho-stratigraphy of the tidal flat and lagoonal zone provides unequivocal evidence of marine transgressive and regressive overlaps. The overlap sequences have analogues in the perimarine zone, and lithostratigraphic and biostratigraphic data from Martin Mere in south-west Lancashire demonstrate the links between these two landscapes. The study of the palaeohydrology of the perimarine and the tidal flat and lagoonal zone and of the coastal sand dunes could be developed further in North-west

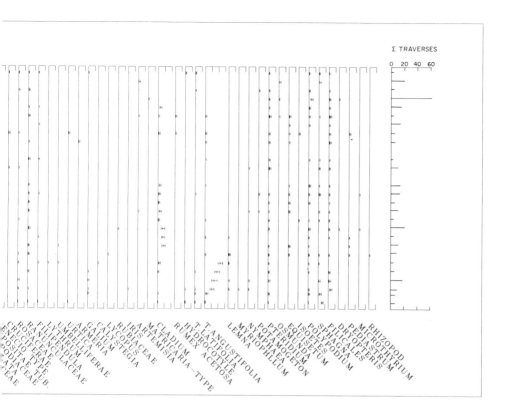

England to provide additional index points showing tendencies of sea-level movement.

In 1913 Clement Reid wrote that on the shores and under the marshes of the estuaries of the Ribble, Mersey and Dee 'are found some of the most extensive submerged land-surfaces now traceable in Britain . . . We have in these strongly marked alternations of peat and warp an ideal series of deposits for the study of successive stages. In them the geologist should be able to study ancient changes of sea-level under such favourable conditions as to leave no doubt as to the reality and exact amount of these changes'. There is no doubt about the richness of the stratigraphic record in North-west England, but considerably more empirical data will need to be assembled from field and laboratory investigations before it is possible to quantify sea-level changes using material from the coastal dunes and barriers, perimarine zone and tidal flat and lagoonal zone during the Flandrian

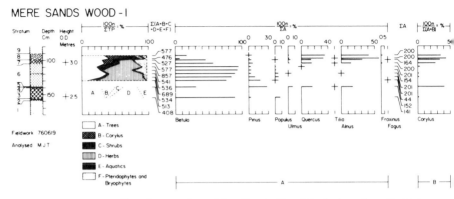

Figure 6.10 A percentage pollen diagram from Mere Sands Wood-1, the location of which is shown on Figure 6.7. Pollen percentage frequencies are based on the formula shown for each life-form or taxonomic class. A plus (+) sign indicates the record of a single pollen grain or spore identified during traversing.

Age, and a start only has been made on the study of pre-Flandrian sea-level changes in North-west England and the east Irish Sea.

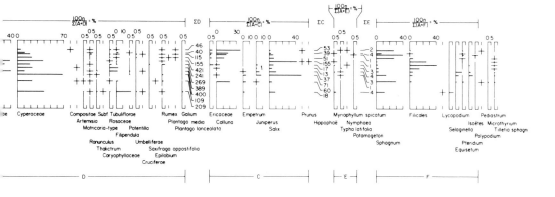

7 *A. M. Harvey*

The river systems of North-west England

Introduction

North-west England is drained by five main river systems, the Mersey/Weaver, Ribble, Lune, Eden and the radial Lake District systems (Fig. 7.1). The modern rivers have developed since the retreat of the Devensian ice sheet and are therefore of late Pleistocene and Holocene age. In the early Holocene cold Late Glacial climates gave way to post-glacial temperate climates and deciduous woodland colonised the land surface, by *c.*7000–5000 BP covering all but the highest and most rugged surfaces. During the later Holocene this woodland was replaced first on the upland surfaces, by open moorland vegetation types, then it disappeared from the steeper valley sides and valley floors. Finally in the lowlands rich oak woodlands were replaced by agricultural land during historic times (Pennington, 1970, 1974). Geomorphologically the early Holocene saw the stabilisation of an unstable landscape and the later Holocene, under the increasing influence of man saw a partial de-stabilisation (Brown, 1979).

Holocene river development has resulted in incision into Pleistocene glacial and periglacial deposits. Sediments have been transported to the major estuaries (Lewin, 1981), themselves formed as sea level rose to near its present level by *c.*5000 BP (Tooley, Chapter 6). The overall incision cannot be related to base level changes except possibly in the early Holocene to some glacio-isostatic uplift, but must largely be due to long-term adjustment to Holocene water and sediment regimes (Ferguson, 1981).

Modern channel morphology reflects sediment type and supply both from Pleistocene deposits and from pre-Pleistocene bedrock. Lower Palaeozoic fine-grained but resistant sedimentaries, low grade metamorphics and igneous rocks occur in the Lake District and Howgill Fells (1 on Fig. 7.1). Lower Carboniferous limestones outcrop around the Lake District and in the Pennines, and Carboniferous clastics with important coarse sandstones and grits occur in the Bowland Fells (2 on Fig. 7.1), and Pennine areas south of the Ribble. Soft Triassic sandstones and marls underlie the Cheshire Plain, coastal Lancashire and the Vale of Eden.

This chapter deals with Holocene river development and the modern fluvial system but first it is necessary to consider the longer-term context.

Origins and the glacial legacy

It is generally accepted that the major drainage alignments developed during the Tertiary, perhaps after early Tertiary uplift and retreat of late Cretaceous seas,

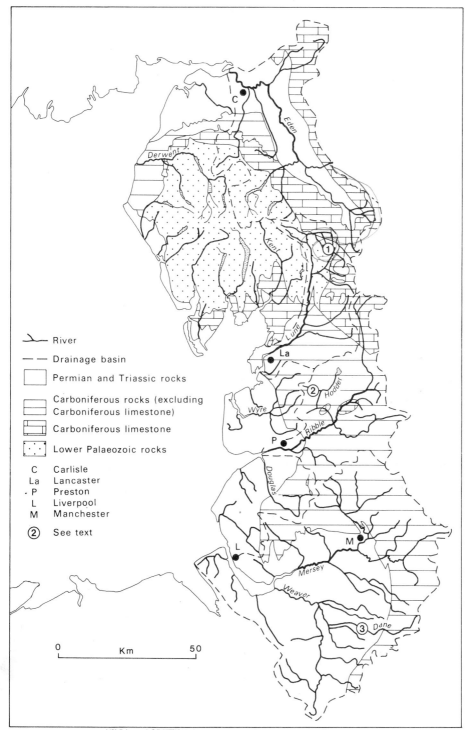

Fig. 7.1   Main river systems of North-west England. (1. Howgill Fells, 2. Bowland Fells, 3. River Dane).

and were modified by mid-Tertiary tectonics (Brown, 1979). The original pattern of easterly consequents, and over the Lake District, a radial pattern (Marr, 1916), was superimposed onto the underlying Palaeozoic rock. Aggressive west coast streams captured the headwaters of the easterly drainage and competed with each other. Late Tertiary systems are suggested by regional slopes of high level erosion surfaces with the eastern part of Bowland, now in the Ribble basin but then draining east to the Aire (Moseley, 1961), and the northern Howgills, now in the Lune basin, then draining north to the Eden (Marr and Fearnsides, 1909; Hollingworth, 1929; McConnell, 1939; King, 1966, 1976).

The aggression of the west coast rivers continued into the Peistocene, locally accentuated by glacial or meltwater erosion. The upper Eden gained at the expense of the Ure, and the Ribble gained its headwaters from the Aire. Lower down the Ribble the Hodder draining west through the Loud valley was diverted south directly into the lower Ribble. The most impressive changes occurred in the Lune headwaters. Pre-glacially the northern Howgill streams (Fig. 7.3) drained to the Eden, first directly across the Carboniferous limestone outcrop to the north then, after the development of the upper Lune subsequent valley, via Scandal Beck. To the south of the Howgills the fault-guided lower Lune appears to have worked headwards through the Lune gorge to capture the upper Lune at Tebay. Here, as the Lune cuts across one of the former major divides of the north of England and right across the major regional structures it seems likely that glacial ice or meltwater, from an early glaciation, may have been responsible for this diversion (King, 1976). The aggression of the lower Lune at the expense of northward drainage has continued on a smaller scale whereby Carlingill (Fig. 7.3) in the western Howgills had captured the headwaters of Uldale Beck through a spectacular gorge, and glacial erosion at Cautley captured the Bowderdale headwaters south into the Rawthey/Lune system (King, 1976). Meltwater modified the Lune/Eden divide near Sunbiggin and lower down the Lune ice and meltwater caused diversions in the Lancaster/Condor valley area. There are many local instances of glacial or meltwater diversions in the Lake District, for example at Aira Beck, north of Ullswater, and in Mungrisdale, north of Troutbeck.

New Pleistocene drainage patterns were established in central Lancashire and the Cheshire plain. The Douglas developed by linkage of meltwater channels and in Cheshire, the Mersey/Weaver system was incised into the sand- and till-covered plain by meltwater streams (Evans *et al.*, 1968). Meltwater modified the Dane/Rudyard valley watershed. The southern watershed of the Weaver is entirely Pleistocene, developing irregularly on the north side of the Devensian Ellesmere/Whitchurch morainic complex (Poole and Whiteman, 1966).

The modern rivers developed from meltwater streams as the Devensian ice sheets retreated, from *c.*16 000 BP in Cheshire to *c.*14 000 BP in the north and from 10–11 000 BP within the valleys of the Lake District (Thomas, Chapter 8). Within the uplands the new river systems generally followed pre-glacial lines modified by glacial and meltwater erosion and in the Lake District interrupted by new lakes occupying deeply scoured rock basins. In the lowlands an entirely new system developed on the drift cover.

Holocene river system development

The Late Glacial was characterised by a periglacial environment with little vegetation on the hillslopes, active slope processes and high rates of sediment supply to the river system. In the Lake District this period was only of short duration and periglacial forms are limited to screes on the steeper slopes (King, 1976). Glacial erosional and depositional forms remain relatively fresh. Elsewhere periglacial conditions persisted for 4000–6000 years allowing considerable modification of valley side slopes. Locally bedrock slopes developed a thin mantle of angular material, and some true screes were formed. On drift-covered slopes considerable modification of the till took place by solifluction, with the upper layers developing downslope fabric and stratification. Gently sloping solifluction terraces were formed on the sides of the main valleys of the Howgills (Plates 7.1a, 7.1b, Fig. 7.2a), the northern Pennines and Bowland (Fig. 7.2b). Locally landslips occurred especially in the Pennine valleys, in Mallerstang (King, 1976) and Longdendale and in the Langden valley in Bowland. Most appear to be Late Glacial but Tallis and Johnson (1980) demonstrate Holocene ages for most of the Longdendale slips.

   Since the development of the solifluction terraces in the upland valleys and throughout the Late Glacial and the Holocene in the lowlands incision has been the dominant geomorphological trend. This appears to reflect the general reduction in sediment supply to the river systems. The incision has not been continuous and is marked by successive river terrace development.

River terrace sequences

On the major rivers several post-glacial terraces may be recognised, up to three on the Lune, at least two on the Eden (Arthurton and Wadge, 1981), and three on the Ribble (Price *et al.*, 1963). Usually terraces may be traced upstream to merge with the valley floor as on the Goyt, Etherow, Bollin/Dean and Dane in the Mersey system (Rice, 1957; Stevenson and Gaunt, 1971; Johnson, 1969c). Elsewhere terraces converge downstream as on the Eden (Arthurton and Wadge, 1981) and in the Derwent/Ellen system (Eastwood *et al.*, 1968).

   Two questions may be asked of the terrace sequences; to what extent may correlations be established throughout the region, and which forms relate to the Late Glacial/early Holocene 'pre-stabilisation' period and which to the later Holocene 'woodland removal' period? At present complete answers to these questions cannot be given, but this chapter will focus discussion on three areas characteristic of the range of geology and environment in the north-west: the Howgills, Bowland and the Dane valley (Areas 1, 2, 3 on Fig. 7.1).

   Along the upper Lune and in many Howgill valleys at least two terrace stages post-date the solifluction slopes and pre-date the modern valley floors. The older, upper terraces, everywhere composed of coarse cobbles, appear to have been deposited by high-energy possibly braided streams, but the lower terraces only show this character in some of the upland valleys. Along the Lune and in the Langdale valley (Fig. 7.3) lensed gravels and silts suggest meandering rather than braided

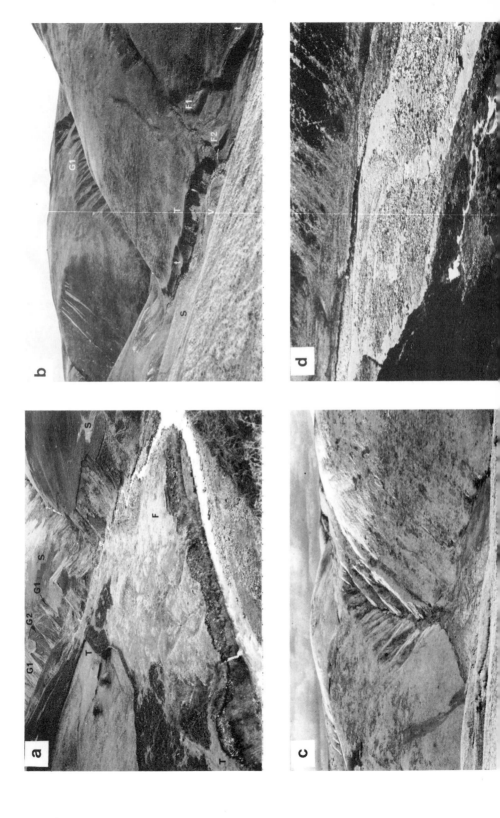

Plate 7.1 (a) Grains Gill from above Carlingill, Howgill Fells. S Solifluction terrace, T High terrace, F Grains Gill fan and Carlingill low terraces. G1 stabilised fossil gullies. G2 Modern active gullies. (b) Middle Carlingill (see also Figs. 7.2, 7.6). S Solifluction terrace. T high terrace, t low terrace, F1 and F2 Tributary fan segments. G1 Stabilised fossil gullies. V Modern valley floor. (c) Blakethwaite, central Howgills. Fossil gully system and fan. (d) Langden Brook, braided channel. (e) River Dane near Swettenham, meandering channel. Valley is trenched into glacial Cheshire Plain (P). T High terrace, t low terrace, f modern floodplain. (f) Langdale Beck, Howgill Fells. Geomorphic effects of the 1982 flood. Note: fresh debris on cone, fresh gravel spreads on valley floor, channel cutoff indicated by arrow.

stream environments (Harvey *et al.*, 1981).

In the Carlingill valley (Figs. 7.2a, 7.5a, Plate 7.1b) the upper terrace trims the solifluction deposits but pre-dates tributary fan formation. The lower terrace is contemporaneous with an early fan phase. Set below is a stable section of valley floor itself a little above that part of the valley recently active. Further downstream (Plate 7.1a) the high terrace is present and can be traced up the Grains Gill tributary. Grains Gill fan, set below this terrace, may be the equivalent of the Carlingill low terrace. It can be traced up Grains Gill and related to the development of now stabilised hillslope gullies, which in turn are being dissected by modern gullies (Harvey.*et al.*, 1981). Buried by the fan was an abandoned peat-filled channel of Carlingill which has yielded a carbon-14 date (Gak-4532) of 2290±80 BP (Cundill, 1976). This indicates that aggradation of the youngest part of the fan and by association the lower terraces of Carlingill took place at some time after that date.

In the Langdale valley (Fig. 7.3) a low terrace is capped by an organic soil horizon which in turn is buried by colluvial deposits in a cone at the base of now stabilised gullies (Harvey *et al.*, 1981). This horizon has yielded carbon-14 dates of 2580±55 BP (UB-2212) to 940±95 BP (UB-2213) for the duration of a stable period when the terrace formed the valley floor, prior to being buried by the debris cone.

The low terraces in the Howgills date from late Prehistoric to early Historic times and appear to have formed at a time of slope erosion associated with hillslope vegetation changes. The upper terraces are considerably older and may be Late Glacial rather than Holocene. In the Carlingill valley only the solifluction surfaces and the high terrace show mature podsolic soil development (Harvey *et al.*, 1984) with only weak podsolisation of the lower terrace and younger surfaces.

In Bowland post-glacial terraces occur along the Hodder and its tributaries and although some may be related to local base level controls their occurrence both up- and downstream of rock-controlled river sections indicates their wider significance. At Burholme (Fig. 7.2b) three terrace groups can be identified, including a high terrace trimming solifluction deposits which may be of Late Glacial age. An extensive main terrace, trimming a till surface, can be traced up- and down-valley and into the Dunsop and Langden tributaries. At Langden (Fig. 7.5b) what appears to be the equivalent of this terrace is buried by peat at the base of which are abundant tree remains. The terrace pre-dates the peat formation which may be assumed to have occurred between 7500 BP and 3500 BP. This terrace also pre-dates the formation of large alluvial cones and fans related to phases of gullying and tributary incision, perhaps as in the Howgills, as a result of valley-side deforestation. These appear to be the age equivalent of the low terraces. Within the low terrace at Burholme (Fig. 7.2b) is a peat-filled channel with pollen indicative of Zone VIIb (A. Mannion, University of Reading, personal communication) suggesting that the lower terrace is late Holocene, the others late Pleistocene or early Holocene.

In the Dane valley Johnson (1969c) has identified five post-glacial terrace phases, which between Congleton and Holmes Chapel (Fig. 7.2c) can be grouped into three: fragmentary upper terraces, an extensive main terrace and a low terrace. The lower

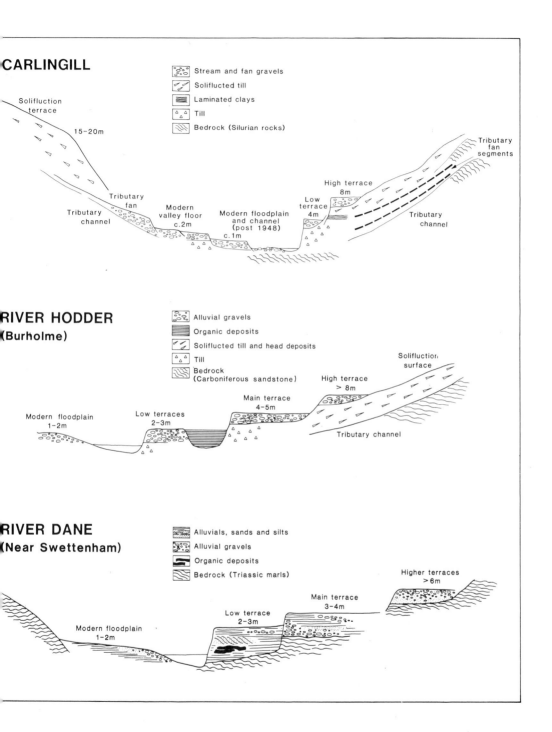

CARLINGILL

Stream and fan gravels
Soliflucted till
Laminated clays
Till
Bedrock (Silurian rocks)

Solifluction terrace

15–20m

Tributary fan segments

High terrace 8m

Low terrace 4m

Tributary fan

Tributary channel

Modern valley floor c.2m

Modern floodplain and channel (post 1948) c.1m

Tributary channel

RIVER HODDER
(Burholme)

Alluvial gravels
Organic deposits
Soliflucted till and head deposits
Till
Bedrock (Carboniferous sandstone)

Solifluction surface

High terrace > 8m

Main terrace 4–5m

Low terraces 2–3m

Modern floodplain 1–2m

Tributary channel

RIVER DANE
(Near Swettenham)

Alluvials, sands and silts
Alluvial gravels
Organic deposits
Bedrock (Triassic marls)

Higher terraces > 6m

Main terrace 3–4m

Low terrace 2–3m

Modern floodplain 1–2m

Fig. 7.2 Selected terrace sequences.

remnants of the upper terraces appear to be those which Johnson (1969c) traces downstream to be equivalent to the high terrace of the Mersey, which developed from Late Glacial to early Holocene (to c.7000 BP). The main and lower terraces therefore are mid to late Holocene in age. The low terrace, earlier described as floodplain (Evans *et al.*, 1968), was formed after dissection of the main terrace, then aggradation to a thickness greater than that of the modern floodplain. Included within the deposits are macro-organic remains which when dated will give this sequence some context. Since aggradation of the low terrace, dissection has taken place and a modern floodplain has been constructed by meander migration, well below the low terrace surface. Hooke and Harvey (1983) suggest that the modern floodplain is little older than 140 years (see below and Fig. 7.5c). Interestingly both upstream of Congleton and downstream of Middlewich the modern channel has trenched into the low terrace but not yet migrated sufficiently to form much of a floodplain.

In all three areas described, high terraces of late Pleistocene/early Holocene age and low terraces of late Holocene age can be identified. The high terraces may reflect water and sediment fluctuations during the pre-stabilisation phases and the low terraces fluctuations during the man-induced deforestation period.

Hillslope gullying

A relationship between alluvial fans or debris cones and the terrace sequence has been suggested, the fans and cones resulting from tributary incision and hillslope gullying. Gully networks ranging from linear to dendritic patterns occurring throughout the Howgills (Fig. 7.3, Plates 7.1a, 7.1b, 7.1c) have since stabilised into 'fossil' gullies and in some places (e.g. at Grains Gill, Plate 7.1a) are now being dissected by a new generation of active gullies (Harvey, 1974, 1977). The fossil gullies, cut in soliflucted till, represented a great increase in drainage density by network extension of about two orders. They appear to be related to a major phase of slope de-stabilisation resulting from the destruction of valley-side woodland. Evidence for vegetation change comes from pollen studies at Carlingill (Cundill, 1976) and in Langdale (Harvey *et al.*, 1981). Where gullies issue into a main valley, fans and debris cones were deposited (Fig. 7.3); small steep debris flow cones at the base of small steep gully systems and larger gentler alluvial fans at the base of the larger valleys. Gullying and fan formation can be related to the terrace sequence with the earliest fan segments grading into the low terraces, as at Carlingill and Grains Gill (Figs. 7.2a, 7.5a, Plates 7.1a, 7.1b) and the main fan segments resting on the low terrace, as in the Lune valley and at Langdale, or grading to the older valley floor surfaces as at Carlingill (Figs. 7.2a, 7.5a, Plate 7.1b). At Blakethwaite on the Carlingill/Uldale watershed (Plate 7.1c) a large fan buries a well developed podsol formed on the underlying solifluced till prior to burial. The carbon-14 dates from Grains Gill and Langdale (see above) indicate that the main gullying phase occurred after c.2000 BP at Grains Gill and in the tenth century AD in Langdale, possibly associated with Viking settlement and the introduction of extensive sheep grazing on the margins of the Lake District/Howgill region (Harvey *et al.*, 1981).

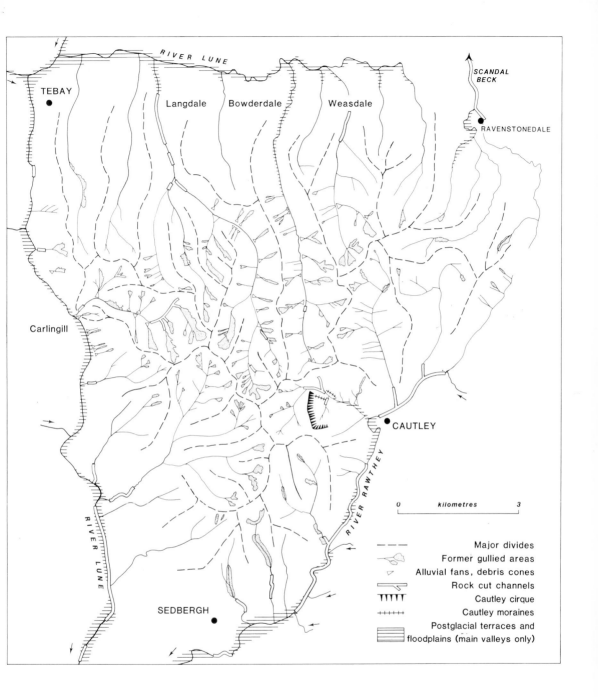

Fig. 7.3 Howgill Fells, Holocene landforms.

In Bowland there is a similar sequence of fan deposition as a result of tributary incision and hillslope gullying. In the Dunsop and Langden valleys, fans post-date the main terraces and grade into lower terraces. They contain as yet undated tree remains, indicating that gullying and fan formation occurred before the valley woodland had totally disappeared. Most of the larger valley-side gullies in Bowland have now stabilised but as in the Howgills there are smaller active modern gullies cutting into the hillslopes.

Hillslope gullying resulting from man-induced vegetation changes, while likely to occur in late Prehistoric to early Historic times need not be synchronous from area to area. Indeed the Howgill phases occurred considerably later than a similar phase of erosion-related valley alluviation identified by Richards (1981) in east Yorkshire, where settlement occurred more intensively at an earlier date.

Before leaving the question of Holocene erosion and deposition mention must be made of peat dissection. Blanket peat began to form on the flatter uplands, especially in Bowland and the Pennines with the degeneration of the woodland from *c.*7000 BP to *c.*5000 BP (Moore, 1975; Tallis and Switsur, 1983).

Peat erosion is common today (Bower, 1960a, 1961; Tallis, 1965, 1973a, and this volume, Chapter 17) and although there is some doubt about the origins and causes of this erosion, it is likely to date back at least 1000 years (Mayfield and Pearson, 1972). Erosion takes place by gullying (Moseley, 1972, Burt and Gardiner, 1982) and piping but often peat debris is deposited on the gully floors allowing stabilisation by renewed vegetation growth.

The modern fluvial system

Hydrology and sediment supply

The north-west experiences all-year precipitation with a slight autumn maximum and totals ranging from *c.*750 mm per year in the Cheshire plain to over 1000 mm in the upland areas above 300 m, and over 2500 mm in the highest parts of the Lake District. Some falls as snow with an average annual snow cover of less than 5 days at sea level rising to *c.*30 days at 300 m and *c.*70 days at 600 m. Even at height snow normally only persists for a short while on any one occasion and there may be several discrete periods of snow cover each winter rather than one longer period culminating in a major snowmelt flood. Potential evapotranspiration is low when compared to precipitation, ranging from *c.*550 mm in Cheshire to less than 450 mm in the uplands (Ward, 1981). This combination produces winter maximum river flows but without extreme mean monthly values. For instance on the Lune the mean monthly flow for December, the maximum north, is only 3 times that for June, the minimum month, and only 1.5 times the mean annual flow (Ward, 1981). Thus variability over the year is low but daily variability, as on other upland 'flashy' rivers (Newson, 1981) is high (Fig. 7.4), and floods can occur at any time of the year. The main causes of floods are: (a) long duration heavy rain resulting from deep depressions, (b) the occasional heavy snowmelt, usually also associated with rainfall and (c) highly localised summer thunderstorms, such as that which caused the 1967 Dunsop floods (Duckworth and Seed, 1969) or that which caused the 1982 Howgill floods (see below and Harvey, in press).

General hydrological characteristics for selected streams are shown in Table 7.1. Mean and flood flows are lowest in Cheshire. Floods are more extreme on the upland streams, with the exception of the Eamont where the regulatory effect of Ullswater is apparent. In part of the region, not represented in Table 7.1, especially in the Mersey basin, the natural hydrology has been considerably modified by water resource developments.

Sediment supply reflects drainage basin geology. Most northwestern rivers are cobble and gravel bed rivers where the calibre of the bedload influences channel processes, slopes and morphology (Wilcock, 1967, 1971). The Carboniferous sandstones of Bowland and the southern Pennines yield a coarse cobble bedload which diminishes rapidly in size away from the supply area (Knighton, 1975b). The lower Palaeozoics generally yield smaller, more platy, but more durable clasts. The finer fractions of sediment loads also reflect basin geology. Sands derived from bedrock, solifluction deposits and glacial outwash are common in Bowland streams and locally in Cheshire, while silts are common from lower Palaeozoic areas, and silts and clays are common in glacial till areas.

Contemporary river sediment appears to be largely derived from channel margins, either directly from bedrock in incised rock-controlled reaches or from Pleistocene glacial and periglacial or Holocene fluvial deposits. Sediment supply from agricultural land is probably not important largely because much of the improved land is pasture and arable lands are mostly on low slopes on the coastal plain. In the uplands, however, sediment supply from the hillslopes is important. Peat debris is common in Bowland and southern Pennine streams and in the Howgills active gullies, cut in soliflucted till, dominate the sediment regime.

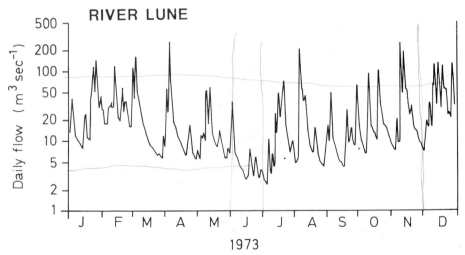

Fig. 7.4 River Lune daily flow hydrograph 1973 (Data from Dept. of the Environment, Water data unit).

Table 7.1 Hydrological characteristics of selected rivers

River	Basin characteristics	DA (km²)	Mean annual Ppt. (P) (mm)	Mean annual River flow (F) (mm)	$\bar{Q}$ (m³ sec⁻¹)	Qf (m³ sec⁻¹)	$\dfrac{Qf}{DA}$	$\dfrac{Qf}{Q}$
U. Weaver (Cheshire)	Lowland Triassic, (Drift cover)	207	768	250	1.64	25	0.12	15.4
U. Dane (Cheshire)	Upland, Carb. sst. Lowland, Triassic	150	1044	492	2.34	63	0.42	26.8
Wincham Brook (Cheshire)	Lowland, Triassic (Drift cover)	148	808	441	2.07	25	0.17	12.2
Ribble	Large complex upland basin	1140	1329	890	32.2	713	0.64	22.1
Croasdale (Bowland)	Upland, Carb. sst.	10.4	(e)1900	1183	0.39	15	1.43	38.2
Lune	Large complex upland basin	995	1498	1069	33.7	672	0.68	19.9
U. Eden	Complex, N. Pennines and V. of Eden	616	1233	684	13.4	304	0.49	22.7
Eamont (Lake District)	Includes Ullswater	396	1813	1105	13.9	166	0.42	11.9

Mean flow [handwritten annotation above $\bar{Q}$ column]

DA is drainage area, P is 30 year means, F is for period of record, $\bar{Q}$ is mean annual discharge.
Qf is mean annual flood, e is estimated value.
Data from *Surface Water: United Kingdom 1977–1980*, HMSO.

At Grains Gill (Plate 7.1a) there is a close relationship between gully and stream processes (Harvey, 1974, 1977, 1984). Sediment production on the gullied slopes takes place by a combination of surface erosion and mass movement processes with winter sediment production dominated by wet mudflows accounting for up to three times summer yields. The fines get washed directly into Grains Gill as suspended load: concentrations of up to 25 g l^{-1} have been measured in rillflow, and in Grain Gill downstream of the gullies may be at least five times those upstream. The clasts accumulate at the gully bases as debris cones, to be removed later by major floods. Such floods, with a return period of *c*.2 years (Harvey *et al.*, 1979), allow continued incision of the gullies by removal of the basal debris cones, and contribute the major coarse sediment input to the stream channel. Channel morphology responds to this sediment input with a wide partly braided unstable boulder and cobble channel downstream from the gullies contrasting with a stable narrow channel upstream (Harvey, 1977). Similar localised channel response to high sediment input is also evident on Langden Brook in Bowland (Hitchcock, 1977a). Total sediment movement on streams with supply from active scars can be high. Wilkinson (1971) estimated the long-term yield of Langden Brook to be *c*.230 t km $^{-2}$ yr $^{-1}$.

Channel morphology and adjustment

Channel morphology responds to water and sediment regime in the context of valley-floor slopes inherited from the Pleistocene and modified by Holocene incision. Several types of channel may be recognised on the basis of channel pattern and stability (Ferguson, 1981).

Three types of stable pattern are evident. Bedrock channels occur either where Pleistocene glacial erosion created rock steps or where Holocene incision has caused trenching into bedrock. The first kind are common in the Lake District, in extreme cases forming waterfalls, either in hanging valley positions as above Borrowdale, or on valley-floor rock steps as in the Kentmere valley. The second kind occurs where a hitherto mobile channel became locked into bedrock during Holocene incision, resulting sometimes in entrenched meanders as on the Rawthey, or simply in a rock-bound channel set below high terraces, as on parts of the Lune, the Greta and the upper Dane.

A second common stable type occurs where the channel margins are non-erodible or protected by boulders and tree roots, as commonly occurs in the lower reaches of many Lake District streams. The third type is common in the lowland areas (e.g. at Crowton Brook and on the River Loud, see Table 7.2), and occurs where low-energy low-slope streams draining till areas have insufficient power to erode clayey banks or where such erosion is so slow as to be undetectable over the approximately 150-year period for which accurate map evidence for channel configuration is available (Hooke and Kain, 1982).

Active channels range from meandering, through low-sinuosity single channels, to braided patterns. Most of the major rivers include the Eden, the Ribble and the Dane (Plate 7.1e, Table 7.2) have gravel beds, well developed pool and riffle sequences and locally active meandering patterns. Some main river reaches, such as on the Lune near Archolme (Thompson, 1984) show a tendency to braid. Smaller lowland streams, where active, have low-sinuosity to meandering channels and active upland streams in the Lake District, the Howgills and Bowland show channels near the single/braided threshold or wide shallow unstable cobble-bed braided patterns (Plate 7.1d).

Of the mobile channels those with silty banks, and lower valley gradients tend to meander whereas those with a coarse cobble bedload and steep valley gradients tend to wide shallow channels and braided patterns (Table 7.2).

Studies of channel adjustment have demonstrated the relationships between sediment calibre, channel slope, sediment transport and hydraulic geometry (Wilcock, 1971; Knighton, 1972, 1973, 1975a, 1975b; Hitchcock, 1977a). Short-term variations in channel form suggest a dynamic equilibrium maintained by a balance between the effects of major channel-forming floods, recurring once every 1–2 years, and moderate more-frequent, essentially channel-modifying events recurring several times a year (Harvey *et al.*, 1979). However, the relationship of these events to longer-term progressive change is uncertain, as is the role of extreme events.

In June 1982 an extreme flood occurred in the Howgills, with a return period over Langdale and Bowderdale estimated to be in excess of 70 years (Harvey, in press). This event caused massive hillslope erosion with the scouring of previously stabilised gullies and the cutting of new gullies. There was a massive production of sediment creating new cones and fans at the base of gully systems, or partly or wholly burying previously stable forms (Plate 7.1f) and in some cases completely diverting the river on the valley floor. There was a massive input of coarse sediment to the stream system, locally creating spreads of boulders and coarse cobbles on the floodplain surface. The effect on the channels of Langdale and Bowderdale becks was to kick channels previously on the meandering/braided threshold clearly into the braided regime, if only temporarily, by channel widening, avulsion and bend cutoff, and slope steepening by path shortening. Slow recovery is beginning but even when it is complete, there will be permanent changes brought about by this event.

Studies of channel change suggest long-term progressive changes in meander morphology as opposed to equilibrium adjustments and short-term changes. Both Hooke and Harvey (1983) on the Dane (Fig. 7.5) and Thompson (1984) on Skirden Beck have identified progressive increases in sinuosity, in the first case linked to the development of the modern floodplain, and in the second as a complex response to upstream hydrological changes. Mechanisms of channel change are seen in two examples (Fig. 7.5). On Langden Brook, a braided channel, changes occur through channel widening and within-channel central shoal formation, but major changes take place during floods by avulsion and secondary braiding whereby the new channel re-activates an old abandoned channel line in the floodplain (Hitchcock, 1977b; Thompson, 1984). The sequence shown in Fig. 7.5 illustrates how the channel taken by the 1975 avulsion, described by Hitchcock (1977b), had by 1983

Table 7.2 Channel characteristics – selected rivers

	Main sediment source	Bed or bank sediments	DA (km²)	Valley gradient	Channel slope	Width (m)	Depth (m)	W/D
Stable channels								
Crowton Brook (Cheshire Plain)	Glacial till	Banks – silt & clay Bed – sand & fine gravel	23	.0045	.0032	3.4	1.2	2.8
River Loud (Hodder valley)	Glacial till	Banks – silt & clay Bed – sand & fine gravel	15	.0030	.0018	6	2.6	2.3
Active meandering channels								
River Dane (Cheshire)	Carb. sst. Triassic sst. Glacial seds.	Banks – silty Bed – sand & gravel	150	.0031	.0016	16	1.6	10.0
Skirden Beck (E. Bowland)	Carb. sst. & glacial seds.	Banks – silty Bed – gravel	41	.0070	.0048	9	1.2	7.5
Active braided channels								
Carlingill (Howgills)	Silurian siltstones & glacial seds.	Banks – loose gravels Bed – cobbles	2.6	.044	.038	6	0.4	15.0
Langden Brook (Bowland)	Carb. sst. & glacial seds.	Banks – loose sand & gravel Bed – cobbles	14	.024	.021	10	0.6	16.7

Sources: Data for River Loud, Skirden Beck from Thompson (1984), for Dane from Hooke and Harvey (1983), for Crowton Brook and Carlingill from survey by author.
Channel data are mean values from sample sites; on Langden and Carlingill for undivided channel reaches.

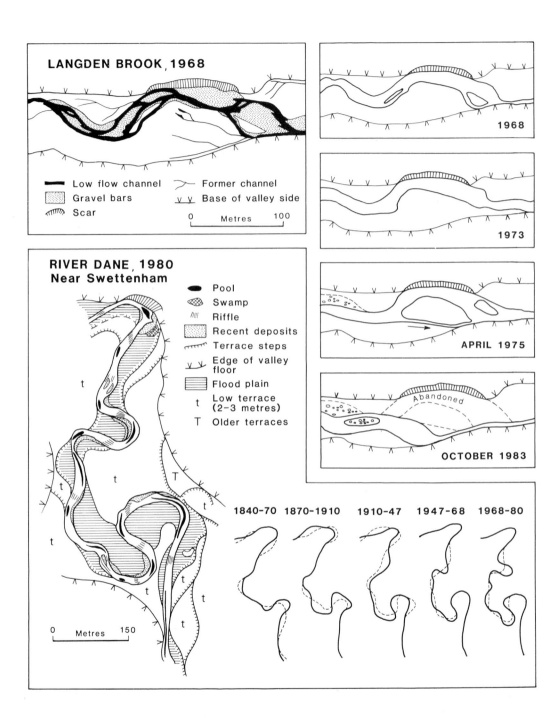

LANGDEN BROOK, 1968

— Low flow channel
⌇ Gravel bars
〰 Scar

⤳ Former channel
ⱽⱽ Base of valley side

0 Metres 100

1968

1973

APRIL 1975

OCTOBER 1983 — Abandoned

RIVER DANE, 1980
Near Swettenham

● Pool
⬡ Swamp
⫼ Riffle
⬚ Recent deposits
⌒ Terrace steps
ⱽ Edge of valley floor
▤ Flood plain
t Low terrace (2–3 metres)
T Older terraces

0 Metres 150

1840–70 1870–1910 1910–47 1947–68 1968–80

Fig. 7.5 Channel pattern and channel changes, Langden Brook and River Dane. (Langden Brook using data in Hitchcock, 1977a; Thompson, 1984; River Dane after Hooke and Harvey, 1983).

itself been abandoned by a further avulsion. On the River Dane the tendency to generate new meanders by 'double heading' is evident. Bends grow too large to sustain a single pool riffle unit and associated secondary flows. They then develop additional pool riffle units which through influencing the secondary flows initiate new bends (Hooke and Harvey, 1983). Most channel change on the Dane takes place by meander growth and migration rather than by cutoff: cutoffs occur (see also Fig. 7.6) as on other meandering Cheshire rivers (Moseley, 1975b) but on the Dane they occur less frequently than the generation of new bends.

The rates of channel change on active streams are such that the modern flood-plains are very recent surfaces (Fig. 7.6). Sample reaches of Langden Brook (Hitchcock, 1977a) and the River Dane (Hooke and Harvey, 1983) where flood-plain ages have been estimated using map evidence, and Carlingill (Harvey *et al.*, 1984) where floodplain gravel-bar surface ages have been estimated by lichenometry all show high proportions of the floodplain to have developed in the last 100 years, a similar rate of floodplain renewal to that observed on Welsh upland rivers (Lewin, 1981). On the Dane this appears to be new floodplain formation as the river cuts into the low terrace, indicating that the whole of the floodplain has developed in perhaps 200 years.

The influence of man

The adjustments of the later Holocene and the contemporary channel changes must be seen in the context of a hydrological environment increasingly influenced by man's activities, but there are certain specific influences which must be emphasised. Although large parts of the region are rural with only low population densities, other parts, particularly the Mersey basin, including the two major conurbations around Liverpool and Manchester, form some of the most heavily urbanised and industrialised land in Europe, with significant influences on the hydrological and fluvial systems.

Major water resource developments have radically influenced the river flows. Most of the small Mersey headwaters around the Manchester basin have been dammed for water supplies as have the headwaters of the Douglas and the southern tributaries of the Ribble in the Rossendale upland. In Bowland the headwaters of the Hodder are either dammed or used directly for water supply and in the Lake District major reservoirs have been created by dams at Thirlmere and Haweswater and some of the larger natural lakes are regulated and used for water supplies. The overall effect of water abstraction is a reduction of river flow and a modification of flow regime. Channel response to reservoir construction has been identified on the Hodder downstream of Stocks reservoir (Petts, 1980a, b; Petts and Lewin, 1975) whereby scour immediately downstream gives way to a marked reduction in channel size over several kilometres of river channel.

Water returned to the river systems as sewage effluent increases the flow down-stream of urban areas especially where interbasin transfers take place. Lake District water enters the Mersey system in this way. Water quality is affected by urban and industrial effluent. The main stem of the Mersey is classified as grossly polluted

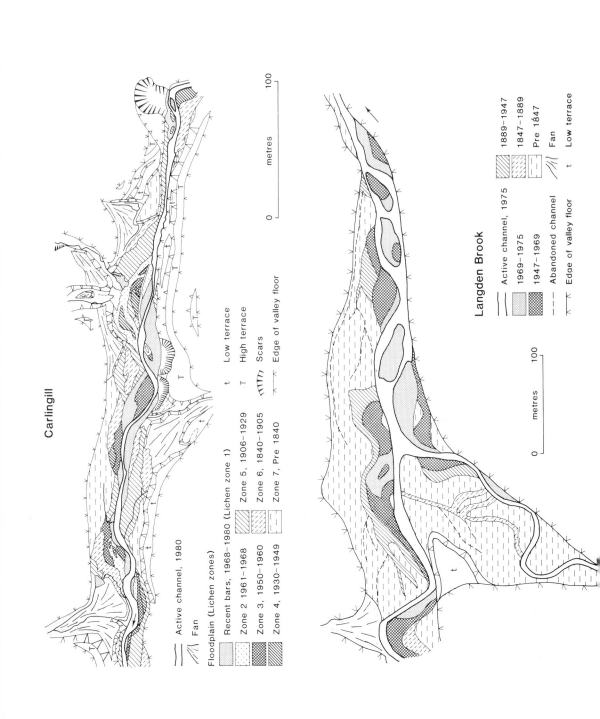

Carlingill

Active channel, 1980

Fan

Floodplain (Lichen zones)

Recent bars, 1968–1980 (Lichen zone 1)

Zone 2 1961–1968

Zone 3, 1950–1960

Zone 4, 1930–1949

Zone 5, 1906–1929

Zone 6, 1840–1905

Zone 7, Pre 1840

t   Low terrace

T   High terrace

⟨⟨⟨⟩⟩⟩   Scars

—⫣—   Edge of valley floor

0   100

metres

Langden Brook

Active channel, 1975

1969–1975

1947–1969

1889–1947

1847–1889

Pre 1847

Abandoned channel

Fan

—⫣—   Edge of valley floor

t   Low terrace

0   100

metres

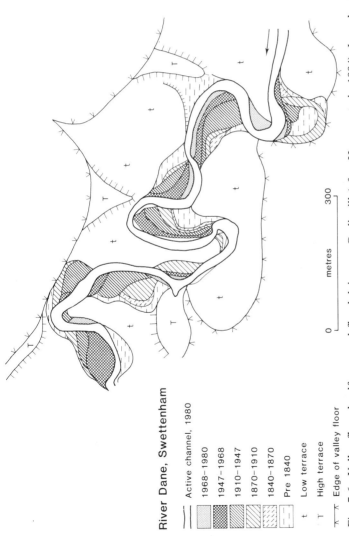

River Dane, Swettenham

	Active channel, 1980
	1968–1980
	1947–1968
	1910–1947
	1870–1910
	1840–1870
	Pre 1840
t	Low terrace
T	High terrace
⊥	Edge of valley floor

0   metres   300

Fig. 7.6 Valley floor landforms and floodplain age. Carlingill (after Harvey *et al.*, 1984); Langden Brook (after Hitchcock, 1977a) and River Dane (after Hooke and Harvey, 1983).

by the Department of the Environment (Walling and Webb, 1981) and the lower reaches of the Weaver, Douglas and Ribble are classified as poor. As well as urban and general industrial waste the chemical industry in Cheshire affects the quality of the lower Weaver and acid mine drainage occurs in coal mining areas in the Mersey and Douglas basins.

Urban development increases flood runoff from paved surfaces which may in turn cause channel enlargement downstream. The affects of the nineteenth century urbanisation on channel morphology are unknown but recent channel changes on the Bollin have been attributed to recent urban development upstream (Moseley, 1975a) and gross channel enlargement has been identified on the River Tawd downstream of Skelmersdale new town (Knight, 1979). Other types of channel modification are more obvious and include channel straightening for navigation, such as on the lower Weaver, bank protection and effectively canalisation of many rivers in urban areas. Even the River Lune, which has escaped much of the worst of manmade modifications has had its upper reaches converted from a wild gravel-bed partly braided stream to a straight rock-lined canal in conjunction with a road improvement scheme!

In summary, the river systems of North-west England, although in places considerably modified by man's intervention, are dynamic systems continually changing and adjusting within the context of the contemporary environment. These adjustments represent part of the longer term progressive changes that can be traced back through the Holocene to the Late Glacial origins of the modern river system.

The Quaternary of the
northern Irish Sea basin

Introduction

With the exception of the narrow North Channel separating Ulster from Scotland, and the restricted channel separating Ireland from Wales, the northern Irish Sea basin is virtually an enclosed, inland shallow sea. Over most of its eastern portion, east of a line from Anglesey through the Isle of Man to the Mull of Galloway, water depths are shallow and average less than 50 m (Fig. 8.1A). To the west, however, a deep trough runs from the North Channel southwards to midway between Anglesey and the Irish coast. Water depths in this trough average 100–140 m but extend to more than 180 m in a series of narrow troughs off the Mull.

Geologically the basin is developed in a strong Caledonian structural setting manifest in both the Lower Palaeozoic massifs that surround it and in the younger sedimentary basins that fill it (Fig. 8.1B). Thus, Carboniferous rocks outcrop over much of the western floor in two NE–SW trending basins that link extensive outcrops of rock of similar age in the Pennines and central Ireland. These basins are divided by a structural arch connecting Lower Palaeozoic rocks of the Isle of Man with those of the Lake District. In the eastern portion of the basin a thick trough of Permo-Triassic sediment overlies the Caboniferous as a north-westward extension of the fault-bounded graben that separates the Welsh massif from the Pennines in Lancashire and Cheshire. Narrow Permo-Triassic basins of similar structural trend occur through the North Channel and Luce Bay but are divided from the main basin by a deep north-easterly trending trough occupying the floor of the Solway Firth.

In recognising the role of the Irish Sea basin as a great reservoir and routeway of ice from the fringing mountain source areas, G. W. Lamplugh (1903) termed the glaciation of western Britain the 'Irish Sea glaciation'. He postulated that, during the last glaciation, and probably most of the preceding ones, ice flowed from the source areas to the north into the Irish Sea basin whence it moved south and south-eastwards to impinge upon the coasts of west Lancashire, to pass through the lowlands of Cheshire and thrust itself high against the mountain massif of North Wales. It also escaped southwards into the southern Irish Sea basin where it coalesced in the west with an independent Irish ice sheet. In its passage it carried large quantities of debris derived from the floor of the basin itself and a significant component of far-travelled erratic rock types brought from the mountain source areas. These characteristics allow the ready differentiation of Irish Sea drift from that derived from the more localised source areas in Wales and the Pennines.

In this chapter we shall concentrate on the glacial record within the basin itself,

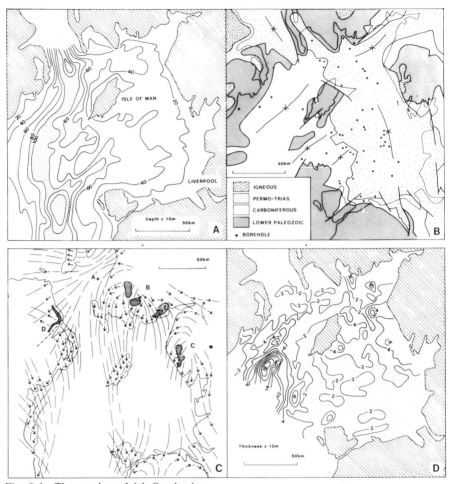

Fig. 8.1 The northern Irish Sea basin.
  A: Bathymetry. Water depths greater than 100 m shown stippled.
  B: Geology (based on IGS offshore maps). Dots show borehole locations.
  C: Generalised ice-flow directions and locations of selected erratic sources. Based on striae, till fabric, and other indicators of ice direction from a variety of sources – *A*: Ailsa Craig riebeckite–eurite. *B*: Southern Uplands granites and granodiorites. *C*: Shap and Eskdale granites. *D*: Ulster flint and chert.
  D: Thickness of Pleistocene drift (after Wright *et al.*, 1971).

including the Isle of Man, and sketch out in particular what is known of the changing environments that occurred during the last, Devensian glacial stage. Details of the influence of Irish Sea ice upon the geomorphology of adjacent areas of North-west England, and the effect of the Holocene rise in sea level upon its coasts, will be found in Chapters 6, 9, 10 and 11.

Ice directions

Assessment of ice-flow directions within the basin is, of necessity, restricted to data obtained from the onshore margins and is based on erratic distribution, striae, till fabric and the orientation of drumlins and other ice-moulded landforms (Fig. 8.1C). None of these indicators, however, can readily distinguish between ice flows during different phases of the same glaciation or, indeed, between different glaciations, and thus they can only show overall patterns.

Far-travelled erratic clasts are found commonly within Irish Sea drifts, especially in North Wales, Lancashire and Cheshire, the Isle of Man and eastern Ireland. Significant indicators include Ailsa Craig riebeckite–eurite from the Firth of Clyde; granite and granodiorite from the plutonic masses of Dalbeatie, Criffel and Cairnsmore in the Southern Uplands; Eskdale and Shap granite from the Lake District; and flint and chert from the Chalk of Ulster. These identify the source or near-source areas and show that the predominant flow of ice into the basin was from western Scotland via the Firth of Clyde and the North Channel. Subsidiary flows radiated outwards from the Southern Uplands, the Lake District, eastern and northern Ireland and, to a lesser extent, Wales.

In some areas evidence shows conflicting flows, related either to separate phases of movement or to competition between ice sheets (Fig. 8.1C). Thus, in Ulster, during an early episode of glaciation Clyde ice passed south-westwards across the province. In a later phase, however, this flow was weaker and a radial pattern of movement was imposed by a more strongly independent Irish ice sheet (Stephens *et al.*, 1975; Stephens and McCabe, 1977). Similar phasing is evident in eastern Ireland (McCabe, 1973), Anglesey (Saunders, 1968a), and the Solway area (Hollingworth, 1931; King, 1976), but in all cases the chronology is uncertain.

Drift thickness and bedrock relief

Information concerning the thickness and distribution of Quaternary drift on the floor of the basin, and of the nature of the underlying bedrock relief, is limited and is derived from the results of offshore seismic and drilling programmes (Wright *et al.*, 1971; Wilkinson and Halliwell, 1979). In general, drift thickness, including both Pleistocene till and outwash and Holocene marine sediment, averages less than 20 m over most of the basin but there are few areas of any extent that show bedrock exposed on the sea floor (Fig. 8.1D). Very thick local accumulations occur immediately onshore in the northern Isle of Man (Lamplugh, 1903; Smith, 1930; Thomas, 1976, 1977) and in an extensive offshore area to the south-west of that island. In this latter area thicknesses reach 180 m but the basal portion may be of Neogene rather than Quaternary age (Dobson, 1977b).

More detailed investigations, using closely-spaced geophysical traverses, have revealed very considerable local variation in both drift thickness and bedrock topography in certain areas. Thus, in the area between the Isle of Man and the Lake District, drift thicknesses locally reaches 120 m in the base of channels cut into underlying bedrock. These channels, occurring in closely-spaced groups and filled with till or other glacigenic sediment, have an onshore analogue in the extensive tunnel valley networks hidden beneath thick drift in Merseyside and north

Cheshire (Howell, 1973). Other channel systems occur in the Kish Bank area off Dublin (Whittington, 1977) but in this case they are cut into till rather than bedrock and are filled with fluviatile and marine sediment. These channels form a dendritic system extending outwards from the mouths of existing rivers and were cut either by meltwater streams draining the retreating ice cap or sub-aerial river systems during a period of low sea level immediately following the last glaciation.

It seems likely that as further detailed information on the nature of the sea floor is revealed much of the bed of the Irish Sea will be seen to display a complex sub- and supra-drift relief related to sub-glacial or sub-glacial fluviatile erosion during ice advance, to incision by proglacial outwash streams during retreat, and to down-cutting by the extension of onshore rivers across the dry sea floor prior to Holocene flooding.

Offshore Quaternary succession

Borehole records (Wright *et al.*, 1971; Wilkinson and Halliwell, 1979) and acoustic profiles (Belderton, 1964; Pantin, 1979) reveal that the Quaternary succession underlying the sea floor is relatively simple (Fig. 8.2). Although there are variations from area to area the succession generally consists of an upward sequence of till, proglacial sediment and marine sediment, each separated by a marked unconformity (Pantin, 1978). The till is relatively uniform lithologically and comprises a stiff, unbedded reddish-brown mud with erratic pebbles of both intra- and extra-basinal type. Lamination occurs locally, but is interrupted and often highly distorted. In these respects the till is very similar to the lower tills of the Cumbrian and Manx successions (Huddart *et al.*, 1977; Thomas, 1977). Only a few boreholes show multiple till sequences and the classic tripartite division of Upper Boulder Clay, Middle Sands and Lower Boulder Clay of much of Cheshire, Lancashire and coastal Cumbria is not evident. In contrast to the southern Irish Sea basin (Garrard, 1977), no interglacial marine sediments are seen underlying the till. Thus, Wright *et al.* (1977) report no chronostratigraphic evidence to identify separate glacial or interglacial episodes.

The proglacial sediments occur up to some tens of metres thick in the area east of the Isle of Man and are widespread to the south. They are almost wholly absent from the area west of the Isle of Man, however, and thick marine deposits directly overlie the till. The proglacial sediments are characteristically muddy but contain thin sand and coarse silt laminae and occasional beds of coarser sediment up to cobble size. Dropstones are common and thin intercalated till units, similar to iceberg dump structures or bergmounds (Thomas, 1984a), also occur. The sediments are not bioturbated and macro-fossils are virtually absent. Foraminifera and organic-walled micro-plankton are locally present, however, in a number of cores from the south-east of the Isle of Man (Pantin, 1978).

The marine sediments vary widely, depending upon locality, with the sectors north and east of the Isle of Man predominantly mud or sand, Liverpool Bay dominantly sand and the area towards Anglesey largely gravel. The area north-

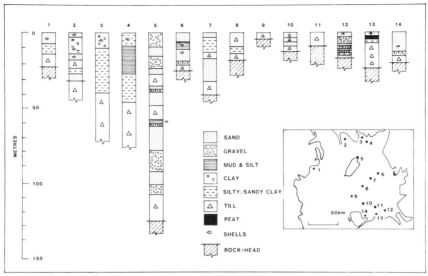

Fig. 8.2 Selected borehole records from the northern Irish Sea basin. (After Wright *et al.*, 1971.)

west, west and south-west of the Isle of Man consists almost entirely of mud. Bioturbation is widespread in these deposits and micro-fauna and macro-fauna, restricted almost entirely to bivalves and gastropods, are abundant.

The Manx Quaternary succession

Commensurate with its location in the north central portion of the basin, astride successive Pleistocene ice streams emanating from source areas to the north, the Isle of Man displays a sequence of Quaternary deposits of extraordinarily diverse character (Fig. 8.3). The greater part of the island, especially that occupied by the Palaeozoic Manx upland massif, has only a thin drift cover. The northern part, however, shows Quaternary strata to at least 145 m below O.D. (Lamplugh, 1903; Smith, 1930) and the maximum thickness, including that exposed above sea level, probably extends to as much as 250 m.

Traditionally, the Quaternary deposits of the island have been divided into two great and mutually exclusive suites (Kendall, 1894; Lamplugh, 1903). A high-level suite (Thomas, 1977) consists entirely of deposits of local composition and is restricted largely to the upland areas. It includes thick sequences of stony clay, interbedded with thinner units of washed gravel, underlain by scree. Clast fabric is everywhere orientated downslope and the deposits commonly thicken in this direction to form pronounced terraces with a marked preference for facing directions in the sectors north-west to north-east. Moraine forms, cirques and other glacial landforms are conspicuously absent (Lamplugh, 1903) and slopes display a zonal character reflecting weathering, transport and deposition in separate and successive downslope elements.

F

A low-level suite of Quaternary deposits consists almost entirely of material of foreign or northerly origin and is restricted to the lowland of the north of the island and to parts of the insular margin. It is associated with a distinctive topography of moraine ridges, kettle basins and sloping sandur surfaces (Mitchell, 1965; Thomas, 1976) cut at the northern extremity of the island by a prominent raised marine cliff fronted by beach ridges, dune sands and lagoonal slacks (Phillips, 1967; Ward, 1970; Carter in Tooley, 1977) (Fig. 8.3). The stratigraphic succession beneath this topography is extremely complicated (Thomas, 1977) but in that part exposed above sea level two major glacigenic formations are identified (Fig. 8.4).

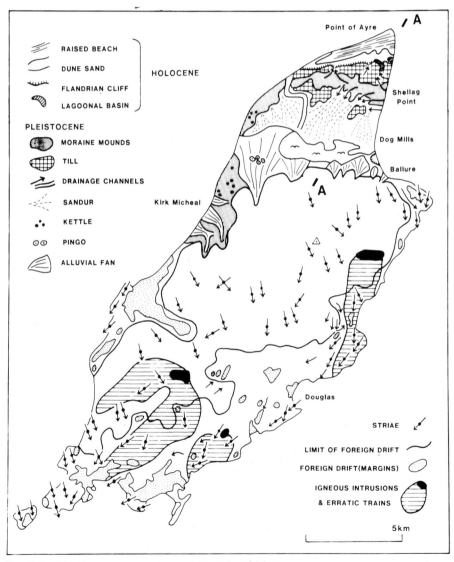

Fig. 8.3 The Quaternary geology of the Isle of Man.

The lower, the Shellag Formation, consists of a basal clay-rich till succeeded by proglacial outwash members. Included amongst the latter is a distinctive suite of sands containing an abundant macro-fauna (Lamplugh, 1903; Mitchell, 1965). This formation extends across the lower parts of the Palaeozoic rock core of the island to a height of some 150 m and intercalates with the local suite of deposits. Its limit of penetration is marked by a distinctive series of moraines in the north-west of the island around Kirk Michael (Mitchell, 1965).

The uppermost formation, the Orrisdale, comprises a complex multi-till sequence and an associated diverse assemblage of proglacial deposits. The till members terminate to the rear of the Bride Moraine, a pronounced ridge rising to more than 100 m above surrounding ground and extending westwards from Shellag Point towards Kirk Michael (Fig. 8.3). At the former locality, the underlying members of the Shellag Formation are disturbed by large-scale glaciodynamic structures (Thomas, 1984b). Large sandur sequences occur forward of the moraine and, around Kirk Michael, deep kettle basins, containing rich organic sediments, pit their surfaces (Erdtmann, 1925; Mitchell, 1965; Dickson *et al.*, 1970; Joachim in Tooley, 1977). At the Dog Mills the sandur sediments pass distally into laminated clays containing an abundant micro-fauna (Thomas, 1977; Thomas in Tooley, 1977). Towards the rock core of the island the proglacial members of the Orrisdale Formation are buried by thick alluvial fan deposits derived from the uplands. Pingos, containing organic sediments in their cores, occur on the outer parts of the fans (Watson, 1971).

At Ballure, the two foreign formations intercalate with the local suite of deposits and bury a fossil coastline showing a rock cliff, sea-stacks and caves. This is the only site in the Isle of Man, and indeed along the whole of the northern Irish Sea coastline, that provides indication of a high, probably interglacial sea level. No datable deposits are found associated with this coastline however.

Underlying the greater part of the foreign drift in the northern area of the island is a remarkably level rock platform at between 41 and 53 m below O.D. (Smith, 1930) (Fig. 8.4). North of the Bride Moraine, however, this platform drops rapidly

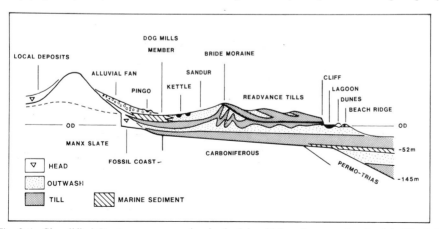

Fig. 8.4 Simplified Quaternary succession in the Isle of Man along section A–A in Fig. 8.3.

away to a depth of at least 145 m at Point of Ayre. Details of the strata resting on this rock platform are known only from boreholes and Smith (1930) recognised three major divisions. The uppermost is equivalent to the foreign glacigenic formations described and is underlain by a further complex of tills and outwash deposits some 40–50 m thick. Below this, at depths between 65 and 73 m below sea level, there is a series of richly fossiliferous silts and sands that Lamplugh (1903) described as marine deposit. Underlying them, and encompassing Smith's lowest division, is a further assemblage of till and outwash deposits of the order of 70 m thickness (Fig. 8.4).

Events preceding the Devensian glaciation

Of events preceding the last, Devensian glaciation in the area of the northern Irish Sea we know very little. Indeed, with the exception of the tills lying deeply buried below the northern Isle of Man, which probably date to at least the penultimate glaciation, basal erosion resulting from ice movement during the last glaciation seems to have stripped away all pre-existing sediment. This picture contrasts with the pattern that has emerged from the southern Irish Sea basin (Garrard, 1977) where two major Irish Sea till units occur over much of the sea floor, separated by interglacial marine sediment. This, however, is perhaps only to be expected for it is characteristic of large ice sheets that, at least in their advancing phase, deposition predominates towards the outer margin and erosion in the interior. Thus, pre-Devensian glacial deposits seem likely to have survived only in localised, overdeepened rock basins on the floor of the sea. Few of these basins have been identified, however, and none explored in any detail.

The record of pre-Devensian interglacial sediments is equally sparse. So far, no Ipswichian marine sediments comparable to the great thickness seen in Cardigan Bay (Garrard, 1977) have been observed beneath tills in marine boreholes and it seems probable that these too have been removed by Devensian glacial erosion. The extent of this erosion is indicated by the widespread occurrence in Irish Sea glacigenic sediments of an erratic shell fauna. Part of this undoubtedly derives from reworking of pre-glacial, Neogene deposits but much results from the destruction of interglacial sea-floor sediment which was covered by ice and incorporated into the basal till.

Around the margins of the basin a number of coastal sites show organic sediment beneath till. Thus, at Wigton in the Solway Lowland (Eastwood *et al.*, 1968), a borehole proved clay containing marine fauna beneath till and gravel while in Lower Furness up to 8 m of organic sediment lie between two tills (Kendall, 1881). These sites, plus the fossiliferous silts at depth below the Point of Ayre (Lamplugh, 1903), have all been provisionally assigned to the Ipswichian interglacial (Evans and Arthurton, 1973; Huddart *et al.*, 1977) but the dating is uncertain. It is also not known if these deposits are *in situ* or erratic blocks. Pollen from organic silts beneath till in Scandal Beck, Cumbria, however, does seem to be Ipswichian (Carter, unpublished) and carbon-14 analyses give minimum ages of 32 500 and

42 000 BP (Shotton and Williams, 1971, 1973).

Raised beach deposits are widespread around the margins of the southern Irish Sea basin and they have been used as an interglacial datum for the lithostratigraphic successions that occur above and below them (Bowen, 1973). This approach cannot be adopted in the northern basin for the coastlines of the Southern Uplands, the Lake District, North Wales east of the Lleyn peninsula, and the whole of the Irish coast north of Co. Wicklow, are devoid of raised beaches overlain by glacigenic deposits. In the southern basin, at least, the apparent parallelism in height of the beaches has implied structural stability (Kidson, 1977) and has been taken to indicate the return of successive interglacial seas to levels only marginally different from those of today. This does not appear to be the case in the northern basin for the evidence, fragmentary and negative though it is, suggests that the last interglacial marine datum is currently below sea level. Thus at Ballure, in the Isle of Man, the only observed fossil coastal feature in the northern basin passes beneath sea level and is fronted by a rock platform, detected by seismic interpretation, at a depth of -4 m O.D. This marine limit is considered to be of Ipswichian interglacial age for nowhere in the complex of cold climate deposits that occur above it is there any evidence for a substantive climatic break.

Two explanations could account for the apparent failure of the Ipswichian datum to attain the same relative level as in the southern Irish Sea. The first is that isostatic rebound, arising from depression of the sea floor during the Devensian glaciation, has not yet achieved equilibrium. Although it is now generally assumed that isostatic recovery is complete, there is very little hard evidence to confirm or deny this (Kidson, 1977), and the explanation remains speculative. The second explanation concerns tectonic instability in the basement. Despite the fact that the Pliocene–Pleistocene appears to have been a tectonically quiet period in western Britain, Dobson (1977) has drawn attention to the probability of movement during the Quaternary along the boundary faults that define the graben structure of the Irish Sea basin. Loading by ice during successive glaciations may have reactivated or accentuated these movements thus complicating the relative vertical movements of sea level between the glaciations.

The Devensian glacial expansion

Although the downturn into cold-climate conditions at the beginning of the last glacial stage began about 80 000 BP, it has been generally accepted that in Britain, at least, major expansion of ice from mountain sources into lowland areas did not occur on a major scale until after 25 000 BP. Of events and environments in the basin prior to this expansion we know little. Evidence from Cheshire (see Worsley, Chapter 11) and elsewhere indicated a generally cold, tundra climate punctuated by complex interstadial fluctuations. Whether the sea penetrated the northern Irish Sea basin during any of these episodes is unknown for the global fluctuations of sea level at this time are unclear and any record in the basin has probably been removed by subsequent glaciation. Some authors, however, have postulated Early or Mid-Devensian expansion into the basin, accompanied by high sea levels (Synge 1977a), and others have demonstrated marine transgression immediately prior to and consequent upon the loading of the crust by the build-up of the last Scottish ice sheet (Sutherland, 1981). The latter author has suggested that the build-up might have occurred as early as *c*.75 000 BP, implying that parts of the northernmost areas of the Irish Sea may have had both a fluctuating glacial cover and an isostatically raised sea level for much of the time interval up to major expansion after 25 000 BP.

Although there has, in the past, been some considerable dispute concerning the maximum limits attained by the Late Devensian Irish Sea ice sheet, a consensus view now concludes that it extended at least as far south as a line from Co. Wexford to Pembrokeshire (Bowen, 1973). This maximum occurred some little time after 18 000 BP for Irish Sea till seals fossiliferous cave deposits in Clwyd (Rowlands, 1971). During this advance it was confluent on its western and eastern margins with semi-independent Irish and Welsh ice caps and a lobe, also confluent with Welsh ice, passed south as far as Wolverhampton.

Using these and other limits, Boulton *et al.* (1977) have constructed a model for the Late Devensian British ice sheet at its maximum stage. For the Irish sea area, this postulates an ice-shed running approximately south-westwards from a maximum surface elevation of 1800 m over the eastern Southern Uplands. Although useful as a first approximation this model can be criticised on a number of grounds. Very briefly stated these include, first, the assumption that the ice sheet was in steady-state equilibrium. This seems unlikely for the stated equilibrium generation time of 15 000 years is too long and expansion may have been stopped short by rapid climatic amelioration after *c*.15 000 BP. Secondly, the model assumes advance across a terrestrial floor. This has by no means been proved and many authors, dating back as far as Lewis (1894), have raised the possibility that the ice sheet may have advanced into a marine margin as an ice-shelf.

Thirdly, the model is based on a modern rigid-bed analogue. Boulton and Jones (1979), however, have shown that when ice sheets move over deformable sediment much lower equilibrium profiles result. The Irish Sea basin seems an ideal candidate for such a profile for prior to the Devensian expansion it can be assumed to have been underlain, as now, by a thick sequence of soft glacial and interglacial marine sediments. A final criticism concerns the transfer rate of rock eroded from beneath the centre of the postulated ice-dome. According to the model some 19 000 years would be required to transport this debris 200 km to North Wales. Thus one would not expect to find rock eroded from beneath the centre of the ice sheet to be very widely distributed. The geological evidence indicates the converse, however, and Southern Upland granitic erratics occur abudantly in Devensian Irish Sea glacial deposits not only in North Wales but also as far south as Co. Wexford and Pembrokeshire. Some of these certainly have been recycled through successive glacial stages but very doubtfully all. Some faster rate is, hence, implied.

If these criticisms are valid it is possible to construct a 'minimum' model for ice geometry somewhat different from the 'maximum' model proposed by Boulton *et al.* This would have three components. Over the rocky and mountainous fringe of southern Scotland and much of Ireland the ice sheet would have a gradient and thickness commensurate with a source-area, rigid-bed analogue. Thus, over most of the northern part of the northern Irish Sea basin ice thicknesses would have probably been in excess of 1500 m. To the south, however, as ice passed out into the basin proper, a lower-gradient thin ice sheet would develop as a function of passage over a floor of deformable sediment. Further to the south, this relatively thin, low-profile sheet may have passed out into the eustatically lowered sea level as a floating ice-shelf. Such a reconstruction would explain the apparent paradox that whilst there is abundant geological evidence for ice in the southern part of the Irish Sea basin, there is no evidence for a heavy load on the crust. Thus strong subsidence would have been restricted to the thick ice conditions in the northern fringe with the thin ice of the south generating no more than slight subsidence.

The Devensian deglaciation

Subsequent to its peak around 18 000 BP, the ice sheet retreated rapidly. The deglacial chronology is poorly known but the ice margin had cleared the Isle of Man by at least 15 150 BP (Bowen in Tooley, 1977), the Lleyn peninsula by 14 680 BP (Coope and Brophy, 1972) and the Solway by c.13 000 (Jardine, 1975). Conventionally, it has been assumed that the retreat was terrestrial and that the retreating ice margin was not in contact with the eustatically lowered sea level. Only during the very late stages of retreat, after 12 000 BP, when eustatic rise and isostatic response were accelerating did the sea follow the retreating ice margin to cut a series of littoral shorelines in front of it in north-east Ireland (Carter, 1982) and south-west Scotland (Jardine, 1975).

Although a number of workers have suggested high glacial sea levels during the period 18 000 to 13 000 BP (Stephens *et al.*, 1975; Stephens and McCabe, 1977; Synge, 1977a, 1980), the evidence has been primarily geomorphological and not founded on the recognition of *in situ* glacio-marine sediment. Recent offshore work by Pantin (1975, 1977, 1978) in the area between the Isle of Man and Cumbria, however, has provided very important evidence for relatively high sea-level during this critical period. He demonstrated that the proglacial deposits of this area, which lie above till and beneath Flandrian marine sediment, were deposited in a cold-water, low-salinity proglacial lagoon, termed Vannin Sound, connected to the open sea. This sea lay to the west, between the Isle of Man and Ireland, where the Irish Sea is deepest. This sea was partially separated from the lagoon by some sort of barrier and Pantin suggests that this may have been a floating ice-shelf possibly divided into two by the Isle of Man. The deposits of this lagoon were sub-tidal rather than inter-tidal, and were fed by sediment-laden meltwater streams. Dropstones and dump structures within the deposits indicate a covering of icebergs and sea ice.

Independent of Pantin, the present author (Thomas, 1976, 1977, and Thomas in Tooley, 1977) has described a similar suite of deposits from the Dog Mills, Isle of Man, some 20 km north-west of Pantin's Vannin Sound. These deposits occur at between 2 and 9 m O.D. and comprise alternating sequences of sand and mud, passing northwards into outwash sandur sediments fronting the Bride Moraine. The deposits contain a rich micro-fauna, including Foraminifera and Ostracoda, in an assemblage very similar to that described by Hughes (Appendix 1, in Pantin, 1978) from the offshore lagoonal deposits. The environment indicated at Dog Mills is cold, low-salinity, estuarine–intertidal with an open-water, current-swept element. The Dog Mills deposits were formed contemporaneously with the Bride Moraine, which was itself created around 15 150 BP.

If we unite the deposits of Vannin Sound with those of the Dog Mills, on the assumption that the latter represent either the shallow-water equivalent of the former, or its northern transgression, we can conceive a picture of glacial conditions in this part of the basin at approximately 15 500 BP (Fig. 8.5). To the east of the Isle of Man a narrow arm of the sea extended northwards to abut directly against an ice margin running from the Bride Moraine eastwards towards Cumbria. As water depths in this area would have been relatively shallow, between 20 and 40 m, it is doubtful if an ice-shelf formed along this margin and it was probably grounded. To the east and west, where the ice margin rose against the bedrock highs of the Isle of Man and Cumbria, the margin was terrestrial and large volumes of proglacial outwash accumulated. As the present coastal areas of North Wales and Lancashire were free of ice at this time, periglacial river systems extended outwards from the position of their present mouths across the floor of the eastern Irish Sea.

To the west of the Isle of Man the ice margin probably extended to the Kells

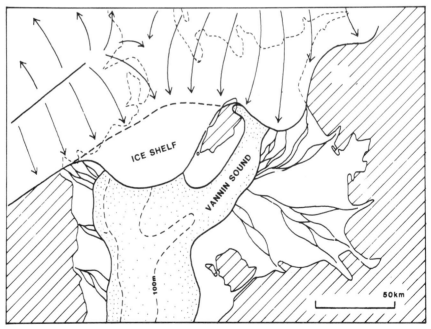

Fig. 8.5   Tentative reconstruction of the northern Irish Sea around 15 000 BP.

moraine of eastern Ireland, with which the Bride Moraine has been commonly correlated (Charlesworth, 1939; Synge, 1952; Penny, 1964; Sissons, 1964; Gresswell, 1967a; Mitchell, 1972; Stephens *et al.*, 1975; Stephens and McCabe, 1977). As water depths along this margin would have been much greater, extending perhaps to 120–140 m, much of it probably stood as a floating ice-shelf fed from the retreating but coeval, Irish and Scottish caps. It is significant that in this area proglacial sediments are missing and Flandrian marine sediments rest directly upon till (Belderton, 1964; Pantin, 1978). This implies, as many have pondered (Synge, 1977a; Thomas, 1977), that the lithologically uniform and widespread shelly Irish Sea till, seen with little variation from Cumbria to Wexford, may, in part at least, be a glacio-marine deposit formed beneath this floating ice-shelf. As in the eastern part of the basin periglacial rivers systems and proglacial drainage from the margins of the Irish Cap extended out across the present sea floor to cut channels into this sediment off Dublin and Kish Bank area.

Two implications arise from this reconstruction. The first is that if marine conditions accompanied the ice margin well north in the basin at *c.*15 000 BP, it may well have done so throughout retreat. If so, this alters radically our current conception of the nature of glacigenic sedimentation in much of the basin but also provides an explanation for the apparently rapid disintegration of the ice sheet between *c.*18 000 and *c.*15 000, for retreat rates in water would have been substantially higher than on land.

The second implication concerns the relationship between isostatic and eustatic sea level needed to explain a sea at a height only a little below that of the modern during a glacial stage. The interplay between an isostatically rebounding crust, a eustatically rising sea level and possible hydro-isostatic effects is yet to be determined and is probably further complicated by local factors such as tectonic activity, differential ice loading and forebulge effects. Take isostatic effects first. The 'maximum' model of Boulton *et al.* (1977) implies that substantial isostatic depression, probably of the order of 400 to 500 m, could have occurred beneath an ice sheet up to 1800 m thick extending as far south as Wexford. This would place much of the basin floor well below even the lowest estimate for eustatic fall and allow for basin reflooding on retreat. It is very doubtful, however, if this amount of depression was achieved due to the short time scale available for advance and retreat. The 'minimum' model, proposed earlier, would restrict significant isostatic depression to parts of the northern basin only. This accords with the observations of Carter (1982) who argued that isostatic depression was limited to Ulster and the periphery of southern Scotland, with much of the central part of the Irish sea rising as a forebulge and the southern basin stable. The maximum amount of isostatic depression has been variously estimated as between 150 and 250 m.

The critical factor is the level of the eustatic sea for if it was very low, and below the maximum amount of isostatic depression, the sea could not gain entry. Even if relatively high it would have to be high enough to counteract any forebulge effect in its transgression northwards through the basin. Estimates of the eustatic sea level at the critical time range from −40 m (Cronin, 1982) to −110 m (Milliman and Emery, 1963), but Kidson (1977) has argued that the weight of evidence now demonstrates that the eustatic sea level during the later stages of the last glaciation was towards the lower end of this range and may, at certain times, have been only a little below that of today. If so, then the basin could have been readily flooded. Although it is difficult to estimate the amount of forebulge rise, a water level as low as probably − 50 m below the present would be sufficient to allow the sea to penetrate the deep-water channel that occurs down the centre of the Irish Sea basin. Differential isostatic loading near the ice margin may have depressed the sea floor to such an extent along part of the Irish coast that the current coastline may have been partially inundated. Consequently, the site of glacio-marine sedimentation at Tullyallen, Co. Louth (Colhoun and McCabe, 1973) may relate to this sea (Synge, 1977b).

It is not known for how long after 15 000 BP the basin was flooded or how far north the sea transgressed. The unconformity that separates the Vannin Sound lagoonal deposits from overlying Flandrian marine sediment, however, demonstrates that a relative fall must have occurred after the deposition of the lagoonal, glacio-marine sequence. This was probably caused by accelerated isostatic uplift outstripping any concomitant eustatic rise. This fall resulted in at least the partial evacuation of the basin. The timing of this event is unknown, but relatively low sea levels are indicated at 12 290 BP in the Solway (Bishop and Coope, 1977) and at a similar date in Ulster (Morrison and Stephens, 1965).

Subsequently, around the Late Devensian/Holocene boundary, as isostatic rebound diminished and major eustatic transgression began, the basin was reflooded. This caused vigorous reworking of the Vannin Sound glacio-marine sequence and the deposition of intra-tidal and sub-tidal sediments upon them (Pantin, 1978). As the sea level continued to rise in the Flandrian it was interrupted by a number of major eustatic fluctuations (see Tooley, Chapter 6), but its return in Northern Ireland and the Solway was complicated by isostatic movements that did not cease until 5000 BP (Andrews *et al.*, 1973; Bishop and Coope 1977; Carter, 1982).

In the Isle of Man there is no evidence for further ice advance after 15 000 BP. Following this time, much of the glacial detritus in the uplands was soliflucted and a generally cold, periglacial climate prevailed. As the regional permafrost declined much of this was reworked by fluviatile action and a series of large alluvial fans built up around the island margin during Late-Glacial times.

A recurrent theme in the Devensian glaciation of the northern Irish Sea basin is the concept of a major, so-called 'Scottish' readvance episode that occurred at a late stage in deglaciation. Its first report was by Trotter (1929) who recognised an 'Upper Boulder Clay' above 'Middle Sands' in Cumberland. He regarded this upper till as a product of a readvance of Scottish ice quite separate from deposits of the main glaciation. This concept was refined by Trotter and Hollingworth (1932) and extended by Charlesworth (1939) as the 'north-east Ireland–Isle of Man–Cumberland' moraine (Line 1, Fig. 8.6). Subsequently, Charlesworth (1939) revised this line to correlate the Bride Moraine with the Carlingford Moraines of Ireland (Line 2, Fig. 8.6). Since this time, numerous workers have drawn many different lines for the marginal limits of this supposed readvance (Fig. 8.6), but others do not recognise it at all (Pennington, 1970; Evans and Arthurton, 1973; Sissons, 1974a). In a review of the concept Huddart *et al.*, (1977) argue that contrary to these latter views there is substantive morphological and stratigraphic evidence to support a broadly correlative readvance limit between Ireland, the Isle of Man and Cumbria at some time between *c.*13 000 and *c.*15 000 BP.

Morphological evidence is unreliable (Richmond, 1959; Flint, 1970), for a large morainic structure may mark a readvance, a recessional stillstand or even a terminal maximum. Amongst the stratigraphic criteria used the presence of one till upon another is the one most commonly adopted. Boulton (1977) has demonstrated from work in north-west Wales, however, that multiple till successions may arise within a single glacial episode without the necessity for invoking readvance of the snout. Even where evidence for override exists this does not necessarily indicate a readvance of wide extent. Thus recent work by the author in the Isle of Man (Thomas, 1976, 1977, 1984b), in Ireland (Thomas and Summers, 1983, 1984) and in Cumbria (unpublished) demonstrate that marginal readvance during the retreat of the Devensian Irish Sea ice sheet was almost endemic. In all cases, however, these readvances represented short-distance, localised oscillations of the snout. They consequently have no significant stratigraphic importance and do not mark ice-sheet fluctuations driven by climatic change. This implies that moraine structures

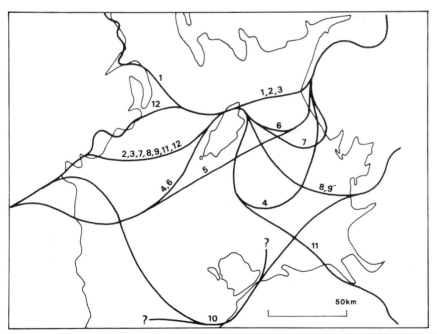

Fig. 8.6 Postulated limits for the 'Scottish Readvance' and correlations with the Bride Moraine. *1.* Charlesworth (1926); *2.* Charlesworth (1939); *3.* Synge (1952); *4.* Cubbon (1957); *5.* Mitchell (1960); *6.* Mitchell (1963); *7.* Penny (1964); *8.* Sissons (1964); *9.* Gresswell (1967a); *10.* Saunders (1968b); *11.* Mitchell (1972); *12.* Stephens *et al.* (1975), Stephens and McCabe (1977).

built during such oscillations cannot be used for long-distance morphological correlation as they are not time-parallel events.

In the author's view the concept of a major 'Scottish Readvance' episode during the Devensian deglaciation is largely illusory, is devoid of stratigraphic or chronological foundation and cannot be used as substance for stratigraphic classifications based on an assumed response to climatic change. The present evidence suggests that the deglaciation was not punctuated by major readvance but was characterised by rapid retreat accompanied by minor snout oscillation caused by essentially local controls.

Quaternary geomorphology of the southern Lake District and Morecambe Bay area

Introduction

The southern Lake District and the fringing hills of Morecambe Bay comprise a compact area of about 1800 square kilometres of exceptional beauty and variety. Much of this variety is a direct consequence of the interplay of two factors: the intensity of glacial erosion and the resistance of the rocks to denudation.

To the north of the region the dominant rock types are the lavas and tuffs of the Borrowdale Volcanic Series of Ordovician age. These outcrops form the topographic centre of the Lake District and here we find a classic glaciated highland landscape with its assemblage of corries, troughs, moraines, and so on. South of a line running from Broughton-in-Furness (SD 212875) to Ambleside (SD 376045) and across to the Shap Fells, the resistant volcanic rocks give way to less resistant Silurian slates, grits and flags. This transition is clearly marked in the topography which is subdued and mostly below 300 m (Fig. 9.1). Nevertheless the presence of deep lake basins such as Windermere and Coniston, extensive till deposits in the valleys and glacially smoothed rock knolls, all point to a widespread and, in places, intense glaciation. Fringing the Silurian outcrops and within the immediate coastal zone are horst-like outcrops of Carboniferous limestones which form an important area of karst topography.

In examining the Quaternary evolution of the region it is worth bearing in mind the growing body of evidence which supports the view that, overall, glacial modification of the Tertiary landscapes in northern England was more limited than previously thought. For example, in Kingsdale, in the nearby Yorkshire Dales, isotopically dated speleothems some 130 000 years old are preserved in drained phreatic tubes only 13 m or so above the present thalweg. This implies that in the glacial phase of the Devensian cold stage the valley floor was lowered by less than 13 m (Waltham, personal communication). If such valley-deepening estimates were applied to the Lake District they would fail hopelessly to account for the many deep glacial troughs. Clearly glacial erosion has been effective, but it is equally evident that it was very selective.

Chronology

Wolstonian cold stage

Some buried tills below organic sediments in Low Furness may pre-date the Late

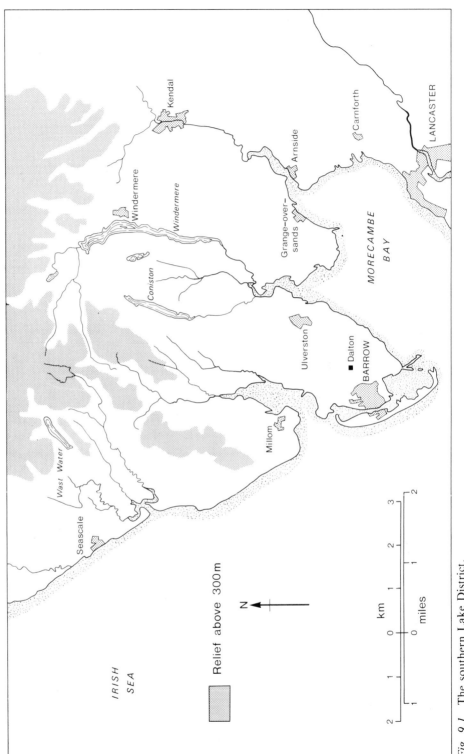

Fig. 9.1 The southern Lake District.

Devensian glaciations and are tentatively placed in the Wolstonian (Huddart *et al.*, 1977; Evans and Arthurton, 1973).

Ipswichian temperate stage

At Dalton-in-Furness (SD 247746) organic deposits were discovered more than a century ago between two till units (Bolton, 1862; Hodgson, 1862). The deposits were not subject to any rigorous examination but the leaves and fruit receptacles of beech (*Fagus*) were identified. If confirmed, the deposit is definitely interglacial. More recently, attempts to relocate these deposits have failed. Hodgson (1867) also described a quarry section near Ulverston (SD 295783) in which a griked limestone surface was capped with up to 3.5 m of till; the grikes themselves were filled with a stone-free tenacious yellow clay. This clay is probably an Ipswichian weathering deposit similar to that described by Pigott in Derbyshire (Pigott, 1965).

The cave systems of the Morecambe Bay karst frequently contain passages choked with glacial deposits (Ashmead, 1974) and their formation must, therefore, date to at least the Ipswichian. The larger cave systems presumably date back several interglacials.

Devensian cold stage

In the southern Lake District there is little evidence of geomorphic activity in the pre-glacial Devensian – that is, the Early and Middle Devensian. During the Late Devensian the whole area was inundated by ice and there is evidence for a complex sequence of glacial oscillations. The lower boundary of the Late Devensian is put at 26 000 BP and the upper boundary at 10 000 BP (Shotton, 1973). Throughout much of the region one glacial till unit is associated with the Late Devensian ice sheet that was centred over the Lake District. This unit is often a grey to reddish-brown tenacious till and in Morecambe Bay is up to 55 m thick (Knight, 1977). In south-west Cumbria the picture is complicated due to the incursions of ice which occupied the Irish Sea basin. Detailed sedimentological work by Huddart (1971b) suggests the following sequence of events.

Phase 1   At an early stage in the Late Devensian valleys draining westward from the central Lake District mountains became filled with proglacial sands and gravels laid down in front of advancing valley glaciers. These glaciers then deposited a grey basal till over the fluvioglacial sediments.

Phase 2   The local valley glaciers were then overrun by a combined Scottish–north Lake District ice sheet which deposited a basal red till above the local grey till and formed the extensive drumlin fields of south-west Furness. At the end of this phase much ice decayed *in situ* and sub-glacial meltwater activity produced complex sequences of meltwater channels and fluvioglacial deposits on the fells around Black Coombe (Smith, 1967, 1977).

Phase 3   Two tills are present on Walney Island and the mainland as far east as a line from Roose (SD 222694) to Roosebeck (SD 256678). The lower till unit has a high percentage of Borrowdale Volcanic rocks, Silurian grits and Carboniferous limestones, whilst the upper till contains erratics from the western Lake District and Irish Sea Basin. The upper till is thought to have been deposited by a readvance of the Irish Sea ice. Huddart named this advance the Low Furness Readvance.

Phase 4   The Scottish Readvance (see also Thomas, Chapter 8; Gale, Chapter 15), is represented by a second readvance of the Irish Sea ice. This ice impinged on the coastal plain north of Black Coombe and its deposits are confined to altitudes below 60 m. At the Annaside–Gutterby Spa moraine (SD 093854) a complex sequence of sand, gravels and tills record the incursion and wastage of this readvance. The basal units of the sequence comprise the Annaside Sands and Gravels which are thought by Huddart to indicate a distal sandur environment. Above these units are tills, sands and gravels of the Gutterby Spa Complex which are interpreted as deposition from basal layers of the Irish Sea ice which advanced with little disturbance over the Annaside Sands and Gravels. This ice is thought to have decayed *in situ* and produced the kettle-holed and hummocky topography.

Late Devensian, Late-Glacial sub-stage

The Late-Glacial sub-stage lasted from about 15 000–10 000 BP and in the southern Lake District its details have been carefully examined by Pennington in numerous cores of littoral sediments especially from Lake Windermere (Pennington, 1978; Coope and Pennington, 1977). Three horizons are recognised in the cores (Table 9.1):

(i) *Lower Laminated Clay*   Paired laminations of silt and clay overlying cobbles and gravels are interpreted by Pennington as marking the retreat of active ice from the Windermere basin. Plant fossils are absent but chironomid larvae and Trichoptera have been recovered (Coope and Pennington, 1977). The upper boundary of the Lower Laminated Clay is regarded as the end of the main Devensian glaciation of the region. Silts and clays not trapped in the lake basins were swept out into Morecambe Bay. During periods of intense periglacial desiccation strong winds winnowed the silts from these outwash deposits to form thick loess which was blown onto the surrounding ice-free hills (Vincent and Lee, 1981).

(ii) *Windermere Interstadial sediments*   Fossiliferous grey-brown organic silts about 50 cm thick in the littoral zones of Windermere show that seasonal melt from active ice was no longer reaching the lake by 14 500 BP. This climatic amelioration and its associated ice wastage was a much earlier event in the southern Lake District than its counterparts in north-west Europe and probably reflects the influence of the warming of the eastern North Atlantic (Ruddiman *et al.*, 1977).

Plate 1. View south-east from slopes of Great Gable to Broad Crag (centre left peak). Note the geological features of the scenery developed in the Borrowdale Volcanics Group. The distinctive bedding in sediments (formed from volcanic ashes) is well seen below the snowline. Fractured rock in fault zones has been eroded to produce gullies of Girta Gill (left) and Piers Gill (right). The valley sides and uplands also show the effects of intense glacial erosion. (photograph – F. M. Broadhurst).

Plate 2. Howgill Fells – Bowderdale. Late Glacial screes and solifluction stages have modified the slopes in this valley following a phase of valley glacier erosion. Holocene debris cones and recent flood deposits are also present (left foreground). (photograph – A. M. Harvey).

Plate 3. The Winnats Pass, near Castleton, Peak District. This deeply incised gorge in the Carboniferous limestone is part of a resurrected Lower Carboniferous sea floor topography. The original submarine gorge was filled by marine muds during the Upper Carboniferous. These muds (now shales) have been easily removed by erosion in very recent geological times with excavation of the feature being dependent upon the removal of the mudrocks in the Hope valley which is located to the north and downstream of the gorge. The Lower Carboniferous topography has, as yet, been little modified by erosion. (photograph – F. M. Broadhurst).

Plate 4. General view of the Peak District limestone plateau. The deeply incised valley of the River Wye is indicated by the presence of woodland in the middle of the photograph. (photograph – John Gunn).

Plate 5. Shale vales and gritstone escarpments near Hayfield, north Derbyshire. The sandstone escarpments are part of a synclinal fold which pitches southwards (to the right). The upper scarp-forming rocks belong to the Lower Coal Measures and Upper Millstone Grit (Carboniferous) series. The valley floor is covered with till. (photograph – R. H. Johnson).

Plate 6. Longdendale. The high plateau areas of the Upland Summit Surface provide a strong contrast in relief to that shown in Plate 5. The gently dipping Kinderscout Grit (sandstone) strata provide strong scarp and bench features overlooking the valley and its reservoirs. To the left of the picture the slopes are covered with landslip debris. (photograph – R. H. Johnson).

Plate 7. The Kinderscout Plateau seen from the west. The slopes beneath this massive sandstone scarp-former were once highly unstable and show much evidence of past landslide activity. The lower slopes have no glacial deposits covering them but were probably covered by ice or snow during the Devensian glacial period. (photograph – Tom Spencer).

Plate 8. View of the Oakwood Quarry looking north-westwards in August 1977. The exposed succession is almost perpendicular to the buried valley trend; the quarry valley floor lies just above the bedrock. The white sands are the Chelford Sands Formation and the black horizon corresponds to the interstadial organic-rich sequence. Here it consists of two separate palaeochannel fills. Above, the red-brown material is the Stockport Formation which is removed prior to quarrying of the white sands. (photograph – Peter Worsley).

Table 9.1 Relationship between North-west European and British Late-Glacial chronologies and the Windermere sediments.

Late Weichselian	¹⁴C years BP	Late Devensian	Windermere sediments	¹⁴C years BP
———	10 000	———		
Younger Dryas		Pollen Zone III		
			———	10 500
			Upper Laminated Clay	
———	11 000	———	———	11 000
Allerød		Pollen Zone II		
———	11 800	———		
Older Dryas			Windermere Interstadial	
	12 000			
Bølling		Pollen Zone I		
———	13 000		———	13 000
			———	c.14 500
			Lower Laminated Clays	
	c.25 000	———		
Middle Weichselian		Middle Devensian		

Source: Coope and Pennington, 1977; Pennington, 1978.

(iii) *Upper Laminated Clay* These sediments, only 50 cm or so thick comprise some 400 paired varves which represent the re-establishment of corrie glaciers in the Windermere catchment between *c.*11 000 and 10 500 BP. This climatic recession is correlated with the Loch Lomond stadial in Scotland (Sissons, 1980). In addition to the formation of moraines, periglacial activity was widespread. Gelifluction and talus formation on many areas of high ground is thought to date from this intense, but short-lived, cold period (Sissons, 1980; Vincent and Lee, 1982). Both Manley (1959) and Sissons have made estimates of temperatures during this period. From his estimates of corrie glacier budgets and his identification of fossil rock glaciers thought to be active at this time, Sissons suggests a mean annual temperature at sea level of no higher than 1 °C and a July mean of slightly below 8 °C.

Glacial erosion

At the maximum of the Late Devensian glaciation it is thought that the Lake District mountains were covered by a dome of ice. Precisely how thick this ice-cap was is a matter of speculation but it must have been thick enough to maintain end-moraines as far south as the Cheshire Basin. Movement of ice away from the centre of the ice-dome was essentially radial (Fig. 9.2). Linton (1957) was of the opinion that repeated radial ice movement fundamentally transformed a simple north–south Tertiary drainage system to produce the now familiar radial drainage pattern. This idea has much merit and there is no doubt that at higher levels in the ice cover

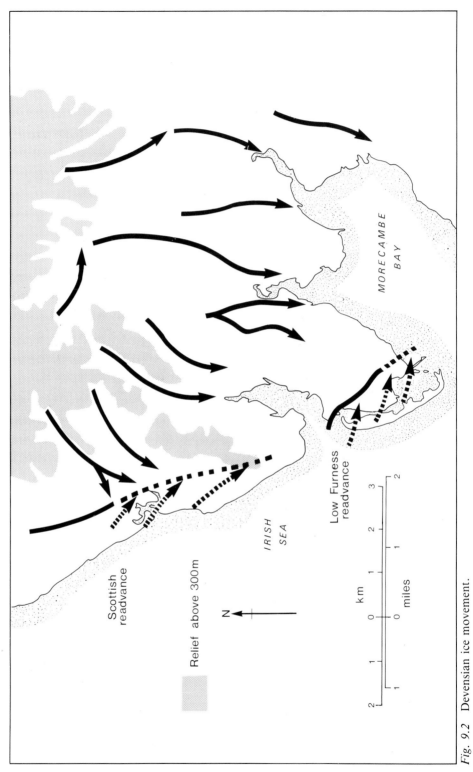

Fig. 9.2 Devensian ice movement.

movement was controlled by the gradient of the ice surface and not by the local relief. It is also true to say that several of the radial alignments now occupied by valleys are in fact structural lineaments and the picture is more complicated than the simple model proposed by Linton.

In the south-western Lake District ice from the Wast Water basin and Eskdale was diverted southward by the combined Irish Sea/Scottish/north Lake District ice sheet. To the east, ice on the Howgill Fells effectively blocked ice movement and Lake District ice was deflected to the north and south. The southerly deflection is well marked by the train of Shap granite erratics which has been traced down the Kent valley and into Morecambe Bay (Marr, 1916).

Corries

These more or less semicircular recesses caused by intense local glacial erosion are amongst the best known glacial landforms in the Lake District. Several researchers have discussed their location, aspect and morphology (Manley, 1959; Temple, 1965; Clough, 1977). Many corries appear to have been initiated in the upper parts of structurally controlled river valleys. The presence of faulting, as at Easedale Tarn (NY 306085) and Stickle Tarn (NY 288078) and high angle jointing, facilitated dilation and plucking, and is characteristic of many Lake District corries. Whether or not a potential site developed into a corrie must have depended very much on the availability of an adequate supply of snow, the protection of the snow from insolation, and altitude. Both Manley and Temple suggest that the dominant winds at the time of corrie glacier activity were south-westerly and associated with cyclonic disturbances. The snow which fell on the pre-glacial upland surfaces would tend to drift into the north-easterly facing corrie sites which offered the best protection against ablation. The efficiency of this process can be judged by the fact that during the Loch Lomond Readvance corrie glaciers accumulated in, and in several case flowed from, their corrie confines in little more than 500 years.

There are still many imponderables associated with the Lake District corries: we have no good evidence as to their age other than that they are post-Tertiary: we cannot assume that they were protected from glacial erosion when the whole area was inundated by ice; we have little knowledge of the balance between glacial and non-glacial geomorphic activity within the corrie basins. In these respects they remain enigmatic features.

Glacial troughs

The glacial troughs of the Lake District form a radiative dispersal system which discharged ice away from a central ice-dome. Many of the troughs are of Linton's alpine type in which a convergent series of corries hang above the main trough (Linton, 1963), as in the case of Great Langdale (NY 295064). In this type of trough ice has used and modified a previously existing valley, even if the modification has transformed it beyond all recognition. Diffluence and transfluent troughs are also found in the southern Lake District. Simple diffluence troughs are created by ice which breaches the divide by branching sideways from an existing valley. A good example can be found at the head of Blea Tarn (NY 289052) where the combined ice from Mickleden and Oxendale breached the southerly divide and flowed south-eastward into Little Langdale (Fig. 9.3a). Transfluent troughs are created at the head of a valley system. In the Lake District the main east–west divide has been breached in several places by north–south transfluent troughs. The necessary conditions for their formation is simply the lack of coincidence between the iceshed and the underlying watershed. Several of the main passes in the central region of the Lake District are probably of this type, for example Kirkstone Pass (NY 403083) and Gatescarth Pass (NY 474094) (Fig. 9.3b).

The rock floors of the major troughs are characterised by overdeepend stretches alternating with rock bars, as at Chapel Stile in Great Langdale (NY 325050) although much post-glacial alluviation has obscured this general characteristic in many valleys. Overdeepened valley floors are often the case at valley junctions and the northern basin of Lake Windermere may be cited as an example of such a confluence basin where ice from Great Langdale was joined by ice from the Rothay valley and Stock Ghyll.

Glacial deposition

Studies of the tills of southern Lakeland and the Morecambe Bay region are very limited and only in the coastal region of south-west Cumbria do we have detailed stratigraphic and sedimentological descriptions. Tenacious lodgement tills are confined to the valley bottoms in the central Lake District valleys but on the coastal lowlands of Furness and Lonsdale they mask much of the landscape.

On the coastal plain of south-west Cumbria two till units, separated by sands and gravels are common and good exposures can be seen on the coast between the Esk valley and Black Coombe and also on Walney Island. An early interpretation of the 'middle sands and gravels' was that they were formed during a waning of the ice which deposited the lower till; a further advance produced the upper till unit. More recent work by Huddart has shown that the sands and gravels are in fact the proglacial sediments of an advancing ice sheet. The upper till unit of complex tripartite sequences is now often interpreted as a flow till, ablation till or even an englacial till (Boulton, 1972). Although such thin, local, tills do exist, as in the lower zones of the Gutterby Spa Complex (Tooley, 1977), the upper tills of south-west Cumbria appear to be of regional importance and contain erratics from Scotland and the Irish Sea basin. Till fabric analysis indicates a west–east or NW–SE direction of ice movement.

Moraines

It is now generally thought that the Late Devensian ice wasted quickly, and often *in situ* (Pennington, 1978). There are, however, several stadial moraines which reflect the pattern of ice wastage back into the central mountains of the Lake

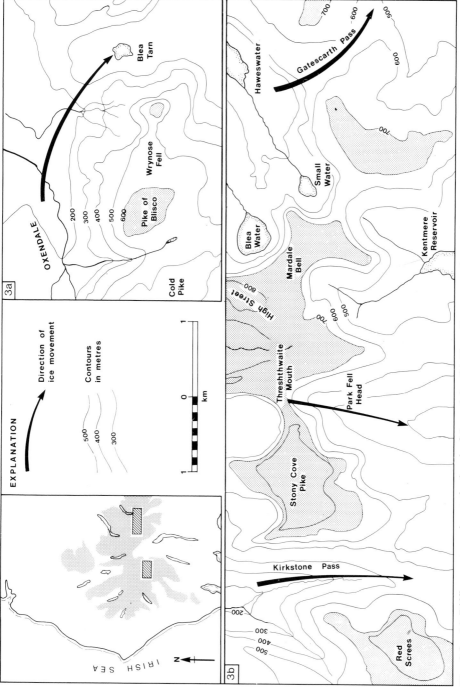

Fig. 9.3 (a) Diffluence troughs at the head of Langdale. (b) Transfluent troughs north of Kentmere.

District and Pennine hills. The detailed mapping by Gresswell (1951, 1962) showed that large stadial moraines exist in the Rusland valley at Haverthwaite (SD 336838), at the southern end of Coniston Water at Nibthwaite (SD 294895), at the southern end of Windermere at Newby Bridge (SD 375864) and at Church Town in the Lyth valley (SD 443905).

At Windermere we know that the lake had become free of ice by 14 500 BP and these outer moraines represent some as yet undated stage which must lie between 18 000 and 14 500 BP. Similarly, the large moraine blocking the upper Conder valley (SD 520645), north-east of Lancaster, must also date from this period. This moraine, which is some 30 m high and 1 km long probably barred the River Lune from an older route down the Conder valley and diverted the river westwards on its present course.

Sometime after 14 500 BP until the end of the Windermere Interstadial the southern Lake District and Morecambe Bay area were free of ice and no varved sediments were formed in the Lakes. But during the 500 years or so of the post Windermere Interstadial climatic recession glaciers grew in the corries and produced a variety or retreat moraines. Recently, Sissons (1980) has produced detailed maps showing the extent of this glacial phase. He recognises several types of moraine associated with this corrie glaciation including marginal moraines at the ice limit, fresh hummocky moraines inside the end-moraines and fluted moraines. The latter are especially well developed in Oxendale (NY 265055). Detailed palynological investigation by Pennington clearly fix the age of the fresh morainic forms. When cored, the corrie tarn sediments do not yield any Late-Glacial pollen, whereas lakes in shallow open basins outside the morainic limits contain a full sequence of Late-Glacial sediments which can be correlated with the Windermere stratigraphic record.

Drumlins

Drumlins are extremely well developed in the region and occur in two major swarms (Fig. 9.4). In the Dalton-in-Furness region the drumlins have been mapped in some detail (Gresswell, 1962; Grieve and Hammersley, 1971). A second impressive swarm can be seen on the low ground between Kendal and Lancaster. Smaller swarms also occur in the Windermere and Coniston basins. Detailed morphometric analyses of these drumlins swarms is still awaited.

Fluvioglacial erosion and deposition

At several sites in the southern Lake District meltwater from the wasting Devensian ice-cap produced complex systems of channels and deposits. In south-west Cumbria Smith (1967) has mapped a sequence of meltwater channels on the western flank of Waberthwaithe Fell (Fig. 9.5a). When examined in detail the channel floors were found to have a variety of gradients and the view that these channels drained water from a glacially dammed lake in Eskdale is clearly untenable.

They are thought by Smith to have formed as a sub-glacial system under a thin wasting, ice margin, the general gradient of which determined the overall

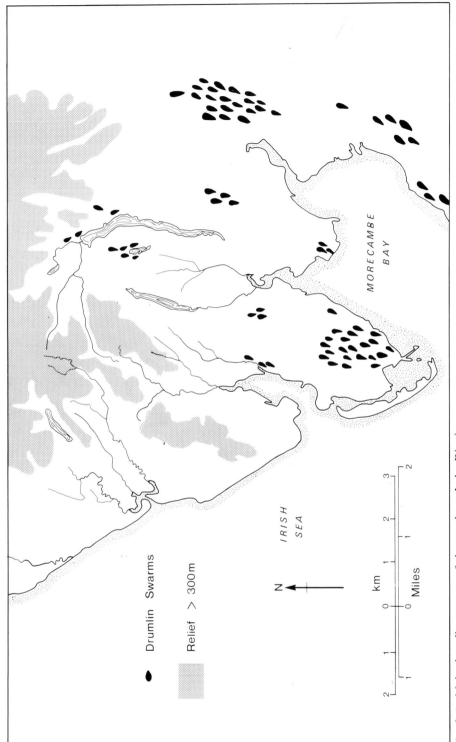

Fig. 9.4 Main drumlin swarms of the southern Lake District.

Drumlin Swarms

Relief > 300m

MORECAMBE BAY

IRISH SEA

N

km
2 1 0 1 2 3

Miles
1 0 1 2

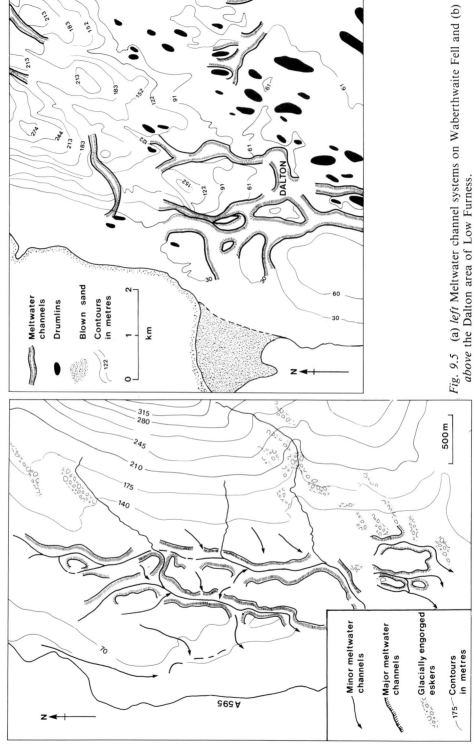

Fig. 9.5 (a) *left* Meltwater channel systems on Waberthwaite Fell and (b) *above* the Dalton area of Low Furness.

north–south slope of the channel floors. At several places, however, the ice was unable to channel meltwater along the fellside and meltwater cascaded downslope to form sub-glacial chutes. Gravels and sands choked many of the chutes to form sub-glacially engorged eskers.

In the Dalton area of Low Furness a complex anastamosing system of meltwater channels developed as ice wasted in the Duddon Estuary. It is not clear if this sub-glacial system related to wastage of the Furness Readvance ice or local ice from the Duddon valley (Fig. 9.5b). In the Carnforth area, north of Lancaster, there are extensive deposits of fluvioglacial gravels immediately to the south of a large drumlin swarm (SD 505715). A complex sequence of sub-glacial rock-cut channels was also formed on the southern flanks of Littledale Fells (SD 535605) as Lonsdale ice wasted (Moseley and Walker, 1952).

As soon as Morecambe Bay became ice-free, fluvioglacial sediments began to accumulate to form a large sandur (Knight, 1977) (Fig. 9.6). Drilling logs from the Morecambe Bay barrage feasibility studies indicate that coarse gravels are absent which presumably indicates that course sediments were trapped in the deeper lake basins.

Periglacial activity

The southern Lake District contains the highest peaks in the region and above

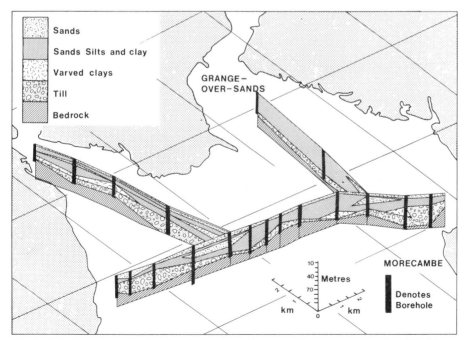

Fig. 9.6   Ribbon diagram showing the glacial sediments of Morecambe Bay (modified from Knight, 1977).

about 850 m one can find much evidence for frost shattering, and here and there, poorly developed patterned ground. In the Late Devensian, however, it is clear that conditions were conducive to extensive periglacial activity. As soon as Morecambe Bay became free of ice the accumulating outwash sediments were winnowed by winds which blew clouds of loess onto the surrounding limestone hills (Vincent and Lee, 1981). At the same time it is thought that extensive talus slopes were forming on the flanks of Whitbarrow (SD 450845) and Arnside Knott. At these sites loess and talus are intimately mixed and bound by calcrete to produce thick cemented screes. As the hills became ice-free extensive snow patches developed and nivation hollows formed on many of the hills (Vincent and Lee, 1982) (Fig. 9.7 and 9.8).

During the Loch Lomond Readvance, periglacial activity was intense. On the west-facing slope of Scafell large gelifluction lobes cover 2 km^2. At High Fell (NY 160080) on the northern side of Wast Water, Sissons has identified a large valley-wall rock glacier 1 km long and some 200 m broad. At lower altitudes conditions were severe enough to produce involution structures in periglacial slope deposits near Dalton-in-Furness (Johnson, 1975).

Karst landforms

Carboniferous limestone crops out over much of the Morecambe Bay region but west of the Cartmel Peninsula it is masked by glacial drift. The Late Tertiary uplift faulted and tilted the outcrops and today they form horst-like blocks up to

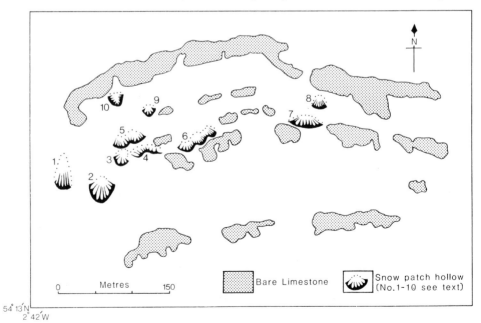

Fig. 9.7 A part of the northern scarp of Farleton Fell showing ten snow patch hollows.

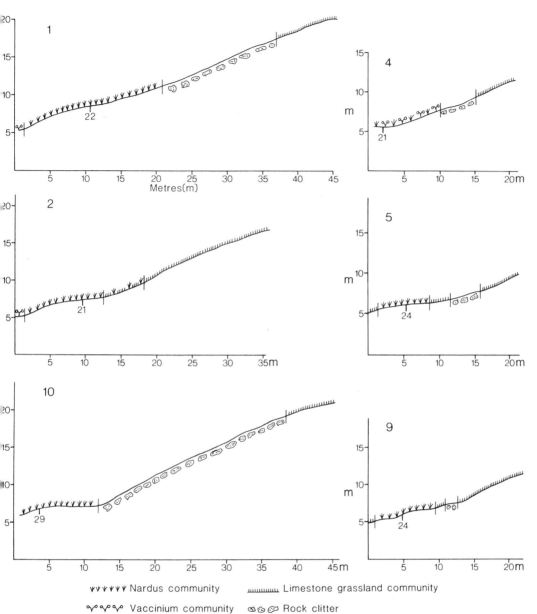

Fig. 9.8   Profiles across six of the snow patches shown in Fig. 9.7. Maximum loess depths in centimetres indicated in each profile.

400 m O.D. From a karst point of view interest in the area stems from the fact that the limestones are at a relatively low altitude as compared with their Craven counterparts 40 km to the east. In addition the limestone outcrops lay directly in the path of emergent ice from the Lake District and the interaction between glacial and karst processes is clearly evident in the landforms.

In spite of the relatively recent uplift of the area, the general conical appearance of hills such as Arnside Knott (SD 455775) and Warton Crag (SD 455775) has led some researchers, notably Corbel (1957) to suggest that the karst of Morecambe Bay is a relict tropical karst. Corbel argues that the reddish soils of the limestone outcrops are a remnant lateritic weathering product. It is now known that the reddish soils are simply stained Late-Glacial loessic sediments. The red staining is derived from ground waters in contact with haematite veins which are common throughout the area (Vincent, 1981).

Cave systems are poorly developed in the Morecambe Bay karst. Ashmead (1974) recognises three types of caves: phreatic networks; abandoned sea caves; abandoned vadose caves. The age of the phreatic and vadose systems is not known but Ashmead indicates that there are deposits of at least one full glacial stage in all of them. The precise origin of the so-called abandoned sea caves is disputed. Both Ashmead (1974) and Tooley (1977) suggest that the sea caves and associated wave-cut notches indicate higher sea levels. Ashmead argues that their altitudinal distribution fits in well with Parry's maps of former Late Tertiary and Pleistocene erosion surfaces in the south-western Lake District. Tooley suggests that those caves found between 12 m and 29 m were formed between 25 000 and 19 000 BP, although more recently he has proposed a Hoxnian or earlier age for these and higher features (Tooley, 1979). Gale (1981a) in a detailed examination of the caves of Morecambe Bay comes to the conclusions that the so-called sea caves are in fact phreatic and their height are not as reported by Ashmead. Gale indicated that Parry's mapping of erosion surfaces was strongly influenced by the contour interval of his base maps. Gale also suggests that the features often interpreted as marine-cut notches are simply the result of glacial plucking of weak bedding planes in the limestones.

Pre-glacial Karst

On the Silverdale Peninsula 10 km north of Lancaster large solutional features resembling poljes and dolines are certainly pre-Devensian (Fig. 9.9). Dolines are particularly well developed on Warton Crag at SD 496730. At this locality the largest doline in more than 100 m in diameter. Clearly such an enormous amount of solution cannot have taken place in the Flandrian.

Poorly drained mosses are a feature of the area and their flat floors and cliffed margins distinguish these polje-like forms from dolines. Good examples of such poljes are New Barns Moss (SD 443715) and Silverdale Moss (SD 474774). Little research has been done on these large solution features and indeed it has only recently been realised that much of the karst landscape in North-west England has been little altered by repeated inundations by ice.

Glacio-karst

When Lake District ice crossed the limestone outcrops it swept away soils and regolith and differentially plucked and abraded the weaker limestone beds. The

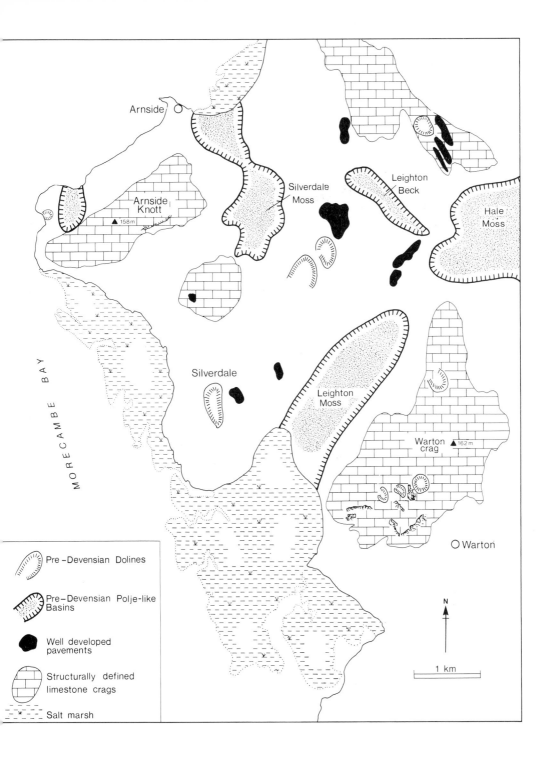

Arnside

MORECAMBE BAY

Arnside Knott
▲ 158m

Silverdale Moss

Leighton Beck

Hale Moss

Silverdale

Leighton Moss

Warton crag ▲ 162m

○ Warton

Pre-Devensian Dolines

Pre-Devensian Polje-like Basins

Well developed pavements

Structurally defined limestone crags

Salt marsh

N

1 km

Fig. 9.9 Karst features of the Silverdale Peninsula.

net effect was to produce a staircase-like landscape with scoured areas, which we call limestone pavements, separated by cliffs. This pavement karst is known by the German term *Schichttreppen* karst (Fig. 9.10). Such an assemblage of steps and rises is typical of the Morecambe Bay karst and good examples can be seen at Farlton Fell and Hutton Roof Crags (SD 555780).

Limestone rocks are usually well jointed and solution dissects the joints to produce blocks or clints (German, *Flachkarren*). The dissolved joint zone becomes a gutter or grike (German, *Kluftkarren*). The precise age of the clints and grikes is not known for certain and whether or not they formed under a soil/peat cover is still a matter of debate. In the Morecambe Bay area we have several lines of evidence to suggest that at least some of the pavements were not plucked down to intact bedding planes by Devensian ice. At Underlaid Wood pavement (SD 484786) several hundred Silurian erratics rest in grikes and only one or two on the clints. Their disposition suggests that the grike system was open when the erratic blocks were released during deglaciation (Rose and Vincent, 1985a). Morphometric analysis of the grikes at this and other pavements in the area strongly indicates the existence of two populations of grikes: a relict set and a post-glacial set (Rose and Vincent, 1985a). Many smaller solution grooves (German, *Karren*) also formed during the post-glacial period. Karst geomorphologists distinguish forms with rounded divides which have formed under a soil or peat cover, from angular forms which are produced directly by the dissolving power of rain. The best known type of covered karren is the rundkarren and some fine examples occur on the sloping pavements of Hutton Roof Crags. The important point about rundkarren is that their presence indicates a former soil/peat cover. Much of this cover has been lost into the developiing grike systems or perhaps removed from the pavements by overland flow (Fig. 9.11).

Several types of bare karst karren are developed in the region including the rill-like rillenkarren and heel-print-like trittkarren (Fig. 9.11). Rillenkarren are locally common at Gait Barrows National Nature Reserve (SD 483775) and at several sites on the coast near Silverdale. Trittkarren occur on the sloping pavements at Farlton Fell in areas which have probably remained free of soil and vegetation (Vincent, 1983).

One of the most common bare karst karren types is the kamenitza or solution pan which is typically developed on flat clints. Kamenitzas are particularly well

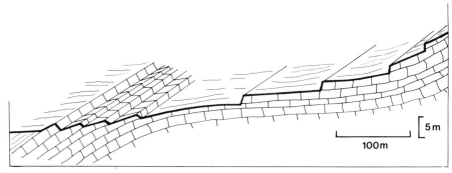

Fig. 9.10 Schematic diagram of a schichttreppen karst landscape.

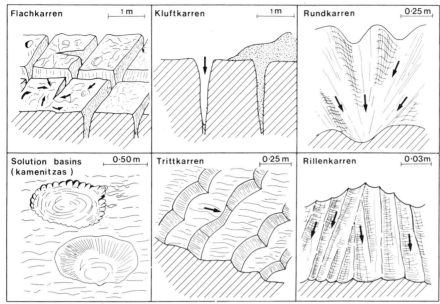

Fig. 9.11   Karren types of the Morecambe Bay karst region.

developed at Gait Barrows where they are up to 1 m in diameter and 10–20 cm deep. Rose and Vincent (1985b) have recently made a detailed study of the kamenitzas at Gait Barrows and, on the basis of simple chemical budgets, conclude that they are at least 3000 years old at this site. Their shape seems to be due to the fact that carbonates are precipitated at the base of the pans during dry periods where it case-hardens the limestone. The walls, however, are not case-hardened and are thus subject to relatively more solution.

There are no general estimates available of the rate of lowering of the karst surfaces of the Morecambe Bay region. A few erratic blocks on pedestals can be seen at Gait Barrows but the surrounding pavement surfaces are far from flat which makes estimates of pedestal height rather dubious. The annual rainfall is about 1100 mm at this site which is considerably less than the karst regions in the nearby Craven uplands and the amount of surface lowering is probably correspondingly less.

The Quaternary History of the Lancashire Plain

Introduction

This chapter is designed to introduce the reader to the major themes of Quaternary history which have been investigated in the Lancashire Plain. There is a notable scarcity of pre-Devensian evidence; hence attention will be focussed on the legacy of a small part of what would normally be considered 'Quaternary history'. The vast bulk of the available evidence within the region is comprised of glacigenic sediments, often in complex vertical and lateral relationships. Quaternary organic sediments, which provide the best evidence of climatic variation and of 'absolute' chronology, are almost entirely restricted to the closing stages of the Late Devensian (Windermere Interstadial and Loch Lomond Readvance, especially). Events leading to the growth of the ice sheet responsible for the sediments and landforms of this region are therefore outlined from evidence outside the region, principally from the Midlands, but also including sea-floor evidence from the North Atlantic. From such evidence, the nature and behaviour of the growing Late Devensian ice sheet is hypothesised. The subsequent landforms and sediments are briefly and selectively reviewed, and interpreted in the light of knowledge derived from present arctic environments. It will be shown that the available evidence points unequivocally to the dominant effects of the *in situ* decay of the ice sheet, followed by modification of resultant surfaces not merely by the post-glacial processes of the Flandrian period, but also by the periglacial processes of the period of the Loch Lomond Readvance.

The physical features and geology of the Lancashire Plain

The Lancashire Plain may be defined as being between the west coast and the 120 m contour from the River Mersey in the south to the River Keer, north of Carnforth (Rodgers, 1962). Its essential character derives from the predominance of glacial and more recent sediments which mask the underlying Triassic and Carboniferous bedrock. To the east of the 120 m contour, Carboniferous sandstones, shales and limestones form the Rossendale Upland and the Bowland Fells, whilst to the north, horst blocks of Carboniferous limestone form the prominent features of Farleton Knott, Hutton Roof Crags and Warton Crag. Triassic sandstones form limited outcrops within the plain itself, most notably in the vicinity of Scarth Hill in south-west Lancashire, and less conspicuously elsewhere, mainly in south Lancashire.

The superficial deposits of the lowland include multiple sequences of till, glaciofluvial and glaciolacustrine sediments, tills of variable character in direct superposition, glacial outwash, aeolian sands, peats and detritus muds, riverine, estuarine and marine alluvium. Drift thickness varies from tens of centimetres to depths in excess of 70 m but natural exposures are few, mainly occurring in river banks and generally partially masked by slumped material and pioneer vegetation.

Fortunately borehole information is copious, mainly being held by the large public undertakings such as the County Highways Department and the North West Water Authority. Such information has thrown much light on the nature of the sediments and has been an invaluable aid in deciphering the legacy of Quaternary events in the area, especially when used in conjunction with civil engineering exposures and geological fieldwork. Boreholes reaching to bedrock, however, have a very patchy distribution, rendering the reproduction of a contoured map of the bedrock surface hazardous.

Much of the bedrock lies below present sea level. Overdeepened enclosed basins are comonly encountered, and many undulations detected by borehole drilling appear to be roughly aligned between NNW–SSE and NW–SE, consistent with ice movement inferred from striations on Harrock Hill, near Parbold (Williams, 1978), between Wavertree and Woolton near Liverpool, and at Daresbury, to the south of the River Mersey. Deviations from the general pattern of south-westerly ice movement are indicated by striations aligned between WNW–ESE and west–east, accordant with the solid rock morphology in the Blackburn area, and by some of unspecified direction in the Douglas valley, near Parbold, where parallelism with local topographic control has been reported (Williams, 1978).

The morphology of the Lancashire Plain may be broadly described as falling into three spatial zones. In the north, around and within the valleys of the Keer and Lune, the terrain is markedly drumlinised. The drumlins range from about 75 to 1600 m in length, have axial ratios roughly between 1.3 and 7.5 and rise up to 150 m above the surrounding terrain. Drumlin orientation is rather variable in the north and north-east of this zone, ranging between north–south and north-east/north-west, probably because of the conditions resulting from the convergence of ice flows passing around the limestone massif of Farleton Knott and Hutton Roof Crags. In the Lancaster area, alignment is much more uniform, trending from NNE–SSW north of the Lune to north–south to the south of it.

South of the River Conder, the drumlinised landforms are less frequent and less clearly developed. The drumlin swarm of the northern zone passes gradually into an area of irregular mounds and ridges often separated by rather flat-floored peaty or alluvial depressions. This area is known as the Fylde. Drumlinoid landforms occur around the mouth of the River Wyre, generally oriented NNW–SSE. Extensive mosslands, notably Cockerham, Pilling, Rawcliffe and Lytham Mosses occupy the low-lying terrain (generally below 15 m O.D.) between the knolls and ridges of glacial and fluvioglacial sediment. They overlie the marine, estuarine and terrestrial sediments resulting from the oscillations in the Flandrian rise of sea level, and thus the nature of the glacial legacy in these areas is much obscured. Several ridges, largely of till, display a general west–east orientation, such as the rather

G

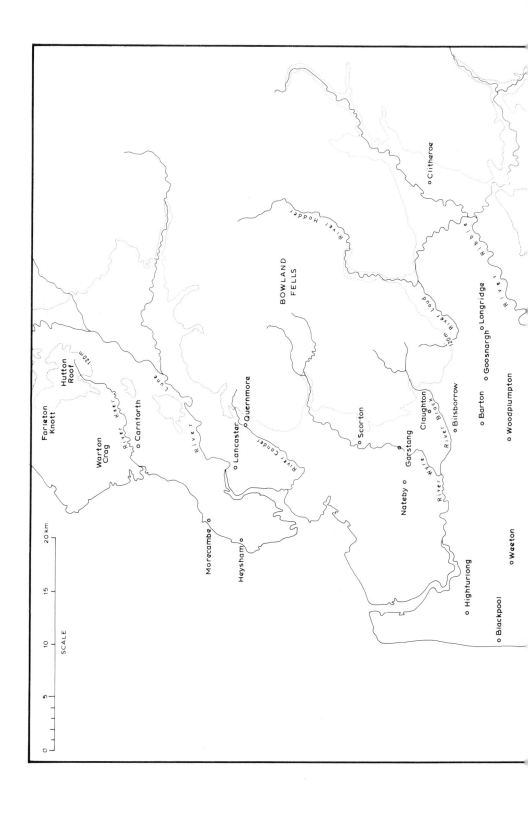

tion of the pre-glacial Devensian evidence for the climatic and environmental changes which took place in Britain and then examines the conditions which led to a particular form of ice-sheet expansion within the Irish Sea area.

For insight into the conditions prior to the Late Devensian ice advance into North-west England, recourse to the now-classic evidence from the Midlands is necessary. The coleopteran assemblage designated Type III by Coope, is found at no less than sixteen sites, mainly in the Midlands, ranging in radiocarbon age from 39 330 years at Queensford in Oxfordshire to 18 240 years at Dimlington (Coope, 1975). This assemblage is indicative of a cold continental climate with average February temperatures below − 20 °C. At some sites, a loss after 25 000 BP of eastern palaearctic species, such as *Agabus clavicornis* and *Helophorus jacutus*, whose present range is confined to Eastern Siberia, led to the tentative postulation of the increasing oceanicity required to build ice sheets (tentative because of the greatly reduced insect frequencies, which make interpretation hazardous). However, recently Denton and Hughes (1981) have proposed a mechanism of sea-ice growth which demands far less precipitation input to the developing Late Devensian ice sheet than has hitherto been envisaged, and this fits more comfortably than expected with Coope's more broadly based hypothesis that the anaglacial climate rapidly approached that of a polar desert in the period roughly from 25 000 to 18 000 BP without the substantial increase in oceanicity. According to Watson (1977), permafrost probably occurred in Britain in the Early Devensian (e.g. at Wretton and Chelford) and certainly in the Middle Devensian, such as is evidenced by fossil ice-wedge structures in datable sequences at Earith (42 000 BP), Four Ashes (42–30 000 BP) and Syston (37 420 BP). This physical evidence is reinforced by palaeoecological indicators of a severe climate, including the Coleoptera discussed earlier and pollen assemblages. The pollen data are consistent with an open, treeless, rather arid landscape and include obligate and facultative halophytes such as *Glaux maritima* and *Juncus gerardii*. They suggest oscillation between tundra and sub-polar desert (Bell, 1969). Coope's re-assessment of the Middle Devensian environment relies heavily on the evidence of the Coleoptera, more mobile and sensitive than plants. At around 45 000 BP from sites at Earith, Four Ashes and Tattershall, the low diversity Type I assemblage unequivocally accords with a climate of arctic severity, whose average July temperature is likely to have been at or below 10 °C and whose winter temperatures were greatly depressed. Apart from a brief, warm, fairly oceanic interlude from about 44–42 000 BP, when the average July temperature reached about 18 °C, the whole interstadial complex is dominated by the Type III assemblage mentioned above. The inference regarding the nature of this data is of a rapid deterioration at about 41 000 BP to a cold, markedly continental regime by 39 000 BP, subsequently a slow decline in temperature and increase in continentality occurred up to about 25 000 BP, and then the temperature fell more rapidly (Coope, 1975). Hence, on the basis of several distinct criteria and a wide range of sites it can be argued that the Late Devensian ice sheet is likely to have advanced across a landscape in which permafrost was well developed.

Given such terrestrial evidence plus the sea-floor evidence of the absence of the

North Atlantic Current from latitudes north of 40 °N by 17 000 BP, and the southward spread of polar waters (Ruddiman and McIntyre, 1973), it is possible to offer a tentative reconstruction of the anaglacial conditions around the Irish Sea margins. Synge (1977b) has already reported glaciomarine sediments below Irish Sea till in the Boyne valley and striations formed during the Late Devensian by shelf ice moving onshore, orthogonal to the coasts of Leinster. Well preserved cold-tolerant foraminifera, such as *Globigerina pachyderma* (s.c.) are found in glaciomarine sediments along the Irish Coast, and are common in sands and clay tills at both coastal and inland locations in the Lancashire Plain, for example at Bare and Quernmore (Reade, 1983, 1904). Growth of shelf ice, along with the continuing fall in sea level, would result in the grounding of the ice necessary for its movement as glacier ice onto the adjoining coastal lands. The shelf-ice hypothesis is supported by an examination of the locations where the present climate corresponds roughly to Coope's estimate for the British climate at around 25 000 BP. As the Late Devensian pleniglacial was approached, it would seem that the polar desert envisaged by Coope is perhaps best approximated by present Antarctic conditions where shelf ice is extensive and has been geomorphologically significant in areas such as those around the McMurdo Sound in Antarctica. What is crucial from the point of view of ice behaviour is the fact that ice movement in such areas as the Irish Sea would tend to be relatively rapid, being wet-based because of the absence of permafrost from the sea floor, whereas any movement onto adjoining land areas would be slowed not only by the regional gradient but also by the compressive flow regime brought about by contact with heavily permafrosted terrain. Hence even in the early stages of such ice-sheet growth, thick imbricate englacial debris bands are likely to have been built above the transition from the wet to the cold-based zone. Furthermore, if the growth of shelf ice resulted in the sealing of landward draining basins, in the initial stages the margins of the sea ice would accumulate (in supra- or sub-glacial positions) substantial quantities of sorted stratified materials, transported there by the brief but relatively high-seasonal-discharge high-sediment-yielding fluvial regimes of that period. (Some confirmation of this is found in maps and satellite photographs of present sea-ice which show that in the melt season debris is carried onto the ice where rivers discharge from continental interiors, but as the season progresses the ice thins or even disappears dumping sediments into sea water.)

From such reasoning, it follows that during the active and *in situ* decay phases of the Late Devensian established by Boulton (1977), from sedimentary evidence in North Wales and Cheshire, large volumes of englacial debris in various states of disarray or preservation would be available for release as lodgement, meltout and flow tills. Furthermore, the coleopteran and palynological evidence from North Wales (Boulton, 1977; Coope and Brophy, 1972), Cumbria (Osborne, 1972) and south-west Scotland, (Coope and Joachim, 1980) supports the application of Boulton's (1972) supraglacial landform and sediment association, by providing a climatic scenario favouring ice stagnating from lack of supply, and melting out slowly because of low temperatures. An examination of the structural, stratigraphic, geotechnical and macrofabric properties of tills of the Lancashire Plain bears out these working hypotheses, and it is to these properties that attention will now be turned.

The Late Devensian glacial stage

The tills of the Lancashire Plain: general survey

By 1967 such borehole information as was then available caused Gresswell (1967b) to doubt the validity of the old 'Lower Boulder Clay/Middle Sands/Upper Boulder Clay' stratigraphy embodied in the Geological Survey Memoirs and embedded in the regional geological consciousness. This conceptual view of the glacial succession was also challenged by Johnson (1971), who defined areas of multiple stratigraphical successions distinguishable from the most general pattern which is of a single till overlying sands and gravels: hence the tripartite succession was said not to be truly representative of the Pleistocene succession in the North West. In 1974 the present author reported that even if strata less than 1 m thick were disregarded, up to three stratigraphically distinct tills and two major bodies of waterlaid sediments were proved in 23 deep boreholes associated with the M55 Preston–Blackpool motorway, whose route traverses one of the areas of multiple strata mapped by Johnson. The section thus produced is given in Fig. 10.2a and is in contrast to the most recent section of the north-western Fylde published by the I.G.S. (shown in Fig. 10.2b). The latter section shows an upper and lower till separated by a substantial body of sands over a distance in excess of 7.5 km, with the lower boulder clay not having been proved in the 2.5 km south of North Shore. The superficial similarity of this section to the old tripartite model is perhaps responsible for the perpetuation of the old model, in spite of the fact that the notes to the section attribute the entire sequence to one phase of glaciation. Unfortunately the nature of the information on which the section is based is not revealed, other than that the stratigraphy has been established by drilling.

Other sections prepared from borehole logs, whilst clearly subject to the problems of complex variability (discussed above), provide ample evidence of the unrepresentative nature of an extensive tripartite succession for North-west England. Figure 10.3 shows a generalised representation of the stratigraphy along the line of the M58 Liverpool–Skelmersdale motorway. Here the dominant pattern is of till lying on bedrock and capped by sands, mainly of the Shirdley Hill type, aeolian in origin. However, near Melling, there are two till strata separated by a substantial sand body, with the lower till resting on sand and gravel. Furthermore, in the Bickerstaffe area there are up to four till-like strata, separated by waterlaid sands and gravels, some of which interpose between the lowermost till and the local bedrock. It should also be borne in mind that from geotechnical criteria it has been established that even where till is present without waterlaid intercalations, a dense, low-plasticity, highly cohesive unit tends to occur towards the base of the till, overlain by softer, less cohesive tills of higher plasticity and less density: hence several till units (e.g. lodgement, meltout, flow) may be present in direct superposition even where the

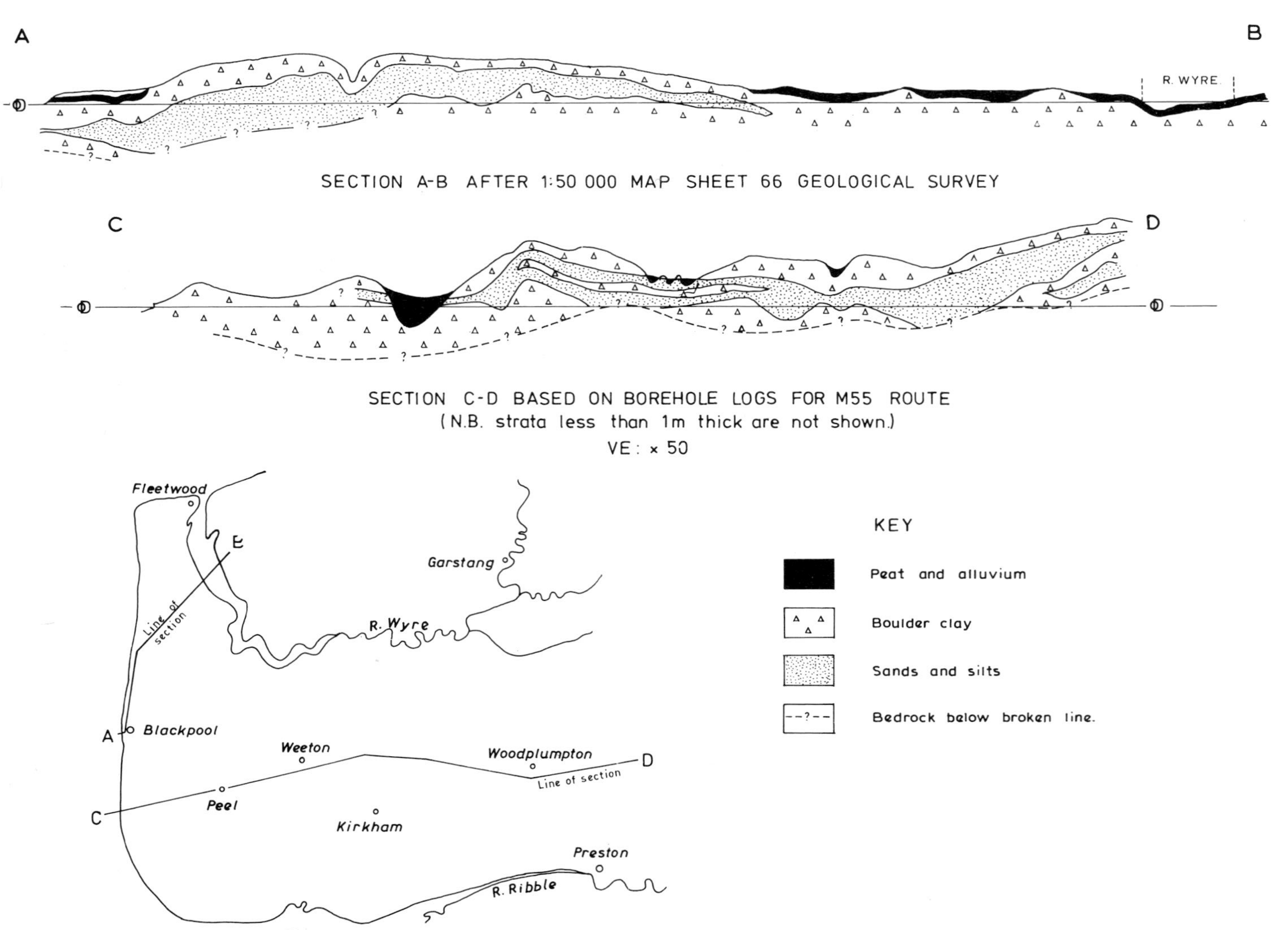

A
B
R. WYRE
SECTION A-B AFTER 1:50 000 MAP SHEET 66 GEOLOGICAL SURVEY
C
D
SECTION C-D BASED ON BOREHOLE LOGS FOR M55 ROUTE
(N.B. strata less than 1m thick are not shown.)
VE: × 50
Fleetwood
E
Line of section
Garstang
R. Wyre
A
Blackpool
Weeton
Woodplumpton
D
Line of section
C
Peel
Kirkham
Preston
R. Ribble
KEY
Peat and alluvium
Boulder clay
Sands and silts
--?--
Bedrock below broken line.

generalised section shows one till (as, for example, at Ox Hey on the line of the M58). This section is also notable for the recognition of an area of multiple successions to the west of Skelmersdale, outside the areas mapped by Johnson. The area concerned extends from Mill Moor to Gillibrands and is remarkable for the rather planar topography at about 67 m altitude. It is part of the feature identified by Gresswell (1964) as the 'Skelmersdale Platform' and interpreted by him to be a relic of a formerly more extensive pre-glacial plain of marine erosion. The M58 survey work shows that for the western part at least, this is clearly not the case with glacial, fluvioglacial and post-glacial sediments forming the surface, and bedrock varying in altitude from about 30 to 57 m O.D. Near Preston, in the Clayton Brook and Goosenargh/Grimsnargh areas new constructional work has proved the presence of a depositional surface at 67–72 m O.D. and the sections have shown that the sediments have the same characteristics as those reported in the Skelmersdale area. These two areas were included by Johnson (1971) in his area of hummocky morainic (i.e. supraglacial drift formation).

The 'Kirkham Moraine' re-examined

Geotechnical data obtained for the M55 motorway surveys has provided much useful information about the glacial deposits in the Kirkham morainic area of the southern Fylde. It indicates that the tills vary greatly in their bulk density, in their undrained cohesion and in their Atterberg limits. Many samples possess the characteristics shown by undisturbed tills described by Boulton and Paul's (1976), 'T-line', but there is no consistent pattern from borehole to borehole, indicating that 'T'-type tills and 'non-T' tills are irregularly distributed in complex inter-digitations. Furthermore it is possible to identify specific areas of topographic inversion such as are produced by the supraglacial meltout system in Spitzbergen, examples of which are set out below.

(i) *Peel Hill, Marton (SD 352336)*   As in much of the data for the M55, immense variability makes the drawing of sections from boreholes hazardous. However, Fig. 10.4 shows clay/stony clay tills draped on water-sorted sediments on either side of the kettle hole about ½ km from the Blackpool end of the motorway, and the irregular distribution of tills of varied geotechnical properties. The association of these sediments with hummocky and kettled topography reflects the presence of dead ice contemporaneous with deposition. Till fabrics at the nearby Peel Brickworks are multimodal and vary in vertical and lateral directions.

(ii) *Church Road, Weeton (SD 385343)*   At this site, during the excavations for the foundations of the M55 overbridge, structures within the tills and sands exposed initially appeared to be consistent with the overriding and shearing processes associated with a readvance of ice. The largest and clearest section available is reproduced in Fig. 10.5, showing the nature of the sand/till contacts. The sands which lie beneath the till bodies range from fine silty sands with silty clay laminations and lenses through medium to coarse sand. Primary structures include small-

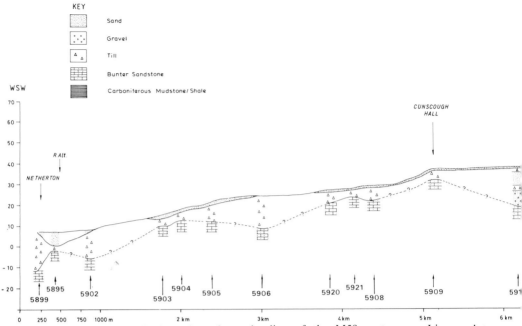

Fig. 10.3 Generalised section along the line of the M58 motorway, Liverpool to Skelmersdale.

scale trough cross-stratification, climbing ripples and tabular beds of almost horizontal laminae, but much is structureless. Post-depositional structures include marked erosional contacts between beds involving tilting of strata prior to erosion, corrugate convolutions in tabular laminae and gentle to overturned folds in sands and silty clays. Folds below till unit C from a face at right angles to the section shown in Fig. 10.5 indicates deformation from a direction of roughly S 20 °W (not corrected for magnetic declination). Till units A to D are therefore interpreted as having been deposited as lobes of flow till, having slumped into a basin dominated by meltout activity. The disturbance of silty clay strata shown in the lower left and lower centre of Fig. 10.5 appears to have been caused by load deformation under pressure from the south whilst the sediments were still wet, and the relatively stone-free character of till unit A may be the result of aqueous deposition. The lobate nature of the 'nose' of till unit B is also suggestive of deposition in a rather fluid environment, whereas the lower portions of till units C and D are very dense and possess well-defined, closely spaced laminae or cleavage planes which may reflect laminar shear as the lower parts of these units were retarded by the greater friction resulting from somewhat drier conditions in the sands beneath, allowing drainage from the base of the flow till units. Till fabrics are consistent with such an interpretation, many being dispersed and variable from horizon to horizon. The uppermost fabric is almost certainly the result of post-depositional gelifluction and is closely associated with an ice-wedge pseudomorph.

(iii) *Whitprick Hill, Weeton (SSD 397344)* This is the highest relief feature in

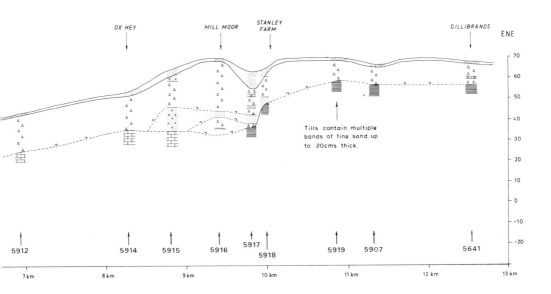

Tills contain multiple
bands of fine sand. up
to 20cms thick.

the Fylde (*c*.45 m O.D.), now capped by a small covered reservoir. Two pits on the south side, at about 39 m O.D. are dry, supporting the extensive presence of well drained sands which are periodically exposed during the ploughing of nearby fields. Borehole logs from the M55 survey record intercalations of sandy clay and clayey sand with bands of silty clay forming the surface materials, resting on till at 1.22 m depth in the west and deepening in excess of 5.5 m to the east over a distance of some 400 m. As the intercalated strata are not continuous from borehole to borehole, the probability is that they have been disturbed by settlement following supraglacial deposition and meltout. There is no evidence of ice-contact morphology, meltwater channels or proglacial lakes in the vicinity which would be consistent with its formation as a kame, *sensu stricto*. The occurrence of well-sorted stratified sediment at the highest point of relief in an area in excess of 600 km^2, must reflect its accumulation within or upon Late Devensian ice and the subsequent creation of a complex, irregular mound once the supporting ice melted away.

(iv) *Pasture Barn Ridge, Medlar (SD 423348)*   A cluster of narrow NNE–SSW ridges form the southern boundary of the Medlar Woods depression and merge with the irregular topography of the broad ridge of the 'Kirkham Moraine'. In addition to the borehole data indicating sands, sandy silts and clayey silts, often capped by till, fieldwork performed on one of the ridges during motorway construction established the draped nature of the surface till, conformably deposited on laminated sands and silts in the central position of one of the ridges and disturbed

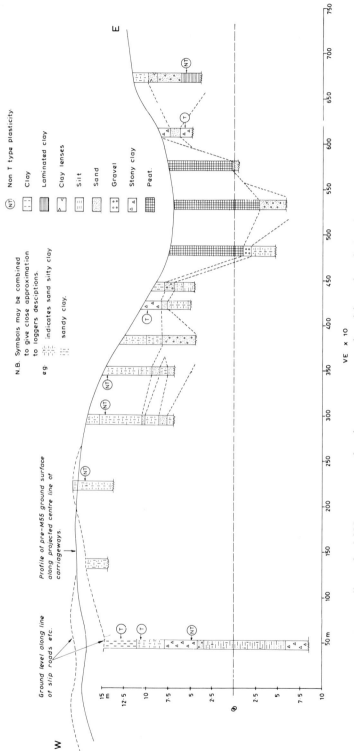

Fig. 10.4 Section at Peel Hall on the M55 motorway showing strong vertical and lateral variation in the drift materials.

by slump and flow on either flank. Till fabrics from this site are consistent with the occurrence of post-depositional gelifluction.

(v) *Bartle Rise (SD 500338)*   This feature forms the southern slope of the valley of Woodplumpton Brook and takes its name from the fact that the motorway rises steadily from about 23 m to the west to 38 m at Broughton. Strictly it is the northern flank of a wedge-shaped ridge extending from Salwick to Fulwood. The motorway crosses this flank at about 32 m, with terrain up to 39 m lying about 500 m to the south. Water-sorted intercalations outcrop in places and extend to more than 5 m below the surface. Sandy clay, silty clay, silts and sands are common, and at Bartle Lane laminated clays are present from 12.2 to 15.8 m below silt and sand strata. Elsewhere, thin till lies on over 22 m of sand and silt, sandy clay lies on stiff, dense till, and alluvial clay lies on soft till. To the east, gravel forms the surface, resting on till below. The morphology of this feature is complicated by small re-entrant valley forms, devoid of surface drainage, but containing alluvium along their floors. Similar features are also present to the north of Woodplumpton Brook. These may reflect the routes by which water drained from the sediments as relief inversion proceeded during meltout, thus accounting for the appearance

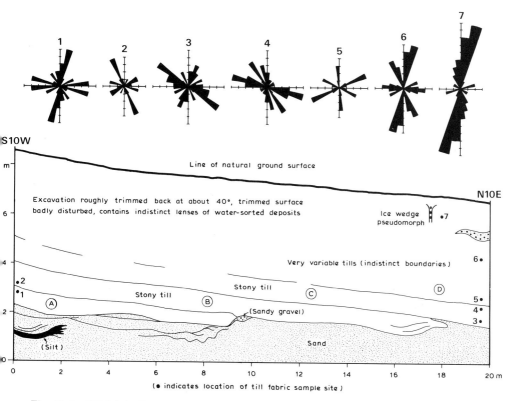

Fig. 10.5   Till fabric diagrams and section on west face of an excavation for the west abutment of the M55 bridge at Church Road, Wetton (SD 385343).

of 'sapping' at the headward ends of the valley forms. (This morphological pattern also occurs along Savick and Sharoe Brooks around northern Preston.)

(vi) *Lightfoot Green Rise (SD 512338)*   Whilst part of the northern flank of the Salwick–Fulwood ridge (and thus of the 'Kirkham Moraine'), this feature is separated from 'Bartle Rise' by a tributary of Woodplumpton Brook. Its topography is markedly tabular, at about 36 m O.D., bounded by gentle to moderate undulations. Ten boreholes drilled in the western half revealed a complex of interstratified sands, silts and clays, dominated by water-sorted deposits. Tills are more common in the eastern portion, but most of these are represented by thin layers of about 0.5 m maximum and are associated with sand and gravel intercalations. Irregular disposition of the beds, coupled with highly variable geotechnical properties and perched water tables made this section, along with the Bartle Rise section, one of the most problematic excavation sites along the whole motorway. Whilst no detailed sections from exposure are available, the testimony of motorway construction workers establishes the existence of post-depositional disturbance structures such as slumps and faults in these sediments.

The lower Ribble valley near Preston

This last feature lies just to the east of the area mapped by Gresswell (1964) as the 'Kirkham Moraine' and slightly south of his 'lateral moraine'. It is here that the topography of the 'Kirkham Moraine' grades imperceptibly into the rather more planar, less undulating relief which marks the eastern extent of the Lancashire Plain and extends into the Ribble embayment, both north and south of the river. Geotechnical similarities have also been noted in these areas, in which high plasticity and low bulk density tills and laminated clays are found, such as at Penwortham, Walton-le-Dale and Clayton-le-Woods. In addition, sand, gravel and silt intercalations are common in these areas, consistent with spatial continuity of both provenance and process.

At Prospect Hill, just south of the M6 crossing of the River Darwen, till forms the upper horizons, varying between 3 m and 8 m in thickness, and beneath that sands and clays (the latter, generally, of rather soft, smooth consistency, high moisture contents and medium to low bulk densities) are intercalated, with only one other till horizon (1.7 m thick) being recorded in a further 18 m of sediment. The abundance of smooth, virtually stoneless clays, commercially exploited at the former London Road Brickworks just north of the Ribble (SD 552393) is recorded in the Geological Survey Memoir for the Preston district (Price *et al.*, 1963). Similar materials and complex intercalations are common to the north of the Ribble, from the edge of the floodplain through Grange Park and Fulwood Row and Broughton, where the M6 skirts the north-eastern suburbs of Preston. There is no record in borehole logs of evidence of post-depositional disturbance in these sediments, but normal fault planes cross-cutting displaced gravel bands within the 'Middle Sands' have been reported by Taylor from the excavation of a cutting for the M6 south of the Ribble near Samlesbury (Taylor, 1958). There is also a record of horizontally bedded sands resting upon a clay surface inclined at between 70° and 90° to the north-east of Fishwick Hall in glacial materials just north of the Ribble, but it is not clear whether this is an erosional or depositional surface: the context could be read as supporting the latter (Price *et al.*, 1963).

North of Broughton is a remarkably planar surface, crossed from south to north by the A6 trunk road and M6 motorway, and known locally as 'Barton Flats'. Apart from its dissection by Barton Brook, the altitude of the surface where traversed by the A6 varies little between 33 and 36 m over a distance of some 4 km. Apart from minor gentle undulations and the much steeper stream valleys, the east–west profile is also rather tabular, falling at an angle of less than ½° over some 3 km. Borehole data from the line of the M6 confirm the repetition of multiple intercalations of tills and various water-sorted sediments, such as at the crossing of Jepps Lane (SD 526380), from which site one borehole log records a succession of clay, till, laminated clays and 'hard boulder clay' resting on water-logged gravel at 8.23 m depth, whilst another borehole for the same bridge site records clay, till, sands and gravels resting on 'hard, stoney boulder clay' unbottomed at 10.68 m depth. Similarly, at Bilsborrow (SD 518398), laminated clays, silts, sands and gravels predominate, with two till horizons identified in one borehole, and none at all in the other. The dominance of waterlaid sediments in this area is clear, and is particularly noticeable in the record of deposits beneath Barton Brook (SD 526378) where only one till stratum, 2.7 m thick, occurs in a total of 15 m of sediment drilled, almost all of the remainder of which is sand and gravel. There is no record of depositional or post-depositional structures in this area, but by analogy with similar complex stratigraphy in adjacent areas, the existence of post-depositional disturbance is highly probable. The 'hard' basement boulder clay identified in boreholes usually possesses low plasticity 'T-type' characteristics, high bulk density and high undrained cohesion. It probably corresponds to the hard, dense, tills which are patchily exposed in local river valleys, and whose fabrics are consistent with a lodgement origin beneath ice moving towards the south-east. In this context it is worth noting that where such tills have been seen in contact with Carboniferous bedrock, the lowest unit is almost entirely Carboniferous in provenance, and a progressive upward change to tills of Irish Sea/Lake District provenance is shown by lithological analysis (see Fig. 10.6). Such occurrences, are consistent with the vertical differentiation of debris reported from modern glaciers (Boulton, 1970), rather than with two phases of ice movement from different directions. The basement till fabrics support this hypothesis, as does the analysis of pebble roundness. Thus, the hypothesis regarding the primary advance of glacier ice down Ribblesdale, laying down Carboniferous-dominated drift east of Preston, followed by the arrival of Irish Sea ice which overrode the 'Ribblesdale' drift and deposited its own distinctive erratic suite (Price *et al.*, 1963) is seriously called into question and the issue warrants further investigation.

East of 'Barton Flats', the rather tabular topography rises with irregular gentle undulations towards the watershed of the eastward-flowing River Loud, where large spreads of fluvioglacial sands form the surface. Streams such as Sparling Brook, Westfield Brook, Blundell Brook and Savick Brook are deeply incised in

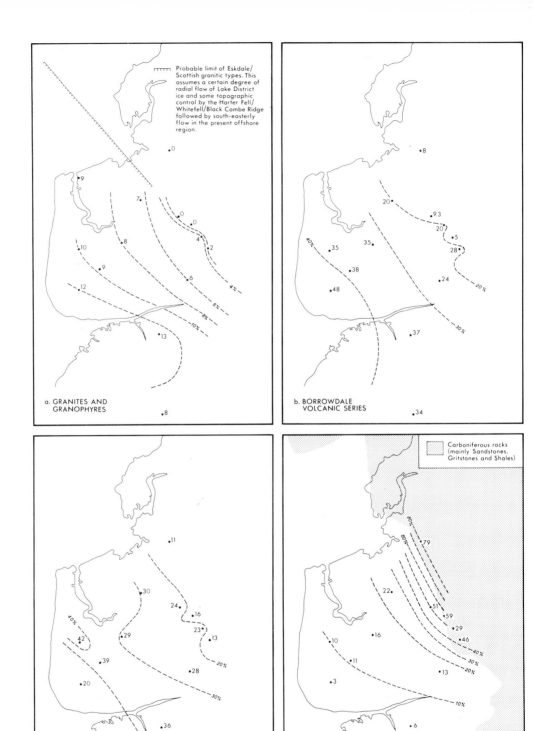

a. GRANITES AND GRANOPHYRES

Probable limit of Eskdale/ Scottish granitic types. This assumes a certain degree of radial flow of Lake District ice and some topographic control by the Harter Fell/ Whitefell/Black Combe Ridge followed by south-easterly flow in the present offshore region.

b. BORROWDALE VOLCANIC SERIES

c. SILURIAN AND ORDOVICIAN GREYWACKES AND SHALES

d. CARBONIFEROUS SANDSTONES AND GRITSTONES

Carboniferous rocks (mainly Sandstones, Gritstones and Shales) misfit valleys with flat floors and steep sides in which large quantities of fluvioglacial sands and gravels, laminated clays and silts are exposed from time to time and from place to place, as stream erosion continues. During the construction of the North West Water Authority's Catterall–Hoghton pipeline, field observation of the exposed waterlaid and glacial sediments was possible. These observations not only confirm the broad extent of the association of tills and waterlaid sediments in often complex intercalations, but also give further weight to the general occurrence of post-depositional disturbance and its association with topography. Where streams had to be crossed by the pipeline, excavation across their valleys provided ample evidence of post-glacial valley-side mass movement and the accumulation of colluvium where meandering channels have migrated away from foot-slope locations. Such manmade exposures in the drift of North-west England provide useful additions to the relatively limited natural exposures, and both confirm the widespread occurrence of the tectonic features referred to by Johnson (1971), such as slumps, sags, faults, and overturned and rolled strata, although his ascription of the former two phenomena to the melting of contained ice, and the latter three to minor ice-front readvances is now questionable. These features all appear to be entirely consistent with the *in situ* decay model propounded by Rich, which Johnson applied to the landforms and sediments of North-west England. At the time of the latter's writing (1971) the model was largely hypothetical, but since that time the principles have been elaborated by Boulton and his colleagues in the light of observation of modern polar glacial environments. Both anticlinal and synclinal settlement structures are common in this region, as are thrust and rolled strata. Distorted strata, particularly in laminated sediments, are frequently recorded in borehole logs, as for example along the line of the M65 at Clayton Brook, and have been seen steeply inclined in a collapsed basin structure transected by a pipeline trench north-west of Langley Lane, Goosnargh where faulted laminated silts and clays dip at angles up to 46°. Rather less severe disturbance occurs in a complex sedimentary sequence exposed during construction of the same Catterall–Hoghton pipeline south-west of Claughton, to the north of the River Brock. Here flow tills, sand blocks and laminated silts are present in generally conformable sequence, with post-depositional disturbance ranging from gentle flexure to an almost recumbent fold. Flexing of the sediments occurs in the laminated silts, in flow tills and in sand bodies, whether lenses or blocks, as shown in Fig. 10.7. Till fabrics sampled from units over 1 m thick in this area show dispersed patterns which possess little modal consistency within or between sample sites. Some variations are consistent with the patterns outlined by Boulton (1971) from lobate flow tills in Spitzbergen. Figure 10.8 shows the topography of the Goosnargh/Inglewhite area, along with the route of the Catterall–Hoghton pipeline, borehole and fieldwork locations, and the outwash sediments as published by the Geological Survey at a scale of 1:63 360 in 1883. Fieldwork between boreholes 23 and 24, just south of Sparling Brook near borehole 26 and between Middleton Wood and Cross House

Fig. 10.6 Percentage contribution of rock types to the lithological assemblage of clasts in the drift deposits of the Lancashire Plain.

in the valley of Westfield Brook established the existence of a dense dark grey till dominated by Carboniferous clasts, resting on Bowland Shale bedrock and overlain by till of Lake District/Irish Sea provenance. Till fabrics are consistent with the basal unit being a lodgement till deposited by ice moving from WNW and capped by the deposition of englacial debris as meltout and flow tills. The dense basement till is also recorded in several boreholes resting on Bowland Shale bedrock and overlain by brown tills and water-sorted sediments. Intercalated water-sorted sediments with occasional till-like horizons are recorded in other boreholes and such materials were seen conformably deposited and subject to topographic inversion during excavations. Lateral erosion in the valley of Westfield Brook near Middleton Wood frequently exposes considerable depths of sands and gravels conformably capped by clays and soft tills. Gravels exposed even within terrace remnants in this valley retain clear striations and possess strong lithological and roundness affinities with the upper tills of Irish Sea provenance. In view of these characteristics, plus their association with post-depositional disturbance structures they are interpreted as fluvioglacial gravels. Both the topography and the sediments of this area bear the hallmarks of deposition in association with dead ice and copious volumes of water. The hummocky, kettled morphology is clearly shown on the 1:25 000 O.S. map, particularly by the 61 m contour. Between SD 56053815 and SD 54603793, the valley of Westfield Brook and the area within 1 km to the south has five mounds enclosed by the 61 m contour, ranging from about 60 m^2 to 1400 m^2 in area, separated by a pattern of valleys and depressions consistent with drainage interruption in the manner proposed by Marcussen (1976) in his discussion of downwasting supraglacial plains. Similar morphology exists in association with the River Darwen between Pleasington and Higher Walton, an area also containing substantial quantities of fluvioglacial deposits.

North Lancashire

Further north, along the margins of the Bowland Fells, kettled and hummocky topography extends from Claughton through Dimples, Gubberford, Scorton, and Shireshead towards the drumlin field of the lower Lune near Lancaster. Most of this area, where traversed by the M6 is characterised by a complex stratigraphy of waterlaid deposits and thin tills in intercalated layers. Till fabrics are variable both within and between sites, and would appear to result from slump and flow processes. Sands and gravels of a torrential fluvioglacial nature are well attested in old workings at Cleveley and in the present Hoveringham workings north of Foxhouses, both these being near Scorton. A dense, Carboniferous-dominated till of possible lodgement origin has been seen in spoil at the latter site, but no good exposures have been seen there by this author. To the west, the glacial legacy of the Over Wyre district mainly lies below peat and alluvium but the Trashy Hill–Skitham ridge has been mapped as boulder clay by the Geological Survey and described as such by Gresswell (1967b) and by Taylor (1961). Exploratory fieldwork by the present author, however, has ascertained that, at least between Trashy Hill and Eskham (SD 443443), fluvioglacial gravels and highly disturbed

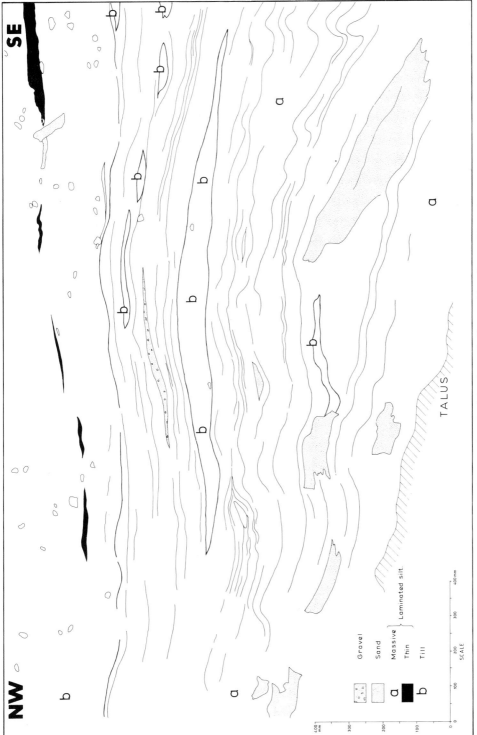

NW

SE

TALUS

Gravel

Sand

Massive ⎱ Laminated silt.
Thin ⎰

Till

SCALE

0   100   200   300   400 mm

Fig. 10.7 Small till lenses and sand blocks within the disturbed laminated silts at Claughton (SD 578416).

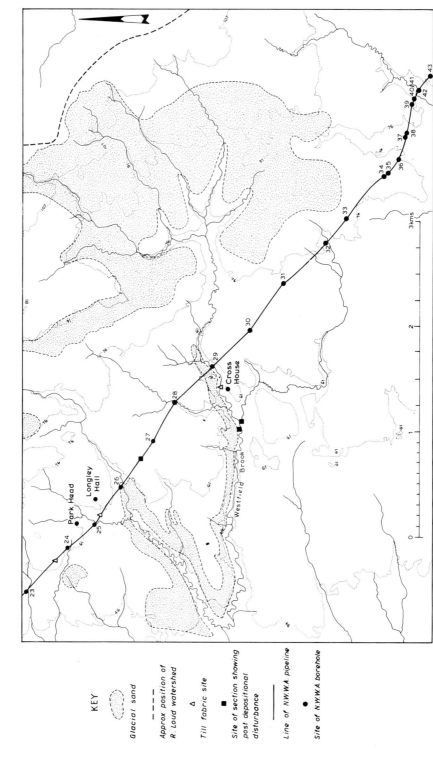

Fig. 10.8 Location map for field sites and boreholes in the Goosnargh/Barton area north-east of Preston.

peat comprise part of the ridge summit, in intimate but as yet undeciphered association with till. North of this, across Cockerham Moss, the landforms take on the appearance of drumlins, with many of them well formed and in a parallel or sub-parallel north–south alignment. However, many of the positive relief features are irregular in form, such as Oxcliffe Hill, Great Fearkla, Byroe Hill, Great Swart Hill, Windmill Hill, and Abraham Heights in the Lancaster area. Unfortunately, no good exposures have been seen in these features during the course of this author's work. Well-formed drumlinoid features north of Morecambe, where truncated by marine erosion, contrast markedly with the Wyre 'drumlins' (which contain waterlaid sediments and slump structures), both in their more classic morphology and in their virtually homogeneous dense till, a reddish matrix-dominated till containing mainly Lake District and Carboniferous limestone clasts. Boreholes commissioned by the CEGB for a pylon line from Heysham to Torrisholme encounter considerable depth and extent of coarse sands, gravels and boulders, possibly of fluvioglacial origin inland of Heysham, and to the east of the drumlinoid form of Torrisholme Barrows two boreholes penetrate sandy tills and sand lenses above a tough, compact boulder clay. Whilst far from conclusive, the available evidence can be interpreted as consistent with the formation of drumlins during the final active phase of the Devensian, followed by *in situ* decay which masked some drumlins and altered the intervening topography by the superposing of englacial and supraglacial debris in the final phase of deglaciation.

The Late Devensian periglacial periods

Fieldwork performed on near-surface till-like materials at sloping sites has established the common occurrence of gelifluction fabrics, indicating that modification of the glacial legacy has occurred under cold-climate conditions. At some sites, such modified tills overlie organic sediments, as for example near Grange Farm, Elswick. At sites below level topography, such as at Peel (Blackpool) and Longton (Preston) tabular stones are commonly found in vertical and subvertical orientation, again consistent with the intervention of freezing processes. Other consistent characteristics include a high proportion of very angular stones, including many with *in situ* fractures, silt cappings on stones, and a considerable systematic variation in the density of the matrix. This latter variation is also identifiable from borehole data from other similar sites. The variation is expressed by two parameters in particular: undrained cohesion and bulk density. When plotted graphically against depth, sharp peaks are seen, with values of both parameters increasing substantially with depth to reach a maximum at depths usually between 1 m and 5 m, followed by substantial decreases in both parameters below the peak density horizon. These variations do not correspond with variation in the composition of the matrix. Given the association of this pattern with the multiple lines of evidence of periglacial conditions, density variation is interpreted as induration caused by freezing processes which caused desiccation of horizons below the freezing plane resulting in increased interparticle packing because of large soil moisture tensions. Horizons having contained ice crystals and/or lenses above the hardened zone might be expected to have a more open texture. Evidence from Late-Glacial kettle-hole infill, such as at High Furlong near Blackpool (Barnes, 1975), indicates that deglaciation was followed by a temperate phase (now known as the Windermere Interstadial) and subsequently by the intense cold period of the Loch Lomond Readvance, during which frost-shattering of stones, and geliflucted slope movement was locally prominent. Such evidence fits comfortably within the regional framework provided by the interpretation of Late-Glacial sediments from Chat Moss (Birks, 1965), Red Moss (Horwich) (Ashworth, 1972), and Windermere (Pennington, 1978). Hence the near-surface evidences of periglacial processes may be ascribed with some certainty to the period of the Loch Lomond Readvance, although precise dating is not available. However it was in this period that the extensive aeolian reworking of fluvioglacial sands in south-west Lancashire took place under periglacial conditions, as indicated by the sediment-ological studies of Wilson *et al.*, (1981) and the palynological and radiometric work reported by Tooley and Kear (1977). Pollen from associated organic sediments in the Ormskirk area is Zone III, and the radiocarbon age is 10 455±100 BP.

Conclusions

In summary, the available evidence indicates that only one glaciation can be recognised in the Lancashire Plain. The glaciation is considered to have been of polar or sub-polar regime, probably involving the formation of sea and shelf ice during the anaglacial period. The ice movement was generally from the North West, but was deflected locally by subglacial relief. The deglaciation period is considered to have been dominated by extensive stagnation and meltout, causing supraglacial sediments to accumulate in considerable thicknesses. These sediments were gradually let down from the ice, being modified both tectonically and sediment-ologically. They were draped onto a surface created by subglacial erosion, lodge-ment and basal meltout. However, supraglacial debris cover may not have been continuous, and certainly varied locally in volume. Consequently, some subglacial landform and sediment characteristics may be detected near the surface in certain areas. Planar surfaces represent those areas where supraglacial sediments had finally lost their extensive ice-core, yet continued to be planed by the lateral migration of low-gradient stream channels, and by the accumulation of sediments in deposi-tional basins. Undulation and kettling of these surfaces resulted from the melting of small amounts of buried ice and the dewatering and consolidation of the various types and thicknesses of sediments. The development of meltwater channels is related in part to drainage along the edge of the Carboniferous upland, in part to subglacial depressions and in part to consequent runoff from planar and irregular depositional surfaces. Organic materials accumulated in kettle holes and channel backwaters during deglaciation and were buried by later slumped tills and stratified sediments. Modification of landforms and sediments by periglacial processes took place during deglaciation and during the period of the Loch Lomond Readvance.

11 *Peter Worsley*

Pleistocene history of the Cheshire–Shropshire Plain

Introduction

The Cheshire–Shropshire Plain has long been regarded as one of the best regional examples of lowland glaciation in Britain, its status being enhanced since it includes the stratotype for the Last Glacial Stage in Britain. The Geological Society of London Quaternary working group originally proposed that the type locality for the Last Glacial Stage should be Chelford in east Cheshire (Shotton and West, 1969) but prior to the publication of the definitive report (Mitchell *et al.*, 1973) a new site had been discovered and investigated at Four Ashes in central Staffordshire. This latter locality proved to contain a more detailed record of environmental change and consequently it was adopted as the type site for the stage in preference to Chelford. Neither Chelford nor Four Ashes lent their names in the coining of the Last Glacial Stage term. Practice up to 1973 was to adopt mainland European terminology, either the north European Weichsel or the Alpine Würm, but the guidelines of the new stratigraphic code insisted that before international correlations could be achieved each country must first erect its own schemes based upon defined sequences. Since terrain is an important element of the late Quaternary sediments it was appropriate that the spatial dimension was reflected in the stage name. Many would have favoured a riverine basis as in Weichsel and Würm but in the event the stage was called Devensian since the region was once peopled by the Devenses, an ancient British tribe. The practice of resuscitating early British stratigraphic conduct by using tribal names is debatable, but be that as it may the reality is that the Devensian stage is now firmly established in national usage.

In ideal circumstances the base of the Devensian would be defined within an unbroken succession spanning the Ipswichian–Devensian transition. An absence of such a sequence in Britain was lamented by Shotton (1977) although subsequent work has identified a sedimentary record bridging this interglacial–glacial boundary at Wing near Rutland in the East Midlands. Unfortunately at this locality only the early part of the stage is present and it is not directly influenced by Devensian glaciation *per se*. Thus Four Ashes currently remains the best stratotype available but rather ironically the type succession proper does not exist since it occurred in a gravel pit which is now worked out. However, the biogenic components are conserved in museum collections.

Virtually the entire Cheshire–Shropshire Plain (here broadly defined so as to include adjacent parts of Clwyd and Staffordshire) was subject to a total cover by land ice in the Devensian stage. The sediments and allied depositional landforms arising from this advance event mantle the surface of a large proportion of the region today. Exceptions arise from either post-deglaciation erosion and deposition or, in some higher areas, from probable non-deposition. Little if any of the exposed surface is directly attributable to glacial erosion.

During the 1950s and 1960s both the timing and extent of Devensian glaciation was a controversial topic and the details of this have already been discussed (Worsley, 1976) and hence need not be repeated in detail. Accordingly the emphasis here will be on findings since that time and appraisals arising from them. Four topics will be selected and taken in approximate stratigraphic order.

The pre-Devensian record

In an earlier overview written in 1967 (Worsley, 1970), data from a borehole at Burland was the sole possible evidence for a pre-Devensian succession but even there the presence of organic clay beneath a 30 m thick till–sand complex was enigmatic since only a basic borehole log had been published. This situation still prevails. In addition the classical till–sand–till tripartite succession which for a century or so had been interpreted by the Geological Survey and some others in terms of two separate glaciations with an intervening major retreat phase was viewed as a complex depositional sequence produced by a single glacial advance event. Knowledge of conditions and events prior to the Last Glaciation was almost non-existent.

Fortunately the present perspective is more encouraging since several newer findings give insights into earlier patterns of environmental change. First, at Four Ashes, evidence has been forthcoming supporting the occurrence of an interglacial record (see Fig. 11.1). Alas, the published stratigraphic detail is meagre but apparently at one location (Site 44) a shallow channel cut into the local Triassic bedrock was filled by a sandy detritus peat and unconformably overlain by 4 m of sand and gravels attributable to the Middle and Late Devensian. A poorly preserved insect faunal assemblage was obtained from the organic detritus. Well-preserved plant macrofossils included wood from *Alnus* and *Taxus*, *Alnus* 'cones', fruit stones, and leaves of *Ilex*, an association consistent with a temperate fen–wood assemblage (A. Morgan, 1973). In addition, pollen showed high arboreal percentages with *Quercus*, *Alnus* and *Corylus*, a spectrum which readily matches that of Ipswichian pollen zone IIb (Andrew and West, 1977). A radiocarbon date on wood from the peat of > 45 ka (Birm-171) is consistent with this. A little further south at Trysull, a locality just within what is mapped as the Late Devensian ice advance limit, some 34 m of glacigenic sands and gravels are unconformably overlain by 1–2 m of silts and clays capped by sands and a thin (< 1 m) sandy till with apparently fresh Irish Sea erratics. The till is the local representative of the Late Devensian glacial event. For present purposes it is sufficient to state that the 1–2 m of silts and clays have yielded a range of animal and plant fossils indicative of an interglacial stage and, in terms of the conventional wisdom, are correlated with the early Hoxnian, (A. V. Morgan, 1973). Although the exact identity of the interglacial stage represented at Trysull may be debatable, there can be

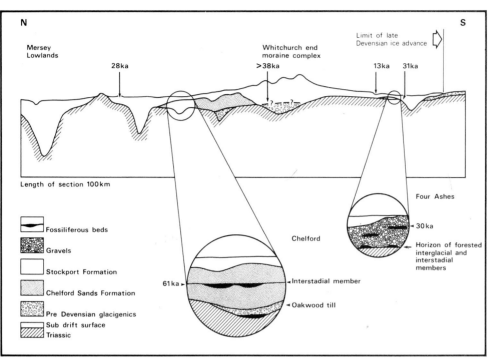

Fig. 11.1 A north–south cross-section through the eastern part of the Cheshire–Shropshire Plain. This shows the southern limit of the Late Devensian glaciation and associated sediments (Stockport Formation) which covers earlier sequences in some areas. Fossiliferous materials are present at Chelford and Four Ashes and the stratigraphic relationships at these localities are shown in the enlargements. The 'absolute' ages correspond to key ^{14}C age determinations

little doubt that it is pre-Devensian in age and consequently the main succession of glacial origin at the site must also antedate the Devensian. It is salutary to note that without the fossil biotic evidence, the entire sequence would have reasonably been assigned to the Late Devensian glacial event. It follows therefore, that in some instances at least, the possibility must be entertained that north of Trysull glacial sediments may occur which belong to a pre-Devensian glacial stage and if exposed beneath Devensian sediments an innocent-looking unconformity might represent a significant hiatus. However, in the absence of either fossiliferous material or sediments which are clearly of non-glacigenic status such a refined discrimination is going to be difficult to achieve.

Observations in east Cheshire (e.g. Evans *et al.*, 1968) have demonstrated that beneath the non-glacigenic Chelford Sands which antedate Late Devensian materials, glacial sediments are present in places. It is very rare to see these at outcrop and often their presence is deduced from inferential evidence such as water seepage zones or on the basis of borehole log descriptions. An example of the latter is afforded by the Brownlow No. 1 Borehole (sited 1 km south-west of Congleton) which records some 10 m of 'clay, red with pebbles' above Triassic bedrock and below over 35 m of sands which, on the basis of new quarry exposures in the vicinity, are mostly if not entirely Chelford Sands. In addition a thin bed (0.15 m thick) described as 'marl silty with plant remains' intervenes at the contact between the potential older till and younger sands. A similar succession has been observed on the floor of the large Oakwood Quarry near Chelford in the same sub-region (this will be elaborated upon later). It is important to recall that in the absence of the distinctive Chelford Sand lithology those pre-Late Devensian glacial sediments would not be recognisable as such and, following the rule that 'the simplest interpretation of the evidence should prevail', they would have been grouped with the overlying spread of Late Devensian glacigenics.

Perhaps the most extensive natural exposure of the glacial succession in the region lies along the south-west Wirral coast at Thurstaston in Merseyside. A reinvestigation of the section has been reported by Brenchley (1968) on behalf of his adult education class. They tackled a project specifically designed to test the validity of the mono- or bipartite glaciation hypothesis as applied to the sequence. The threefold lithological sequence was identified, with each unit having a similar maximum thickness of almost 9 m at outcrop. However, the base of the lowest was always below beach level, with geophysical data suggesting a sub-drift surface at less than 15 m O.D. Detailed studies of erratic shape and type failed to reveal any differences between what may have been two chronologically separate tills. Nevertheless distinctive textural contrasts were apparent and this factor, in conjunction with the relative sparseness of erratics, softness, coloration and presence of bedding structures led the investigators to propose that the so-called 'Upper Boulder Clay' was a lacustrine facies associated with ice retreat. A consequence of the adoption of this inferred depositional environment was that the entire sequence was likely to be the product of 'a single episode of ice retreat'. Subsequently Lee (1979) undertook a detailed analysis of an unstratified silt body some 30 m long and 2 m thick occurring within the middle lithological unit. His results indicated that both the composition and geotechnical behaviour of the silt were consistent with its being true loess and, as such, this bed must signify a sub-aerial depositional environment. This conclusion is clearly at variance with that of Brenchley's group, but since the latter demonstrated an undeniable difference between the lowest and upper units, the possibility re-emerges that the sequence is best interpreted in terms of a bipartite glacial depositional model. When considered along with the evidence for a pre-Devensian glaciation from the localities previously discussed the current drift of opinion would seem to be in favour of perhaps a more extensive survival of glacigenic materials from an earlier glacial stage than has recently been the case. It has to be stressed that this does not mean an endorsement of the interpretative model formerly championed by the majority of Geological Survey workers. Rather, refinement of methodology and increasing knowledge of the sedimentology are enabling a more sophisticated comprehension of the succession.

The final evidence relating to this particular aspect comes from Chelford where recent extensive quarrying and exploration boreholes have enabled a revised understanding of the stratigraphy to emerge (Worsley *et al.*, 1983). It is now possible to identify a buried bedrock-cut palaeovalley trending SE–NW and an exposure affording a reasonably complete cross-section of this valley occurs at the Oakwood Quarry. Close to the valley axis spoil from a sump excavation indicated the presence of a biogenic deposit. Initially it was thought (Worsley, 1980a) that this fossiliferous material, associated with clay and silts in the spoil heap, came from a horizon *above* the previously mentioned glacigenic bed which outcropped on the quarry floor below the Chelford Sands. After excavation of a trench in these deposits, it unexpectedly became clear that the biogenic material occurred *beneath* the glacigenic horizon (see Figure 2 in Worsley *et al.* (1983)). In outline the succession consisted of unweathered bedrock (Triassic Mercia Mudstone), fluvial gravels followed by laminated silts unconformably overlain by a complex of till and stratified clays, silts and sands. The entire sequence was less than 3 m thick. The basal gravels and silts appeared closely related and were probably indicative of a passage from free flowing water to ponded conditions. The occurrence of only a very minor component amongst the clasts of distant material suggests that the sediments are non-glacigenic, a view substantiated by the presence of a range of fossils. This biota includes pollen, plant-macro material, Mollusca, Coleoptera and Ostracoda and it appeared to be concentrated in the silts. From the pollen an open treeless landscape is suggested by the virtual absence of tree pollen and the dominance of terrestrial herbs with a significant aquatic representation. As might be expected from pond sediments the macroplant fossils were dominated by aquatics and similarly the great majority of the Mollusca were compatible with relatively shallow water, a picture collaborated by the Ostracoda. The Coleoptera suggest a similar situation and also throw more light on the immediate area, which was probably marshy in some places and better drained in others. Generally the fossils are tolerant assemblages but with modern ranges which extend into the Arctic. Thus the gravels and silts appear to be a record of sedimentation in an open treeless cool environment, generally comparable in character to these known from the Middle Devensian stadials. However, in this case the stratigraphic setting on bedrock close to the axis of a palaeovalley and beneath a glacigenic horizon which in turn is overlain by Chelford Sand indicates that the age of the sequence is at least Early Devensian and is most likely to be pre-Devensian.

As can be seen in Fig. 11.2 showing the total Oakwood Quarry stratigraphy, the sequence exposed in the trench lay directly below the peats and muds associated with the Chelford Interstadial and hence there is little doubt that a glaciation occurred after the accumulation of the stadial biota and before the commencement of Chelford Sand deposition. When first identified in the mid-1970s the glacigenic material consisted of massive till and hence was informally designated 'Oakwood till'. Subsequent exposures have shown it to be a more complex unit and the sequence in the excavation typifies the variability often displayed by it. Palynomorphs have been obtained from the till matrix and are dominantly Quaternary pollen and freshwater algal spores (Hunt, 1984). They are possibly partially derived from the underlying biogenic sediments. Contrary to Hunt, it is premature to claim that the source ice originated in the Pennines.

The lateral facies changes may in part be a response to post-depositional

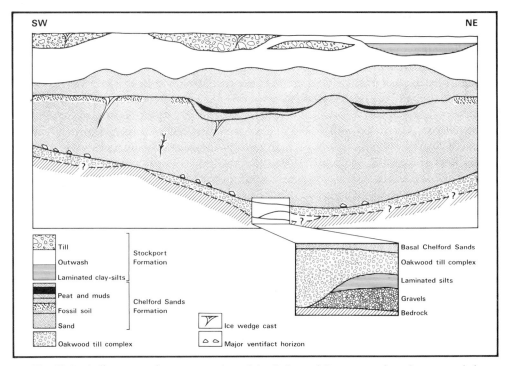

Fig. 11.2   A diagrammatic representation of the Oakwood Quarry stratigraphy as revealed
   between 1973 and 1984. The plane of the section is roughly normal to the bedrock-cut
   valley and the interstadial palaeochannel within the Chelford Sands Formation. Both
   slope to the north-west.

modification allied to periglacial weathering regimes and the discovery of a ventifact clast at a depth of over 1 m in till is consistent with this possibility although
the exposures revealed during excavation showed no evidence of re-deposition.
Normally it is only seen exposed in shallow gullies which develop on the quarry
floor and the effects on its fabric as a result of heavy machinery moving over its
surface must be always borne in mind. Despite this problem, the till was
undoubtedly associated with a cover of ventifacts, some of which developed from
large erratic clasts which protruded above the till's upper surface. A lack of
pedogenetic features suggests a major phase of wind abrasion of the till prior to
burial beneath the aggrading lower Chelford Sands. Although the age of the till
is unknown, the balance of evidence favours a pre-Devensian age. Since recycled
erratics also occur in the basal fluvial gravels, two separate glacial advances can
thus be demonstrated each of which antedates the Chelford Sands. Similarly rare
erratics in the main gravel sequence at Four Ashes testify the probability of a pre-
Ipswichian glacial event in that locality. On the basis of available data, correlations
with the main Trysull outwash succession are considered inappropriate.

The preglacial Devensian

Sediments assumed to be of this age are known from several parts of east Cheshire and in a limited area around Four Ashes. Early Devensian aged materials are identified as such primarily upon the evidence of biogenic-rich beds indicative of a boreal-type forested interstadial. This interstadial, the Chelford Interstadial, was first described by Simpson and West (1958) following the discovery of organic deposits within a thick sand sequence at the Farm Wood Quarry, Chelford. The overall stratigraphic context of the Chelford material was originally misunderstood since its existence was taken to confirm an ice recessional phase occurring between two major glacial advances of regional importance. Thus the interstadial sediments were considered to lie within 'Middle Sands', a sandy outwash component of a claimed regional tripartite till–and–till succession. Later it became evident that the host sands were not glacial outwash *per se* but rather a distinctive suite of locally derived alluvial sands transported by rivers in a periglacial climatic regime (Boulton and Worsley, 1965; Worsley, 1966; Evans *et al.*, 1968).

These alluvial sands – the Chelford Sands – normally overlie the Triassic bedrock, as previously noted. They may occasionally overlie residual beds of older Pleistocene sediments. Natural exposures are rare since the sands are usually blanketed by the Late Devensian glacial deposits and only occasionally has

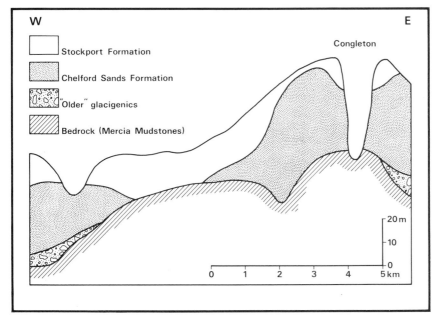

Fig. 11.3 A schematic east–west section through the Pleistocene succession in the Congleton area of east Cheshire, based in part on unpublished data. It shows the irregular sub-drift surface, residual early glacigenic sediments, the Chelford Sands Formation and the scale of erosion prior to the emplacement of the Stockport Formation in the Late Devensian. The deep valley at Congleton represents deglacial and subsequent erosion by the River Dane.

postglacial erosion created outcrops. These are almost exclusively on the sides or bottoms of incised valleys. A generalised east–west cross-section through the main Chelford Sands area from Congleton to near Sandbach shows their mode of occurrence (see Fig. 11.3). The deep valley at Congleton represents that of the River Dane and the depression in the west the Taxmere area. At Congleton the sands attain their maximum thickness and approach 70 m, but post-depositional erosion has been locally severe such that the entire sequence has been removed (Worsley, 1967a). Most data on the sands are derived from a number of quarries which work the sand as raw material for foundry and glass-making purposes.

The Chelford Sands are typically extremely well-sorted medium sands with a relatively low clast content. Their most striking property is a dominance of low angle bedding planes with a general westwards dip in the 1–3° range. Despite the near-horizontal stratification which is most apparent from a distance, in detail some significant variations occur. Locally parts of the sequence may comprise stacks of trough cross-bedded sets, or solitary scour channels filled with simple planar cross-strata. Measurements of these cross-stratifications reveal an overall east–west pattern of palaeocurrent flow. Besides the cross-bedded material, units of evenly laminated sands are frequently associated with what appear to be multiple deflation surfaces. Many of the scattered clasts are highly polished and some display excellently developed facets (Thompson and Worsley, 1967). Studies of clast composition show some variability between exposures but the commonest lithologies are medium to coarse grained sandstones, orthoquartzites and vein quartz. Typically these constitute some 80–90 per cent of the total clast population. Of the exotics, flint, which varies between 1 and 9 per cent is perhaps the most interesting since the nearest outcrops are in Northern Ireland.

In was suggested (Worsley, 1970) that the Chelford Sand depositional environment was that of a series of low angle alluvial fans. This conception envisaged ephemeral streams draining the Pennine uplands transporting sands and locally derived clasts onto the low ground of the Cheshire Plain. Whilst this depositional model is still applicable the increase in Chelford Sands data since that time has emphasised the role of aeolian processes in contributing to the sedimentation (T. R. Good, work in progress).

In a discussion of the Dutch coversands Ruegg (1983) has identified a sequential sedimentological trend within the Weichselian sequence, with a reduction in fluvial influences and a corresponding increase in aeolian effects. Such a pattern appears to be present in parts of the Chelford Sands, e.g. Oakwood Quarry, but in the absence of biogenics it is not currently possible to place most of the exposures in a time-based order. There are undoubted parallels in sedimentological character between the coversands of the Netherlands and the Chelford Sands but advances in the interpretation of both have been dependent upon the recognition of a sand sheet environment capable of producing extensive evenly laminated sands in association with clearly fluvial structures. One possible analogue occurs in the south-west of Banks Island where fluvio-aeolian sedimentation has been recently investigated (Good and Bryant, 1985). After a study of Chelford Sands grain surface textures under the electron microscope Brown (1973), using established criteria, was able to distinguish features indicative of fluvial, aeolian and glacial abrasion. Although she had reservations on the power of environmental discrimination using this technique it is interesting to note that these are the kinds of textural groups which might be anticipated.

Although the provenance of the clastic material is clear, an unresolved question revolves around the original source of the sands. Matters are complicated by colour contrasts within the sands. Normally basal occurrences are reddish brown but in the two westernmost outcrops the lowest part of the succession is white. However, the vast sediment bulk is coloured reddish yellow. The genetic significance of the contact between the white and reddish yellow sands is controversial but in the writer's experience the balance of evidence supports a primarily uncomformable relationship. Concerning sand origin, the two obvious possibilities are the sandstones associated with the local Carboniferous and Permo-Triassic solid outcrops. Although the last major phase of sand movement appears to have been generally westwards if, as suspected, aeolian-powered transport is important, then the final fluvial phase may have simply redistributed sand derived from a different direction.

Exposures created in the Oakwood Quarry in the 1970s have clarified the stratigraphic context of the interstadial deposits within the sands. This has been possible largely because in the new quarry a dry working technique has been employed necessitating continuous pumping of the ground water discharge. As a result the Chelford organics could be seen to lie within the fill of a palaeochannel system which trends SE–NW and a schematic cross-section through this is shown in Fig. 11.2. The significance of the coincidence of the interstadial palaeochannel and the bedrock-cut palaeovalley is currently unclear, except that bedrock geomorphology during the interstadial may have influenced drainage patterns. Borehole data indicates that the Farm Wood exposures lie in the downstream equivalents of the palaeochannel fill.

Dimensionally the palaeochannel is on average 150–175 m wide and 4–5 m deep and the norm consisted of a single channel-system. However, sometimes the sections demonstrated the occurrence of two and rarely three independent channel cross profiles within the main feature. The south-west bounding channel slope (averaging 22°) was much steeper than its counterpart on the north-east side giving an overall asymmetric form to the feature. Within the channel fill considerable complexity was displayed. Usually the south-east margin presented the clearest exposure of internal structure, largely because it consisted almost exclusively of sand. Bedding relationships indicated at least two phases of large scale planar cross-bed construction with dips of up to 32° towards the channel axis. Seemingly the sediment supply came from the direction of the landsurfaces beyond the channel limits suggesting that the main process involved was aeolian lee-side deposition rather than fluvial bar aggradation. The main organic bed (compacted felted peat and wood) accumulated in a period of stability during the emplacement of the south-west marginal fill, for the material occupying the palaeochannel floor could be traced up part of the marginal slope and rarely up the entire feature to pass laterally into a thin organic horizon (palaeosol) on the bounding landsurface. Occasionally *in situ* tree stumps could be found on the slope but were commonest within the main channel floor. Towards the north-east the main organic bed retained its integrity for up to some 75 m but was then truncated by an unconformity indicative of lateral migration by a subsidiary chanel. Beyond the north-east limit of the main organic bed the fill succession consisted in the main of a mixture of drifted clastic tree material, sands, silts and muds with no systematic organisation. Finally the main palaeochannel limit would normally be defined by a gently shelving thin organic bed which then became near horizontal. In places on either side of the main feature, upon what would have been the flanking landsurface, thin organic accumulations were found associated with *in situ* rootlets suggesting a weakly developed palaeosol. For the most part, however, the palaeosol was eroded and its position marked by a minor stone lag with many faceted and polished pebbles and occasional ice wedge casts.

The flora from Farm Wood was described by West (Simpson and West, 1958) and compared with the *Pinus–Betula–Picea* forest of northern Fennoscandia. Coope (1959) examined the Coleoptera and came to a remarkably similar palaeo-environmental conclusion. Later he slightly revised his original material (Coope, 1977b) and after a re-evaluation stated 'for this interstadial at least, complete harmony of flora and fauna existed'. Published work on the Oakwood material is currently restricted to short discussions by Whitehead (1977) and Holyoak (1983) on the spruce. K. A. Moseley (unpublished) has undertaken a detailed study of the contained Coleoptera and found excellent agreement with Coope's pioneer work.

The radiocarbon dating saga of the Farm Wood Quarry organic material has been related in detail by Worsley (1980b). In Fig. 11.4 a plot of the various dating assays is presented. The important point to note is that effectively the interstadial material lies beyond the limits of conventional radiocarbon dating. Accelerator techniques currently under development may ultimately be able to remove the current uncertainty. So far the Oxford laboratory has used wood from Oakwood in experiments evaluating chemical background and incidentally confirmed that the material is not less than 50 ka in age (J. A. J. Gowlett, pers. comm.).

To complete the pre-Devensian record, recourse is necessary again to the Four Ashes locality, where biostratigraphic evidence from both the Early and Middle Devensian is available. Unfortunately the 'earlier' material lies immediately above the basal unconformity occupying the same relative stratigraphic position as the previously mentioned interglacial biota at Site 44. In all some fifty or so separate organic beds were discovered within an area of about 1 ha in the late 1960s. Each of these has been given a separate site number by Anne Morgan (A. Morgan, 1973). At Sites 9 and 10, which were less than 25 metres apart, detritus muds yielded coleopteran faunas including six species which are particularly associated with trees. Anne Morgan concluded that the prevailing vegetation was 'of coniferous forests represented mainly by *Pinus* and *Picea* together with *Betula* and possibly *Salix*'. Later Andrew and West (1977) noted the presence of high tree-pollen percentages represented by *Betula*, *Pinus* and *Picea* in material from Site 10. Thus both the coleopteran and pollen data suggest that a boreal-type forest environment prevailed during the accumulation of detritus muds at Sites 9 and 10. On the basis of current understanding it appears that a correlative of the Chelford Interstadial is present

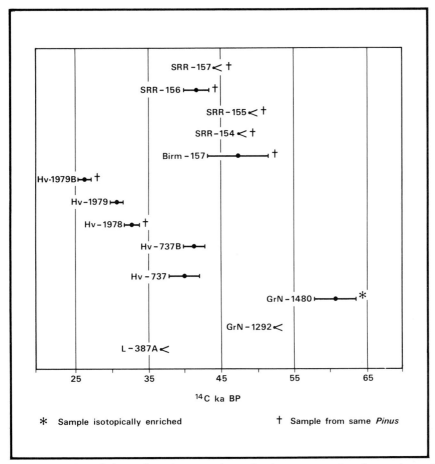

Fig. 11.4 A plot of the radiocarbon age determinations on wood and peat from the interstadial member at the Farm Wood Quarry, Chelford. Despite stratigraphic evidence suggestive of a relatively short duration event of several millennia at the most, the radiometric results embrace a much longer timespan. Note particularly, the spread of dates from the same tree. The oldest date GrN-1480 was obtained after isotopic enrichment. All the finite results probably reflect varying degrees of contamination of 'infinitely old' organics by younger materials and the 'true' age of the interstadial is likely to be greater than 65 ka.

at Four Ashes, but it has to be emphasised that the lithostratigraphy only gives an 'older' limiting age to the superincumbent sands and gravels. Similarly it is only an assumption that Site 9 and 10 biogenics are younger than that from Site 44 and it is of relevance to note than Anne Morgan was aware of the potential stratigraphic ambiguity arising from another basal lens with an 'arctic' fauna.

The sands and gravels which overlie the basal organic-rich channel fills range from some 0.5–4.5 m in thickness and consist almost entirely of material derived from the local Triassic Kidderminster Conglomerate (Bunter Pebble Beds). Comparison with the Chelford Sands reveals parallels, local sediment sources supplying clastic materials from non-glacial outcrops. A major difference is the absence of evidence at Four Ashes for clastic deposition prior to the forested inter-stadial (assuming synchroneity) as was happening in Cheshire. On weathered faces bedding structures in the gravels suggest a low-relief alluvial environment consistent with an intermittently functioning braided river system. Throughout the sand and gravels many lenses rich in detritus muds and clays were present. Anne Morgan's analysis of the coleopteran faunas from 24 different lenses showed a remarkable degree of ecological integrity within each, suggesting no mixing arising from erosion of previously deposited adjacent material. This feature she ascribed to two factors, first vulnerability of insect fossils to mechanical breakdown and fungal attack, and second proximity to the depositional basin's watershed with implied low energy runoff. All the faunas *within* the gravels suggest a persistence of an open treeless landscape whilst the aggradation progressed. Of a number of radiocarbon dates from the lens material, some are infinite but most are finite and define a timespan of 45 ka to about 30 ka. Apart from a period somewhere between 38.5 and 42.5 ka a rigorous continental climate persisted. Within the latter timespan a relatively thermophilous fauna was present and appears to correspond with the Upton Warren Interstadial *sensu stricto*. A period of permafrost presence is indicated by a horizon of ice wedge casts in the middle of the succession and is judged to have occurred after 38 ka (A. V. Morgan, 1973) but his claim that another such event occurred during sedimentation cannot be accepted since the evidence is a sole 'wedge' structure penetrating the bedrock. This is truncated by the basal gravels and thus, on the argument outlined previously, is of indeterminate age.

The Late Devensian glaciation

At Chelford and Four Ashes the stratigraphy clearly shows the earlier Devensian sediments unconformably overlain by glacigenic deposits which form part of the regional surface cover. The extent of this unit in relation to the last glacial event can be seen in the north–south profile of Fig. 11.1. Traditionally the bulk of these glacigenics has been classified into a 'Lower Boulder Clay', 'Middle Sands' and 'Upper Boulder Clay' tripartite lithological succession (as noted in the pre-Devensian till section). The British Geological Survey, having a primary duty to produce geological maps, had inevitably to seek some lithological order and certainly the use of a tripartite lithological succession aided their task. However, it was its extension beyond pragmatic mapping needs which led to difficulties. Detailed examination of various quarry exposures revealed that the till and sand units were often characterised by lateral and vertical impersistence. This sedimento-logical factor, the significance of which had long been appreciated by many independent workers, together with the landform context and increasing knowledge of processes operative in contemporary glacial environments led to the conclusion that a simple lithology–event approach was unrealistic. Hence the tripartite inter-pretative model had to be abandoned and when a formal stratigraphic code came to be applied, the whole succession was given formational status (Worsley, 1967a).

Thus the Stockport Formation embraces the Late Devensian glacigenic deposits and if needed local units may be recognised with member or bed status. Such an approach enabled the reconciliation of both simple and complex successions within a single drift sheet.

As highlighted previously (Worsley, 1975), the Cheshire–Shropshire lowlands played an important role in the development of conceptional models in glacial geology. In current terminology this would be expressed as the 'Supraglacial Land-system' (Boulton and Paul, 1976; Eyles, 1983; Paul, 1983), that is, a landform–sediment association which develops from deposition upon the glacier surface and is subsequently subject to varying degrees of modification as a result of buried ice melting. This kind of situation arises as a response to compressive glacier flow inducing basally transported material to become raised such that surface ablation processes release it from the ice. The induction is usually a consequence of decelerating ice-flow velocities occurring in zones of obstruction. These include (a) outer slow-moving 'frozen-to-the-bed' glacier margins, (b) interaction between different ice sources, and (c) bedrock form. A major belt of supraglacial land-system type terrain extends across the central part of the lowlands, swinging northwards parallel with the bounding uplands to the east and west. This feature was termed the Wrexham–Whitchurch–Bar Hill morine by Boulton and Worsley (1965). Another belt lies just south of the River Severn in central Shropshire. Between the 'Supraglacial Landsystem' zones the more familiar 'Subglacial Land-system' prevails, relatively low relief lodgement till plains with localised drapes of glacio-lacustrine sediments.

The main Late Devensian ice source lay to the north and north-west with a major ice lobe moving out from the northern Irish Sea Basin into the lowlands. Since this ice stream crossed relatively erodible Carboniferous and Triassic bedrock overlain by marine deposits the resultant tills are brown to red-brown in colour and fine-grained and often contain comminuted though occasionally whole, marine shell material. The presence of Quaternary marine dinosysts has been demonstrated by Hunt (1984) confirming an Irish Sea source. No independent ice sheet was nurtured in the southern Pennines but on the western side of the lowland a separate glacier system developed over Wales. A belt of ground along the Welsh borders corresponds with the zone within which the Irish Sea and Welsh derived ice sheets interacted. In some areas such as the Shrewsbury district, the Welsh ice deposits overlie Irish Sea materials (Worsley, 1975).

Most of the western lowlands have been subject to an intensive programme of sand and gravel resource assessment survey during the last decade, with over 500 boreholes being sunk to aid determination of the sub-surface geometry of the aggregate body. Rather disappointingly not a single buried organic horizon was encountered and no revision of the chronostratigraphy has been possible. The timing of the Late Devensian glaciation remains rather loosely defined. At Four Ashes the youngest sub-till organics have yielded a carbon-14 age of some 30 ka (Birm-25). Dates from within the Stockport Formation are restricted to three shell assays: 28 ka (I-1667), 32 ka (I-2939), > 38 ka (Birm-60); and a single date on 'organic silt' of dubious parentage: 26 ka (I-2800). The oldest date from sediments which accumulated after ice retreat comes from Stafford and is 13.5 ka (Birm-150). If speculative correlations outside the region are to be resisted then we must conclude that the glacial maximum has an age somewhere between 25 and 14 ka and no greater precision is possible.

Although the mineral assessment boreholes shed little light on the absolute chronology they did provide a wealth of sub-surface lithological data. Of note was a confirmation of locally complex suites of sediments and extensive sub-regional lacustrine deposits, many indicating impounded water prior to burial by advancing ice. In order to illustrate the kind of debate which is currently being waged over the origin and significance of the Late Devensian glacigenics, the Wrexham area of the North Welsh borderlands will be considered further as a case study.

The Wrexham Plateau – a case study

This area is of interest since, first, it has an extensive and thick suite of glacigenic sediments which rapidly thin westwards against the Welsh Uplands, and second, sediment supply was derived from both northern (Irish Sea) and local Welsh glacial sources. The first comprehensive account of the Wrexham area appeared in the standard Geological Survey memoir (Wedd *et al.*, 1928) describing the 'one inch to the mile' map sheet. Of note is the fact that the bulk of the glacial geological account was written by G. W. Lamplugh, a very experienced worker with first-hand familiarity of modern glacial depositional environments (Lamplugh, 1911). He introduced the term 'Wrexham Delta-terrace' to refer to the 'plateau-like shape' of the main mass of sand and gravel covering some 50 km^2 around the town itself. From a present-day perspective it is easy to question the use of this genetic term but there must be doubt whether Lamplugh contemplated the possibility that his term would be so extensively used subsequently. In order to avoid continued usage of a designation which is no longer appropriate the substitute term 'Wrexham Plateau' is proposed. Lamplugh described the diversity of surface form to the plateau, the northern part being fairly flat but southwards passing into hummocky terrain. The latter character was seen in part as a response to ice meltout producing kettle holes but he also identified a well-defined esker at the southern end of Gwersyllt Park some 3 km north of Wrexham. It seems that the relatively abrupt easterly limit to the plateau was regarded by Lamplugh as a delta front and hence the derivation of his proposed genetic term 'Delta-terrace'.

As part of a wider study of the Alyn River basin Peake (1961) postulated that the plateau feature was the end product of a 'composite prograding delta' built into a series of narrow ice-dammed lakes associated with an easterly-retreating ice margin, with the final easterly limit being regarded as an ice-contact slope. In the same year Poole and Whiteman (1961) came to a somewhat different conclusion by using the term 'morainic outwash' deriving a sediment supply from the east. Brief reference by the writer (Worsley, 1970) to the feature drew attention to the parallels with Flint's (1929) ice marginal terrace model and in accordance with this also viewed the eastern termination as an ice-contact slope. Francis (1978) provided a critical overview of previous work and gave descriptions of two working and two abandoned sand and gravel pits near to Gresford just north of Wrexham itself together with brief reference to a pit in the south-eastern sector. He regarded his sedimentological observations as being compatible with an aggrading sub-aerial alluvial fan depositional environmental model. The fan, he envisaged, was supplied by outwash from a general western source. Interbeds of poorly sorted stony clays were considered to be either debris flows analogous to those in semi-arid areas or flow tills. In the east, the plateau termination was claimed to be constructional in origin 'without resorting to deposition in contact with ice for which no evidence is available'. To him the existing landform corresponded with the final depositional surface of the unconstrained fan. The linearity displayed by parts of the hummocky topography on the plateau surface was ascribed to 'ice-contact borders' but with the margin of the main glacial mass lying westwards rather than on the east as previous workers had invoked. This latter conclusion provoked a critical response by Peake (1978).

Dunkley (1981) supported the concept of a fan body fed from the westward-lying valleys. Within the sedimentary mass he was able to detect both an overall fining pattern from west to east and a coarsening upwards succession. Thus in the central and western parts in particular, the former depositional process was viewed in terms of proximal braided stream systems but eastwards these were thought to be replaced by distal flood deposits overlying sands, silts and clays deposited in standing water. Further it was suggested that in the northern part of the fan an upper gravel sheet lying unconformably on the fan sequence might be a separate outwash body associated with a subsequent ice readvance from the north encroaching upon the fan surface. The till deposited by this advance was the same sedimentary body upon which Peake (1961) had based her concept of the Llay Readvance. However, Francis (1978, p. 51) regarded this lithological unit as a flow till and hence his conclusion 'the invocation of a readvance seems unnecessary'. Finally Wilson *et al.* (1982), attempting to integrate the newly acquired mineral assessment survey data from the western Cheshire–Shropshire Plain, argued that the main Wrexham sand and gravel succession formed part of an extensive prograding outwash plain (sandur) succession. They reasoned that the responsible advancing ice was not a limited readvance but rather a major regional one which subsequently overran the entire feature. In their view, the abrupt eastward termination of the Wrexham Plateau was a glacial erosional feature with the ice having removed the distal toe of the outwash fan. It is clear, therefore, that unanimity over the plateau origin has yet to be achieved!

However, the diversity of interpretation apparent in the preceding outline of published opinion does in part reflect the intrinsic complexity of the main plateau sedimentary body. It is suggested that most, if not all, of the previously inferred depositional processes have contributed to the landform which we see today. Accordingly attempts to invoke a single depositionary process would appear to be unsound. If we take the concept of the alluvial fan, which is the essence of the newer interpretative framework, the following observational factors are difficult to reconcile with it: extensive kettle-topography, occurrence of an esker, flow till units, interbedded Irish Sea type tills and sands along the eastern plateau margin, and crenulate relatively steep slopes. Of course there is a danger of slipping into a semantic debate over the subtleties of discrimination between a delta, alluvial fan and sandur, especially when there is agreement that the sediment flux is powered by glacial meltwater. In this instance most would acknowledge the existence of intergrades. The absent ingredient in all the published discussions to date is the supraglacial landform–sediment association, yet such a concept permits a reconciliation of all the known facts.

The overall character of the Wrexham Plateau can be appreciated by reference to the glacial geological map (Fig. 11.5). Apart from the extensive area of alluvium fanning out from the exit to the Alyn Gorge just north of Gresford in the northeastern sector, other sediments younger than the deglaciation have been omitted. In the main, lithological boundaries follow Ball (1982) and Dunkley (1981), since revision of the Quaternary geology was part of the mineral assessment programme. Limited outcrops of silt have been subsumed within the till distribution, partly for clarity but also because sometimes the till in the east displays lamination – indeed it was initially considered to be of lacustrine origin by the Geological Survey. As can be deduced from the pattern of bedrock exposure there is an overall trend for the glacial succession to thin westwards against the rising ground of the Welsh Uplands and this is also clear from the cross-section (Fig. 11.6).

Along the higher ground forming the eastern flank of the uplands two major rock-cut meltwater channels can be seen near to Brymbo trending NNW–SSE. Both have humped longitudinal profiles although the crest of the westernmost lies just beyond the map border. This latter property indicates that they are likely to be of subglacial origin and probably last functioned when the last ice sheet was close to its maximum. It is possible that the channels originated in an earlier glaciation and were reoccupied by meltwater in the last event. The Alyn Gorge (floored by bedrock) is also likely to be related to glacial meltwater activity and may have tapped the flow of the Brymbo channels once ice wastage had progressed to the extent that water flowage was no longer being confined to a general north–south direction.

The map also shows the extensive distribution of enclosed hollows: they occur on both the sands and gravels and the tills. Of specific note is the way in which some of the depressions transgress the lithological boundaries between the tills and the sands and gravels. In this context the existence of the Gwyersylt esker is also of significance. These relationships indicate that extensive deposition of glacigenic materials occurred over and within glacial ice.

A cross-section through the area, drawn roughly perpendicular to the postulated main ice-contact zone along the eastern plateau margin is shown in Figure 11.6. The sub-surface information is based upon six mineral assessment boreholes and the log from the shaft of the old Wrexham-Acton colliery. As is evident from a comparison of the borehole successions, correlations of lithological units are difficult but one can be reasonably confident that at least the higher till units are irregular in shape and of limited aerial extent. Consequently the portrayed shape of the till bodies below the landsurface are partly conjectural, yet it is obvious that no simple tripartite lithostratigraphy prevails. Similarly in quarry exposures

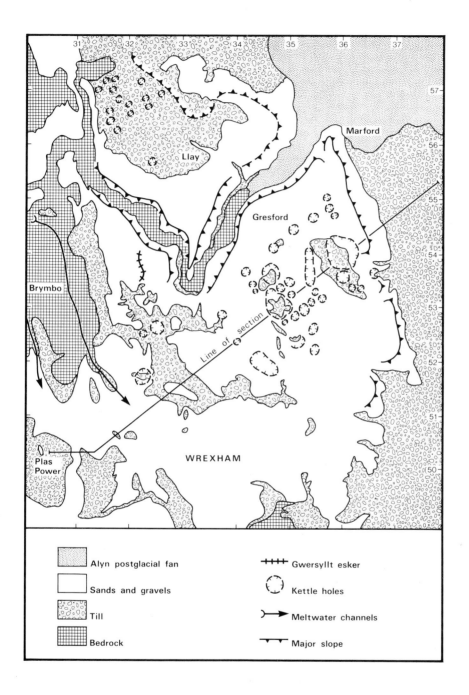

Fig. 11.5 Glacial landforms and sediments in the area around Wrexham. This area forms part of a regional 'Supraglacial Landsystem' belt. Scale is given by the national kilometric grid.

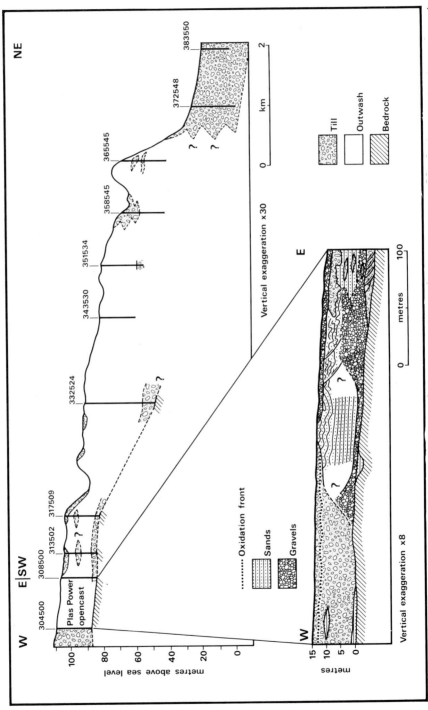

Fig. 11.6 A cross-section through the Wrexham Plateau (see Fig. 11.5 for location). Note the variety of landform irregularities and associated facies changes. The amplified section shows the stratigraphy revealed in the former Plas Power open-cast coal workings. This area coincides with the former interface between local (Welsh) and Irish Sea glacier deposition.

till units are commonly encountered with limited vertical and lateral extents lying within thick dominantly fluvial successions. In a section cut into the eastern plateau limit lateral facies changes occur, from till in the east to cross-stratified sands and silts in the west. The extensive till spread north of Llay was originally interpreted as a readvance feature (Peake, 1961) but later the seeds of doubt became evidence due to subsequent improvements in the understanding of glacial processes (Peake, 1978). Of relevance is a field discussion in 1979 held by most of the British workers concerned with till studies. This was held at an exposure through the Llay till and after examination there was broad agreement that the sediment had been emplaced by a mass flow mechanism.

At the western end of the cross-section the NE–SW orientation deviates to east–west in order to incorporate the sequence revealed by open-cast coal mining in the late 1950s and early 1960s. In Plas Power Park a section some 15 m thick and 350 m long through the glacigenic cover was thereby created. The amplified cross-section in Fig. 11.6 shows a composite schematic diagram of the surficial materials exposed in 1963–6. At the western end the main till mass was rich in cobble and boulder sized clasts of limestones, sandstone and shale which could be matched with local bedrock outcrops. The matrix was relatively coarse, light to dark grey in colour but where affected by groundwater seepage it became 'khaki'. This till is assumed to be of local (Welsh) origin. In contrast most of the till at the eastern end of the section comprised the typical red-brown Irish Sea material with a finer matrix and comminuted shell material and relatively rare clasts which included flints and Lake District rocks amongst the erratic component. These two tills would appear to correspond with what Peake (1961) described as 'yellow' and 'red' boulder clays. The fluvial facies was divisible into two types: (a) poorly sorted material ranging from sand up to boulder size with feebly developed stratification, and (b) well sorted sands with occasional lenses of gravel or silt and prominent small-scale cross-bedding. Since the latter sands were usually directly associated with the Irish Sea till bodies and directly comparable in character with the glacial sands which occur throughout much of the lowland region to the east it appears that this suite is derived from a source lying in the plains area. By contrast the local till and associated poorly sorted coarse outwash materials probably emanated from the higher ground lying to the west and north-west.

Whereas the local till outcropped as a massive body with only minor included washed zones which probably represented subglacial stream courses within an aggrading lodgement till, the Irish Sea till was quite different. It occurred as irregular masses within the well sorted outwash and four separate beds could be identified. Directly above rockhead at the eastern end of the section, beneath the till–outwash complex just noted, a 3–4 m thick till body lay on the bedrock. Close examination revealed that it consisted of imbricate thrusted slices of tough consolidated local and Irish Sea material. The stress field responsible for this was undoubtedly from the east. Unfortunately the quarry limits prevented a fuller understanding of this tectonic unit.

The local till possessed a surface horizon varying from 2 to 6 metres in thickness which was reddened through oxidation (Eyles and Sladen, 1981) and upon first encounter was mistakenly taken to be a separate till sheet. It may be that this phenomenon can account for the extensive upper red till shown in Fig. 4 of Peake (1961).

The Plas Power succession thus represents interactive sedimentation from two distinctive glacier ice masses. Further, the style of sedimentation associated with each may possibly reflect fundamental contrasts in glacier character as determined by thermal regime history. Since the locality appears to be co-extensive with the main Wrexham Plateau we may tentatively conclude that the entire landform is the result of deposition influenced by two glacial sources. An absence of buried biostratigraphic evidence creates difficulties in determining the age of the whole sequence bearing in mind the conclusions derived previously (e.g. Trysull). Nevertheless it appears likely that the bulk of the succession is of Late Devensian age but the lowest units may antedate that time. The association of tills, fluvioglacial units, hummocky terrain with well-defined kettle holes, an esker with a capping of till, and major ice-contact features are indicative of a complex depositional environment. A significant component of the sedimentation appears to have been supraglacial in character since post-depositional meltout has apparently led to deformation of the upper parts of the sequence. This does not deny a role for a major sediment input from the north-west into the interface zone between local and more distant ice masses. The thick outwash and subsidiary tills proved by the boreholes demonstrate the occurrence of a well-defined major sedimentation sink and the marginal interbeds suggest that residual ice masses imposed controls on the outer limits to the area. Hence it is proposed that the Wrexham Plateau ought to be regarded as part of a Supraglacial Landsystem which swings in arcuate fashion across the entire lowland thereby linking the Welsh and Pennine ice marginal sequences.

Conclusion

In the foregoing account, several themes have been selected for discussion and chronologically these group into three: (i) the pre-Devensian, (ii) the preglacial Early and Middle Devensian, and (iii) the Late Devensian glaciation event. All matters concerning landscape evolution after the last deglaciation (c.16 ka) extending through the 'Late-glacial' period (Windermere Interstadial and Loch Lomond Stadial) and the Flandrian 'Post-glacial' must be left to other contributors.

Much of the current chronostratigraphic understanding is based on the assumption that the Chelford Interstadial is of Early Devensian age. This is based on the evidence of stratigraphic position, radiocarbon dating and a *Betula–Pinus–Picea* tree assemblage. Although future work may well confirm this view it is not however possible to categorically deny that it might have a pre-Ipswichian age. For instance the interstadial is absent from the thick sequence above the most continuous Ipswichian–Devensian succession known in Britain at Wing (Hall, 1980). During the Ipswichian low frequencies of *Picea* pollen are present; however the quantity is sufficient to confirm the tree's membership of the Ipswichian forest community from biozone Ip IIb to early Ip IV – after that it seems to disappear. Matters are complicated by the low pollen productivity of *Picea* and an absence of macrofossils from the last interglacial record. West (1980) has drawn attention to the paradox that although *Picea* becomes progressively rarer in successively younger interglacials (interpreted as an overall trend for an eastwards migration of glacial stage refugia with time) it appears in abundance during interstadials dating from the early parts of the cold stages. This pattern he attributes to rapid migration of boreal-type forests after climatic amelioration pre-empting the initiation of the typical temperate-stage forest development cycle. Even so, the behaviour is puzzling and its presence in an Early Devensian boreal-type forest may be an artifact of a misinterpreted stratigraphy. Until a Chelford Interstadial deposit can be shown to lie superimposed upon Ipswichian sediments some caution must be exercised over its precise chronological attribution.

The most promising immediate prospects will be derived from the application of amino acid geochronological techniques to the terrestrial and marine molluscan fossils within the Pleistocene succession. A knowledge of the D-alloisoleucine to L-isoleucene ratios, which are related to protein diagenetic degradation within the shell material, yields a relative chronostratigraphy if a uniform regional thermal history is assumed. It can be anticipated that the problem posed by the differential extent and superimposition of lithologically similar tills will be elucidated in the light of the D : L ratios of their contained derived marine faunas. This approach would utilise the concept that the youngest faunal element within a given till unit will impose a maximum relative age for the glacial event responsible for the till deposition. Thus there is a high probability that significant new results will soon be forthcoming necessitating refinements of the outline presented here.

The Quaternary History of North Staffordshire

Introduction: general physique

The physiography and drainage of the area and the general features of the solid geology appear in Figs. 12.1 and 12.2. Throughout the area, high ground is commonly associated with anticlines and scarps with faulting for the topography is strongly influenced by both structure and lithology. Structurally, the dominant features are: (a) the Western Boundary Anticline which lifts the Namurian and Westphalian rocks in a complex flexure before plunging them down below the Triassic rocks in the west, probably without faulting (Challinor, 1978); and (b) the south-pitching synclinorium of the Potteries coalfield with its associated complex faulting. In general, the clays, shales and mudstones of Namurian and Westphalian age occupy the lower ground, while the grits and sandstones form ridges and scarps. In the case of the Triassic rocks, mudstones and soft sandstones usually form the plateaux or lower ground while conglomerate Pebble Beds constitute the higher land.

Overall, the highest ground in the area lies to the north and along the western margin. To the west lies the hummocky terrain of the Woore–Bar Hill–Wrinehill moraine complexes and the Cheshire–Shropshire plain. To the south and east, the ground slopes with the drainage which is discordant with structure in a number of places, an anomaly explored by Hind (1906), Barke (1920, 1929) and Yates (1956, 1957). Generally, the main drainage divide lies close to, but slightly east of, the highest ground of the western anticline, where it separates streams flowing west or north-west to the Severn or Mersey–Weaver systems, from those flowing south or south-east to the Trent. A number of drainage derangements thought to be the result of Quaternary ice-sheet invasions are examined later.

Yates (1956) suggested that the broad outlines of the drainage pattern have been inherited from an ancient surface which sloped from north-west to south-east. The initiation of cols in the Western Boundary Anticline is attributed by Yates to pre-glacial piracy by west-flowing streams with gradients steeper than those flowing south-east.

Yates (1956) and Lawrence (1967) claim to recognise a number of erosion surfaces in the area and these have been assigned a Tertiary age. Lawrence suggested that the presence of surface remnants indicates that parts of north-west Staffordshire were unglaciated, an untenable hypothesis in view of the universal occurrence of till or individual erratics, but a claim prompted perhaps by under-recording of glacigenic deposits on I.G.S. 'drift' maps of the area.

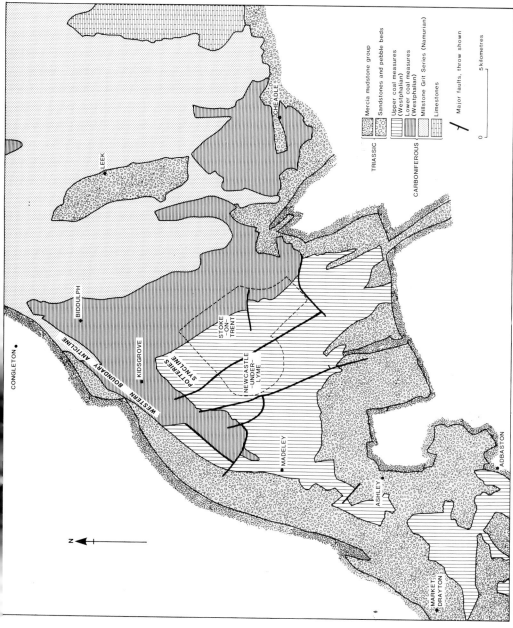

Fig. 12.1 North Staffordshire – geology (solid).

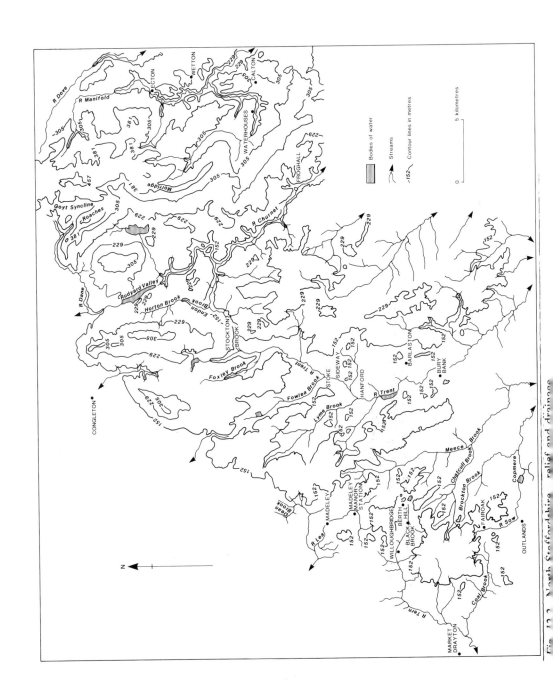

Fig. 12.2 North Staffordshire: relief and drainage

Glaciation

Whatever the character of the landscape on to which the first Quaternary ice sheets advanced, uncertainty remains about both the number of glacial inundations and in some instances the provenance of the ice.

Evidence from mid-Staffordshire suggests that there were at least two glaciations of the north-west Midlands. At Four Ashes, between Stafford and Wolverhampton, Late-Devensian till rests on gravels which, while essentially of local derivation, nevertheless contain 'a few weathered and different erratics which must derive from an earlier glaciation or glaciations' (Morgan and Morgan, 1977). Presumably the degree of weathering and infrequency of these erratics militates against their being pro-glacial outwash from the overlying till.

The glacial history of the Cheshire plain is also relevant to events in North Staffordshire (Worsley, Chapter 11). Past views of that history have shown considerable differences, varying from a single glaciation with retreat phase pulses (Boulton and Worsley, 1965; Worsley, 1970) to multiple glaciations extending to Saalian (Wolstonian) times (Poole and Whiteman, 1966; Poole, 1968).

Until lately, the evidence from the region immediately to the north and north-west of Staffordshire (Worsley, Chapter 11; Johnson, Chapter 13), pointed to one recent event (the Late-Devensian glaciation) but hinted at an earlier event of uncertain age but greater areal extent (Evans *et al.*, 1968; Worsley, 1970, 1980a). However, new and unequivocal evidence has emerged from Chelford in Cheshire, of a third, possibly much earlier, glacial episode (Worsley *et al.*, 1983, Chapter 11).

In North Staffordshire itself, King (1960) and Beaver and Turton (1979) refer to several episodes of advance and retreat and Gemmell and George (1972) suggest that North Staffordshire was invaded by Elster (Anglian), Saalian (Wolstonian) and Weichselian (Devensian) ice. The question of whether Welsh ice ever invaded the Potteries region is highlighted by Wedd's (Gibson *et al.*, 1925) reference to the discovery of felsite erratics with Arennig affinities within the area.

In sum, consideration of adjacent areas leads to the conclusion that ice invaded North Staffordshire, certainly on three occasions and perhaps more, but that within the area itself, there is *clear* evidence for only one such event. Deep-sea and ice-sheet cores suggest that there were other glaciations, the evidence of which may have long since disappeared.

The eastern limits of glaciation in this area are difficult to recognise in detail though there is agreement on their general location. Thus Jowett and Charlesworth (1929) have the ice limit just west of the Roaches and running along the Morridge. King (1960) shows a firm 'limit of Irish Sea ice' in much the same location, while Johnson (1965a) indicates 'limit of boulder clay' and elsewhere, 'altitudinal limit of glacial erratics', both somewhat to the west of the Jowett and Charlesworth margin. The I.G.S. 1:625 000 Quaternary map (south) shows a bold line for the Devensian ice limit running north–south just east of Leek, with a non-glaciated embayment around Cheadle. Francis (1980) recently questioned these previously defined limits and reported formerly unrecognised deposits which ease the margin of the Devensian ice eastwards in places.

Little has been written about the impact of glaciation on the landforms of North Staffordshire. The depositional evidence is meagre compared with adjacent areas. In the last published revision of the Stoke-on-Trent geological memoir (Gibson *et al.*, 1925) only 9 from a total of 107 pages refer to glacial and related themes, little improvement on the 1905 edition. By contrast about 45 per cent of the memoir *The Geology of the Country between Stafford and Market Drayton* (Whitehead *et al.*, 1927), relates to glacial and fluvio-glacial matters. This contrast is perhaps not surprising in view of the overriding economic importance of the solid geology in the Potteries.

The borehole record is partially old or very old in North Staffordshire and logs are commonly unreliable for a number of reasons but, notwithstanding these drawbacks, borehole data, combined with other evidence, usually permit some general observations. Thus over large areas, 'glacial deposits' range in thickness between about 2 m and 6 m. Rural areas to the west (often marked driftless on I.G.S. maps) commonly have only between 1 m and 2 m of stony material with far-travelled erratics. The thinness of the till compared with areas to the west has been attributed by Wedd (Gibson *et al.*, 1925) to the shearing of clean ice over the more debris-laden lower layers as the ice front surmounted the western anticline and high ground to the south. There is no clear pattern to the variations in till thickness, except that the larger valleys show thick, largely fluvio-glacial infills of sand and gravel with, more rarely, the addition of till or peat.

To date no borehole or exposure has yielded organic material which might be of interglacial or interstadial age. Grey clays and 'organic' sands more than 20 m thick at Stockton Brook may partly be weathered Coal Measures strata, but the more recent component is unlikely to pre-date the final withdrawal of ice from the area. Another deposit, probably of Windermere Interstadial affinity, underlies an apparent solifluxion deposit at 'the Bogs', Blackbrook (Fig. 12.4). There is no secure date for this deposit but again, it can hardly be older than the departure of the late-Devensian ice from this area. Beyond this, the age of the till or fluvio-glacial deposits in North Staffordshire must rely on extrapolation of chronologies established in adjacent areas where similar field relationships are encountered but where lithostratigraphic units are bracketed by dated deposits as, for example, at Four Ashes, (A. Morgan, 1973), Stafford (Morgan *et al.*, 1975), Wrinehill (Yates and Moseley, 1967) and Chelford (Simpson and West, 1958; Worsley, 1980a, b, and Chapter 11).

Glaciation and the alignment of drainage in North Staffordshire

Meltwater drainage systems

A number of workers have examined the trunk valleys for guidance on the glacial history of North Staffordshire. The deglaciation model which emerged had water-shed cols, trunk valleys and pro-glacial lake shorelines generated, then abandoned in sequence as an ice sheet wasted. By means of this model, it was considered that

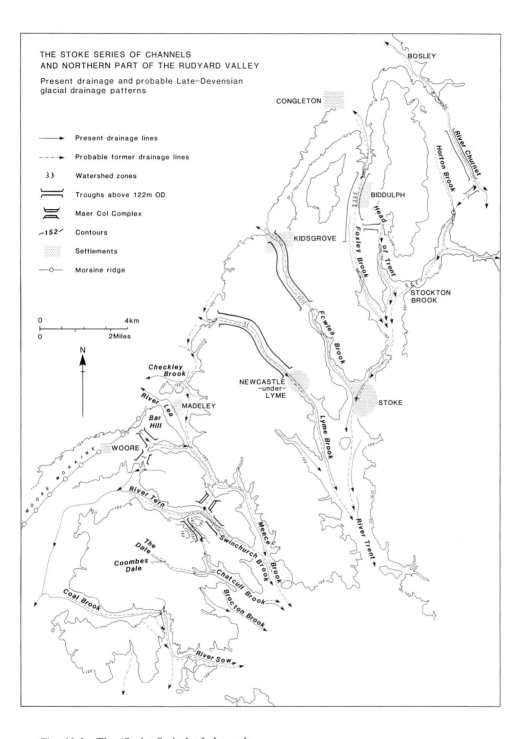

THE STOKE SERIES OF CHANNELS
AND NORTHERN PART OF THE RUDYARD VALLEY

Present drainage and probable Late–Devensian
glacial drainage patterns

Present drainage lines

Probable former drainage lines

Watershed zones

Troughs above 122m OD

Maer Col Complex

152 Contours

Settlements

Moraine ridge

0   4km

0   2Miles

N

BOSLEY

CONGLETON

River Churnet

Horton Brook

BIDDULPH

Head of Trent

Foxley Brook

KIDSGROVE

STOCKTON
BROOK

Fowlea Brook

NEWCASTLE
–under–
LYME

STOKE

Checkley
Brook

River Lea

MADELEY

Bar
Hill

WOORE

Lyme Brook

River Trent

River Tern

WOORE MORAINE

The
Dale

Coombes
Dale

Swinchurch Brook

Meece Brook

Chatcull Brook

Brocton Brook

Coal Brook

River Sow

Fig. 12.3 The 'Stoke Series' of channels.

the history of at least one deglaciation could be reconstructed.

Wedd (Gibson and Wedd, 1905) was the first to propose the idea of water ponded in valleys whose downstream parts were temporarily obstructed by ice. In his opinion, the upper Tern was at one stage blocked by ice whose former presence is manifested by the Woore moraine (Fig. 12.3). He suggested that water first spilled into the Trent system but later escaped along the ice margin as the latter retreated, so cutting a new permanent channel for the middle Tern.

A similar hypothesis for the Rudyard valley was applied by Pocock (1906), and various workers have considered other valleys where glaciation or deglaciation was thought to have wrought permanent drainage modifications (Hind, 1906; Barke, 1920, 1929). However, it was not until Jowett and Charlesworth (1929) published a regional deglaciation model covering an area from north Lancashire to mid-Shropshire, that the idea of relating most of the major valleys in this area to deglaciation was fully developed.

The model had ice entering north-west and central England from the north and north-west and ultimately surmounting the west Pennines and high ground to the south-west. During deglaciation, once the main watershed was exposed, further retreat revealed an increasing area of 'reverse slope'. Suitable embayments, the valleys of former west-flowing streams, were filled by pro-glacial lakes which ultimately spilled over the watershed generating and rapidly enlarging cols and channels. Where embayments were absent, water flowed along the ice margin sometimes from one lake to another. As deglaciation proceeded, these lakes emptied in sequence from south to north. This system of spillways and marginal channels supposedly generated during a deglaciation of North Staffordshire was named the 'Rudyard series' in the north and the 'Stoke series' in the south (Jowett and Charlesworth, 1929).

The model remained broadly intact for more than thirty years. During that time, Yates (1955) examined channels around Kidsgrove, distinguishing between river capture wind gaps and marginal or pro-glacial channels. Yates and Moseley (1958) elaborated the model for the Madeley area, citing evidence suggestive of pro-glacial lakes associated with an ice-marginal drainage system which flowed south from near the Dean Brook to the Whitmore trough (Fig. 12.2). King (1960) reconsidered the evolution of the Rudyard–Churnet valley system, and Walton (1964) evaluated evidence favouring ice-marginal channels and former lakes around Biddulph and Mow Cop. Curiously, Whitehead *et al.* (1927) while explaining part of the Sow–Coal Brook System in terms of the overspill concept, were not disposed to develop the idea.

By the late 1950s, serious doubts had arisen about this deglaciation model, introduced by Kendall (1902) and applied in this area by Jowett and Charlesworth. Perhaps the most conspicuous critic of the model's over-zealous employment was Sissons (1960a, 1961), who demonstrated that many channels in a number of areas of the country did not satisfy criteria which would identify them as spillways.

It was natural therefore, that in a review of the impact of glaciation on the west Pennines, Johnson (1965a) rejected many aspects of the Jowett and Charlesworth model and by implication, their view of the evolution of the 'Stoke Series' came into question. Further work on the Rudyard–Churnet valley (Johnson, 1965b) and its intake area around White Moor (Johnson *et al.*, 1970) demonstrated the application of a new model in which the development of the trunk valley and the majority of the adjacent channels was seen as resulting largely from the work of water flowing marginally, sub-marginally, or sub-glacially under hydrostatic pressure as ice wasted '*in situ*'.

Two pieces of evidence were cited in favour of the new concept. First, little indication of strandlines, glacio-lacustrine sediments or related phenomena could be found to authenticate the former presence of pro-glacial lakes. Second, many of the supposed marginal or pro-glacial channels appeared to have 'humped' or undulating long-profiles, a characteristic difficult to reconcile with the flow of water in open channels.

The two models described above appear capable of a degree of synthesis if modifications are accompanied by the introduction of a new concept. Jowett and Charlesworth were concerned with deglaciation and do not appear to have considered the possibility of pro-glacial lake development during the ice advance phase. Though very much smaller volumes of meltwater were produced during glaciation, seasonal melting of ground and glacier ice and snow produced substantial runoff. If the least favourable palaeoenvironmental reconstructions are accepted, then for the Late-Devensian glaciation, mean July temperatures of 7 °C to 8 °C (Coope, 1977a, 1980; Lamb, 1977b) and an annual precipitation about one-third of today's (Williams, 1975) would still, bearing in mind the long summer day, have allowed considerable meltwater production.

A number of workers have referred to the development of pro-glacial lakes during glaciation (King, 1960; Beaumont, *et al.*, 1969; Merritt, 1982) and Shotton's classic 'Lake Harrison' model, though recently criticised (Sumbler, 1983), depends on the concept (Shotton, 1953). The synthesised model has the following stages.

Advancing ice impinged against west-facing embayments and ponded seasonal meltwater. Ice advance and seasonal variation in meltwater volumes resulted in ephemeral strandlines. Lake sediments consisted of a mosaic of proximal high-energy sands and gravels, distal quiet-water sediments and scattered deposits of till swept randomly forward with detached pieces of ice. Eventually, advancing ice overrode and incorporated some of these sediments.

How long such lakes persisted in unclear. Boulton *et al.* (1977) estimate rates of advance of the Late-Devensian ice in the Midlands at between 150 m and 500 m *per annum*. Though perhaps reasonable for ice advance southwards across the Cheshire–Shropshire plain, this is probably too fast for encroachment onto the high land between Congleton and Adbaston. To the west of this high land the rockhead is masked by a considerable thickness of glacial and fluvio-glacial sediment (Yates and Moseley, 1967) and the ascent of the ice front would have been more difficult than appears today. In practice, the ice was partly deflected south-west by the high ground except where it had access to embayments. Thus thickening of ice in an easterly direction was in part a function of the southward advance of the ice margin. This being so, pro-glacial lakes generated during glaciation must have disappeared first in the north, a reversal of the sequence proposed by Jowett and Charlesworth in their deglaciation model.

The pro-glacial lakes of the glaciation phase grew to the point where they overflowed the watershed at the valley heads. Overflow cols and escape conduits on the distal side of the watershed developed rapidly, but once the ice margin surmounted the watershed, seasonal meltwater escaped supra-glacially, marginally, englacially and sub-glacially down the lee-side valleys which were part of the Tertiary landscape.

During deglaciation, ice east of the watershed and in valleys oblique to the direction of ice flow experienced sluggish forward momentum as it thinned and 'retreated'. Marginal and sub-glacial meltwater now not only occupied and enlarged channels cut by meltwater of the glaciation phase, but cut additional channels in harmony with the changing hydrology of the decaying ice sheet.

Once ice-thinning revealed the main watershed, marginal retreat dominated, though masses of ice still became detached where topography encouraged this. Marginal lakes as envisaged by Jowett and Charlesworth developed, but now, watershed cols and associated channels were already in existence. The size of such lakes was therefore restricted. A mosaic of sediments such as those mapped by Yates and Moseley (1958) in the Madeley basin and described earlier, was deposited and again, as outlined earlier, lake strandlines were of insufficient size to resist subsequent erosion (Sissons, 1981).

Earlier it was suggested that North Staffordshire experienced at least three glaciations. Evidence for the earliest is entirely absent from the area, and is scant for the penultimate event. Two, and perhaps all three, ice sheets crossed the main watershed. The sequence of events occurring during glacial invasion is likely to have been repeated three or more times. However, all glaciations subsequent to the first entered a landscape where the watershed cols and associated channels were already cut. Though possible, it is difficult to imagine that these conduits were repeatedly wholly plugged by till or fluvio-glacial sediment from the previous retreat phase. Consequently, the second and subsequent glacial events must have had a diminishing potential for producing pro-glacial lakes. Therefore at some stage in the sequence of glaciations the 'throughways' of the 'Stoke Series' must simply have allowed pro-glacial meltwater of both the glaciation and deglaciation phases to flow unimpeded through these watershed breaches. Strandlines therefore, are hardly to be expected and any lakes developed during the retreat of the Late-Devensian ice are likely to have been shallow and the result of temporary valley blocking by ice or debris. Lacustrine sediments intermingled at the margins with slope deposits and elsewhere with partly reworked till and the whole suite of deposits was subjected to fluviatile reworking.

The 'Stoke Series' of channels therefore consists of six valleys, four of which slice right through the watershed between Congleton and Adbaston (Fig. 12.3). The present watersheds within most of these through-valleys are simply a function of glacial or fluvio-glacial deposition in the last deglaciation phase and the surface drainage within the valleys has gentle gradients and is weakly organised. The alternation of broad and constricted sections is a reflection of structural and lithological constraints. Most possess continuous buried channels in which the thickness of fill varies (Al Saigh, 1977).

These buried channels are polygenetic. Developed through more than one glacial event, their undulating bedrock character reflects the action of sub-glacial water though there may be a meandering component (Schumm and Shepherd, 1973) which available data is insufficient to determine except for stretches of the upper Coal Brook and Sow valleys. The valley fills date largely from the last deglaciation and include a number of large peat-filled kettle holes.

The glacial derangement of the main drainage system

Barke (1920) refers to 'considerable modification of drainage by glaciers' in this area. Space only permits reference to the more significant of the supposed modifications. First, the River Lea's unusual course was noted by both Barke (1920) and Yates and Moseley (1958). The anomalously large valley compared with stream size, the 120° change of course near Madeley Manor station and the presence of high-level, now dry, tributaries orientated upstream, suggest enlargement by glacial meltwater and ultimate blocking of the former westward course by the Woore moraine at Onneley.

This moraine is also thought to have been responsible for blocking the former westward course of the upper Tern near Willoughbridge and thus for the new, anomalous course which cuts across the grain of the Triassic sandstones at Bearstone, 4 km west of Willoughbridge (Fig. 12.2).

Third, Hind (1906) suggested that the Trent formerly followed a course along the broad valley between Stoke and Barlaston but that moraine blocked this route at Sideway causing the present westward flow between Stoke and Hanford (Figs. 12.2 and 12.3). At Hanford, the river joins the valley of the Lyme Brook and flows south to join the former valley at Barlaston. The course is certainly anomalous but the supposed moraine remnant at Sideway is difficult to recognise in a disturbed industrial area. The former presence of 'dead' ice in the broad valley south of Stoke may provide a more acceptable explanation for this modification.

Fourth, the Coal Brook and upper Sow valleys between Market Drayton and Copmere occupy one trough divided by a low watershed at Fairoak. The upper Sow north of Outlands appears once to have been part of the Coal Brook but the cutting of the gorge at Outlands, first by overflow and subsequently by sub-glacial meltwater reversed the flow between Outlands and Fairoak. The anomalous gorge section east of Outlands runs across the grain of south-orientated dry valleys and must have developed during deglaciation when water flowed marginal to an ice mass which abutted against high ground to the north of the river's present course.

Finally, the Stockton Brook 'gap', a broad east–west orientated channel separating the Trent and Churnet drainage, has received attention from Pocock (1906), Gibson *et al.* (1925), Barke (1929), King (1960) and Johnson (1965). It has been suggested that the Horton Brook once flowed through the 'gap' as a tributary of the Trent at a time when the Churnet north of Froghall formed part of the same catchment. Subsequent glacial episodes, probably more complex than is allowed in the literature, created the south-flowing Churnet and led, following cessation of voluminous meltwater flow, to the abandonment of the Stockton Brook gap and the aquisition by the Churnet of the Horton and Endon Brooks as tributaries.

Periglacial processes and environments

Periglacial environments, almost certainly attended by some form of permafrost, were present during glaciation and probably deglaciation phases in North Staffordshire. Landform development during such phases is variously manifested within the area but the distribution of phenomena is patchy and the evidence is sometimes equivocal. The last periglacial conditions were experienced during Late-Glacial times but their impact is unclear. The landforms developed during periglacial episodes are examined below.

Mass movement phenomena

(i) *Landslides* It is axiomatic that mass movement was accelerated during periglacial episodes, but it has become clear that landslide development was by no means confined to these episodes (Franks and Johnson, 1964; Muller, 1979; Tallis and Johnson, 1980). Most North Staffordshire slides are confined to Namurian and Westphalian rocks of the Dane valley and the uplands north of Leek. Two slides, however, were revealed during road construction close to Stoke-on-Trent (Early and Skempton, 1972; Searle, 1973) and doubtless others remain to be exposed. Only in the case of the Walton's Wood slide, revealed during construction of the M6 motorway, has it been possible to assign a tentative minimum age to the disturbance. Here, basal pollen-bearing sediments of Zone VI age were buried at the toe of the slide, but the investigators concluded that initial movement occurred in Late-Glacial times (Early and Skempton, 1972).

(ii) *Other mass movement phenomena* Much of the upland area in the eastern half of North Staffordshire has a mantle of solifluxion or gelifluxion debris. Where underlain by shales, for example in parts of the upper Churnet catchment, these mantles consist of relatively fine-grained, rudely stratified overburden, up to 3 m thick at the base of slopes. Where grits and sandstones constitute the bedrock, rockfalls, screes, small blockfields and arrested ploughing blocks occur. Examples which probably date from Late-Devensian or Late-Glacial times appear in the upper Dane catchment and the Goyt syncline. It is inconceivable that these phenomena were overridden by ice and thus unless they entirely post-date the Late-Devensian glaciation, which is unlikely, they indicate that the Devensian ice lay to the west.

Possibly the best known feature of this type is the cemented scree at Ecton (Prentice and Morris, 1959). This deposit consists of angular clasts of limestone, cemented together and showing a strong downslope preferred orientation. It has a thickness of at least 3 m, possibly more than 5 m, at the base of the slope.

Cementation is the result of percolation of lime-rich solutions through the deposit.

Though rarely exposed, most other steep-sided limestone valleys of the North Staffordshire 'Peak' district must possess similar deposits, perhaps genuine *grèzes litées* (Warwick, 1977), spread over their lower slopes. None of these phenomena have been dated, but they are likely to be composite features accumulated during both the Devensian and the Late-Glacial episodes.

Dry valleys

Dry valleys are encountered wherever Triassic sandstones or Pebble Beds constitute the bedrock and even where less permeable rocks occur. Such valleys are endemic on the Carboniferous limestones. In a few places underlain by permeable Triassic strata, dry valleys are particularly abundant and display a rich variety of size and shape (Fig. 12.4). While the smaller examples are probably a product of the latest periglacial episode and partly perhaps of post-glacial pluvial episodes or even of anthropogenic activity during the Holocene, the larger examples are polygenetic.

In the Triassic areas, the valleys can be described under three headings; (a) fluvial-type channels, (b) short steep-headed valleys and (c) bowl-like forms. The latter are rare, type (b) more common and type (a) the most common.

Type (a) commonly display a meandering habit, especially the longer ones. Many have intakes cut abruptly into gently sloping plateau surfaces and a few are headed by clusters of anomalously short first-order channels. Some support channelled surface flow along part of their length.

Most of type-(a) valleys are concavo-convex in cross-section, though smaller examples are more V-shaped. No examples of asymmetry have been recorded. Long profiles, without exception, are concave though the character of the buried bedrock profile where substantial fills exist may be more complex. Sedimentary fills are difficult to estimate because, without the clear development of palaeosols, augering cannot discriminate between fill and weathered bedrock if these are lithologically similar. Absence of palaeosols suggests that the majority of the filling occurred prior to pedogenic activity and is thus probably Late-Glacial or early post-glacial.

Some of the larger valleys have floors veneered with erratic-rich material up to 2 m thick in places. While it is impossible to infer from this that these valleys are pre-glacial (though some may be), it suggests that they were at least developing while till, which was later translocated to the valley bottoms, was being deposited on the adjacent interfluves.

Morphology indicates that a number of type-(a) valleys could not have been generated solely in a sub-aerial environment. Fig. 12.4 shows the concentration of dry valleys around the watershed separating the Meece, Lea and Tern headstreams. Spring sapping can be invoked to explain some development, for example in Coombesdale (D) and valley (C) at Shelton-under-Harley. Elsewhere, however, the dry valleys or short channel fragments run parallel to the trunk valleys only to turn abruptly to join them (e.g. SJ 820382–823377, 805401, 780383–788385, 780388–782388 and both north and south of location (G) (Fig. 12.4). Such parallel 'feeders' can only be explained as marginal or more probably, as sub-marginal channels, originating while ice still occupied the trunk valleys at least to the level of the 'feeders' but probably to much greater altitudes (Shreve, 1972).

It is probable that those larger dendritic type-(a) valley systems where intakes are multi-headed and abrupt, are also partly the result of sub-glacial fluvial erosion operating during deglaciation. None the less, like all the valleys of types (a) and (b), these dendritic systems must have experienced considerable development during periglacial episodes. Widespread development of 'derasion' phenomena (dells, '*vallons en berceau*') on a variety of lithologies has been reported from Eastern Europe (Pécsi, 1964; Klatkowa, 1965). In Britain, the dry valleys of the southern Chalk downs (Reid, 1887; Bull, 1940; Kerney *et al.*, 1964) and the West Country (Lewin, 1969; Gregory, 1971) offer environmental analogues except in lithology and the absence of glacial impact. Permafrost rendered formerly permeable strata impermeable and seasonal thawing of the active soil zone allowed rapid valley development through solifluxion and fluvial action. The parallel with the Chalk breaks down when the question of sedimentation is addressed. In the sandstone areas of North Staffordshire there is no equivalent of the chalk rubble fills with their pedogenic horizons. Instead, eroded materials were mobilised as masses of primary sand grains and consequently were dispersed over wide areas.

Type (b) valleys may be embryo type-(a) forms but in many instances appear to be partly the result of nivation or snow-patch erosion. Worsley (1967b) mapped examples north-east of Maer (Fig. 12.4, SJ 794389) as nivation hollows. The group of valleys close to the Lymes Farm and adjacent to the M6 motorway are of this type. Facing south-east, they occur where sandstones rest of mudstones (Exley, 1970). Spring sapping therefore may well have played a part in their evolution and though there is presently no channelled flow, fines may be removed via percolines (Bunting, 1964).

Type (c) is best exemplified by a bowl-shaped depression on a west-facing slope north of Shelton-under-Harley (Fig. 12.4, SJ 813398). This feature is bowl-shaped, has a discernible 'lip' and is cut in sandstone. It is difficult to see it as the product of processes other than snow-patch erosion or nivation.

The post-glacial period

Post-glacial environments have 'softened' and stabilised most landforms. Kettle holes have been infilled, the smaller ones mainly with inorganic sediments, the larger ones with considerable thicknesses of peat. The early anthropogenic impact is unclear but must have been important from the Neolithic onwards in terms of grazing and forest clearance (Evans, 1975; Simmons and Tooley, 1981). The Elder Bush Cave, Wetton (Bramwell, 1964) in the north-east of the county has yielded fauna and artifacts which, with material from Cauldon Low, Thor's Cave and Waterhouses fissure (Jackson, 1953) indicate human occupation of the area from

Fig. 12.4   Dry and misfit stream valley networks within the catchments of the Meece Brook, River Tern and Whitmore trough.

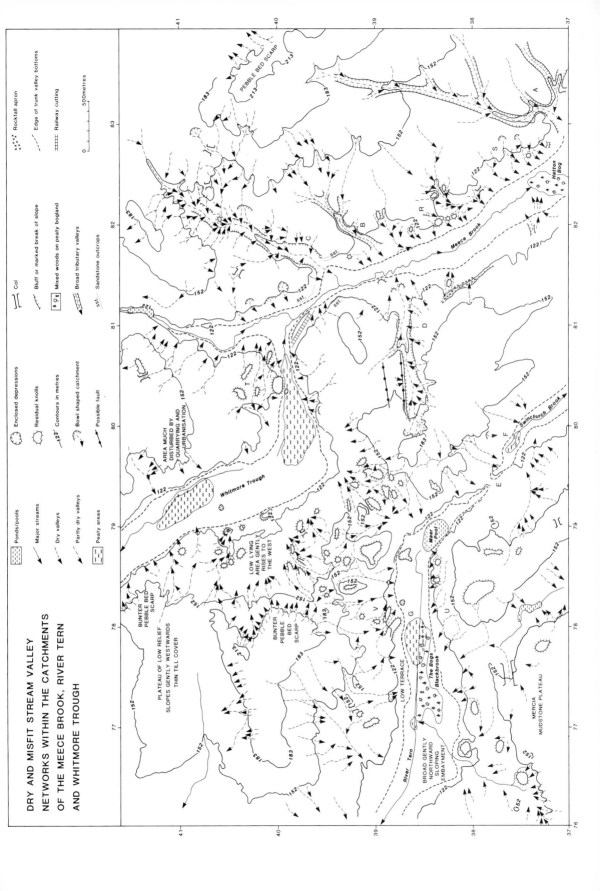

DRY AND MISFIT STREAM VALLEY
NETWORKS WITHIN THE CATCHMENTS
OF THE MEECE BROOK, RIVER TERN
AND WHITMORE TROUGH

Legend

Ponds/pools
Major streams
Dry valleys
Partly dry valleys
Peaty areas

Enclosed depressions
Residual knolls
Contours in metres
Bowl shaped catchment
Possible fault

Col
Bluff or marked break of slope
Mixed woods on peaty bogland
Broad tributary valleys
Sandstone outcrops

Rockfall apron
Edge of trunk valley bottoms
Railway cutting

0   500metres

PEBBLE BED SCARP

BUNTER
PEBBLE BED
SCARP

PLATEAU OF LOW RELIEF
SLOPES GENTLY WESTWARDS
THIN TILL COVER

BUNTER
PEBBLE
BED
SCARP

LOW LYING
AREA GENTLY
RISES TO
THE WEST

AREA MUCH
DISTURBED BY
QUARRYING AND
URBANISATION

Whitmore Trough

Meece Brook

Swinchurch Brook

Meer
Pool

Hatton
Bog

LOW TERRACE

The Bogs
Blackbrook

BROAD GENTLY
NORTHWARD
SLOPING
EMBAYMENT

MERCIA
MUDSTONE PLATEAU

River Tern

the Upper Palaeolithic. Elsewhere, a series of hill forts and enclosures (Berth Hill, Bury Bank) and chambered tombs (north-east Staffordshire) testify to the increasing impact of *Homo sapiens* on the landscape.

Forest clearance led to increased runoff, erosion and sediment transport, but no hiatus or important pedogenic horizon representing such a phase has yet come to light. In early Iron Age times, clearance coincided with the increased coolness and pluviality of the early Sub-Atlantic (Turner, 1981) and the impact on micro-landforms of this double pressure may have been considerable. On the other hand, sediments at Whixall Moss (Shropshire) closely parallel those found in a kettle hole on a gravel terrace at Fairoak in the Sow valley, and suggest rapid rises in the water table accompanied by peat accumulation some time between 5 and 4.5 ka bp. This would be before the Iron Age phase of settlement and possibly coincident with an early Neolithic intrusion.

The impact of later ploughing can be seen in the size of hedgerow 'soil banks' where these run sub-parallel to the slope contours. Sediment thicknesses of more than 1 m have accumulated upslope of hedges, particularly where these are known from other evidence to pre-date nineteenth century enclosure. South of Madeley, the boundary of the old medieval deer park shows an accumulation more than 2 m high though the hedgerow has now gone.

However, the greatest anthropogenic impact on the North Staffordshire landscape has occurred during the past 150 years of the Quaternary era. Industrialisation, including the extraction of minerals, particularly coal, iron, sand, gravel, limestone and clay, and the sedimentation of spoil, have generated completely new landscapes in rural as well as urban areas. Building material exploitation and the employment of open-cast working techniques mean that this process is far from over. Increasingly, the destruction of evidence capable of permitting interpretation of long past events accompanies the benefits of the exposure of new evidence which such work engenders. This is a problem which Quaternary geologists share with archaeologists and which, in North Staffordshire, has rendered certain long-exploited areas virtually sterile as far as further work is concerned.

13 *R. H. Johnson*

The imprint of glaciation on the West Pennine Uplands

Introduction

During the past decade many British Pleistocene landforms and sediments have been re-appraised using analogue models of landform assemblages and sediment associations that were first identified in modern polar ice-margin environments (Boulton, 1972). Although such 'landsystem' models are of great value and of practical application (Fookes *et al.*, 1978; Eyles, 1983), they cannot be exact replicates of past British ice sheets for the broad reasons that these latter were formed in temperate rather than sub-arctic latitudes, were more extensive than the polar analogues and, unlike them, in many areas reached sea level. In some British Pleistocene situations too, ice movements were *towards* areas of moderately high relief which, although covered by ice, greatly influenced the mode of glacial erosion and deposition. Such constraints operated in the West Pennines, and in consequence the glacial landform and sediment pattern differs from that usually presented in textbooks (see Derbyshire *et al.*, 1979; Embleton and Thornes, 1979). This chapter therefore includes a new 'model' for such environments, which, if valid, may usefully be employed for similar areas elsewhere.

The chapter is divided into several parts and subsections. Following this introduction, the study area and its Pleistocene literature will be briefly described, and thereafter the evidence for any pre-Devensian glaciations will be considered. The next and most important section concerns the Devensian glacial period. A review of past Pleistocene events is given for two main areas: first, the Manchester embayment and its surrounding Pennine uplands and, second, the hill terrains that enclose the Craven lowland and the lower Ribblesdale valley. The concluding paragraphs offer a synthesis and formulate conditions for an ice-sheet model in which hills form a partial constraint at the margin of the ice.

The study area

Throughout this section reference is made to Fig. 13.1. If the karst regions of the upper Ribble are excluded, the West Pennine hills constitute a narrow upland tract (II), 15 km wide, which only becomes extensive where the Bowland (IIE) and' Rossendale (IIH) Forests form two major salients of high ground. Although separated from each other by the deeply entrenched valleys of the Mersey, Ribble and Wyre tributaries, the formation of each is very similar. The topography is one of tabular hills, low plateaux and sharp crested cuestas eroded in Carboniferous

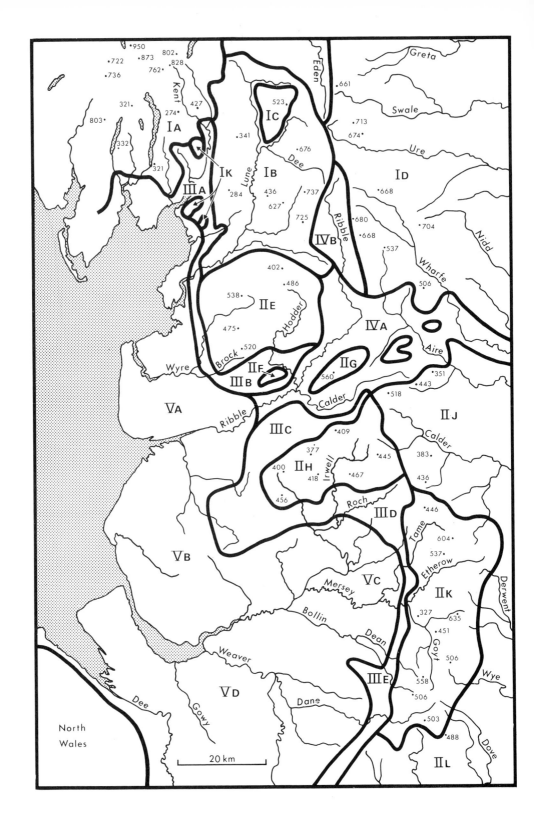

(Namurian and Westphalian) sedimentary rocks, with upland surfaces lying between 300 and 600 m O.D. The distribution of each landform is largely controlled by the local lithology and geological structure, and in the south-west Pennines (IIK) the relief is almost 'Appalachian' in character with a series of inward- or outward-facing, sharp-crested but bevelled escarpments which reveal the dissected fold structures (Johnson and Rice, 1961). There is thus a strong contrast between this region and the other upland areas, for in the Bowlands (IIE), and the Rossendales (IIH), and throughout the eastern part of the study area (IIJ, IIL) the folding is less severe and the cuestas form much broader summit ridges and tabular hills. Here, the extensive plateau surfaces are mostly bevelled by erosional processes, but between the Rossendales and the Bowlands (IVA) there has been intense folding of the north Lancashire Coal Measures and Millstone Grit formations and this has greatly facilitated erosion. Only the Pendle Hill (IIG) and Longridge (IIF) escarpments now remain as outliers of the uplands.

As its western margin the upland surface is terminated abruptly by a series of steep escarpments which can be traced almost continuously from the Lake District southwards to Staffordshire. These scarps mark a major geomorphic and geological discontinuity which is of great antiquity and provides a transition zone between upland and lowland areas. At its foot, the West Pennine scarp has a belt of low hills 100–122 m high formed by Coal Measure or Permian sandstones veneered with till (IIID). A series of step faults mark their outer edges, for the faults reflect the tectonic boundary between the Pennine upland and the Cheshire basin.

However, the variety of landforms present in the study area cannot be described solely in geological terms alone. Many geomorphic and especially glacial processes have been equally important in sculpturing the present scenery. In attempting to assess the efficacy of ice in particular it is useful to begin by examining its role in all the upland areas of North-west England, for although even the highest relief was covered by ice on several occasions, the efficiency as a denudational process has varied from place to place. As an erosive agent it has been most effective in the Lake District (IA) which was a centre of ice-sheet nourishment and growth development in an area of resistant Palaeozoic (Ordovician and Silurian) rocks. Here the ice created a mountain scenery of corries, arête ridges, ice-scoured hill and rock bosses, over-deepened valley troughs, hanging valleys and truncated spurs, nearly all of which occur within a 14.5 km radius of High Raise in the Langdale Fells (Waters, 1976). Outside this zone ice was far less effective as an erosive agent,

---

Fig. 13.1   North-west England: a geomorphological sub-division. **I.** Cumbrian Mountains and North Pennine Uplands: *A.* Lake District; *B.* Lonsdale Fells and the Upper Lune valley; *C.* Howgill Fells; *D.* North Yorkshire Dales; **II.** Central and South Pennine Uplands: *E.* Bowland Forest; *F.* Longridge Hill; *G.* Pendle Hill; *H.* Rossendale Forest; *J.* South Yorkshire Dales; *K.* Southwest Pennines; *L.* Peak District. **III.** The Upland–Lowland Transition Zone: *A.* South Cumbrian and Lancastrian Fells (IK), and adjacent lowland; *B.* Bowland foothills and moraines: *C.* Rossendale foothills and moraines; *D.* South Pennine foothills and moraines; *E.* Alderley Edge and moraines. **IV.** Lowland embayments into the Uplands: *A.* The Craven Lowlands, Ribblesdale and Lower Lonsdale. **V.** Till-covered lowlands: *A.* Fylde lowland and low escarpments. *B.* Southwest Lancashire plain and low escarpments. *C.* The Manchester embayment. *D.* Cheshire Plain and low escarpments of the mid-Cheshire hills.

and although there were large and extensive ice fields on the south-eastern flank of the Lake District, in the Howgills (IC) and other Lonsdale Fells (IB), most erosional work was done by glacier ice streams flowing within trunk valleys. Similar glacier streams were also formed in the Yorkshire dales and supplied from ice sources located on the high plateaux of the North-Central Pennines (ID). The outward flow of ice was not confined to the main valleys, however, and many examples of ice-breached watersheds occur at low cols on the divides (Fig. 13.2; Gresswell, 1952, 1962; Moulson 1967; Clayton, 1981).

Further south, along the margin of the Bowland, Rossendale and South Pennine uplands, the ice-sheet edge was compressed against the hillslopes, and hummocky moraines were formed (Zone III). These moraines form a key element in the glacial landform/sediment pattern of the upland margin. However, in the uplands *per se*, ice covered the hills but its erosive work was done mostly by meltwaters and its principal role was once of deposition. Glacial sediments mantle the lower hillslopes and have infilled the valley floors while on the upper slopes (especially in the east of the study area), periglacial processes replaced ice action as the most effective agent of denudation. Indeed, so strong were the frost processes on the hill summits of the eastern Rossendale hills that Jowett (1914) failed to find any evidence for former ice movements and thought that an unglaciated enclave had persisted here throughout the last glacial episode.

Previous theories and literature

The early theory that the 'drift' sequences of tills, sands or gravels found in North-west England were the sedimentation products of sea ice and icebergs has long been rejected and has been replaced by theories which attribute their deposition to the work of land ice mechanisms (Binney, 1848; Hull, 1864; Tiddeman, 1872). At first the deposits were thought to have been formed during a single glacial episode but this explanation was rejected by Hull who recognised in many parts of the study area a tripartite division in the Pleistocene deposits which included a lower till, an intermediate series of fluvio-glacial outwash, and an upper till which he regarded as the product of a second glaciation. Mapping within the Irwell, Goyt, Tame and Etherow valleys and throughout much of the Lancashire Calder and lower Ribble catchments further north, Hull observed many sections where the succession was present, but he did not appreciate that it only occurs within certain prescribed topographical situations. The deposits Hull mapped are now recognised as a single stratigraphic unit, known as the Stockport (glacigenic) Formation, which post-dates the inter-stadial deposits at Chelford in mid-Cheshire and was deposited during the last major Devensian glacial incursion (Worsley, 1967a). Hull's interpretation is now therefore no longer generally accepted (Gemmell and George,

Fig. 13.2 Glacial geomorphology of the Rossendales *1.* Reservoirs. *2.* Main watersheds. *3.* Meltstream channels. *4.* Limits of Ribblesdale ice sheet tills. *5.* Limits of Irish Sea/Lake District ice sheet tills. *6.* Glacial sands and gravel. *7.* Till. *8.* Till 'outliers' separate from main till depositional area.

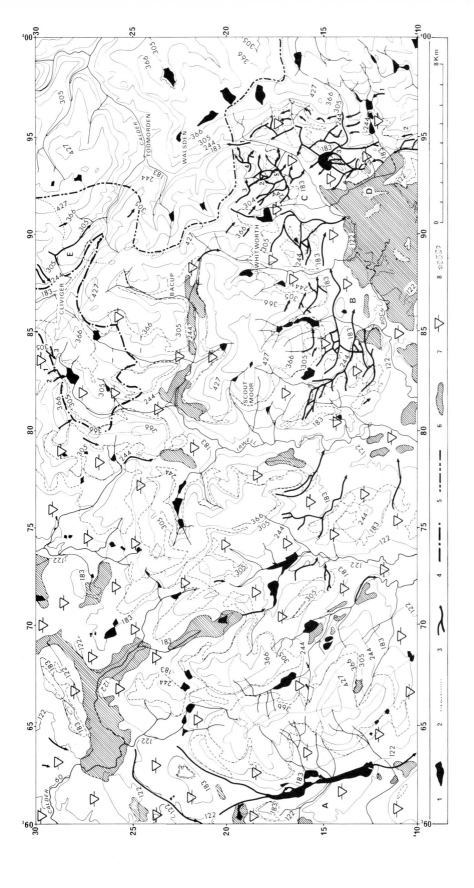

1972), but quite recently it has been shown that not all the glacial formations in Cheshire relate to a single glacial episode and work specifically at Chelford (Worsley *et al.*, 1983) and in south Shropshire (Wilson *et al.*, 1982) clearly demonstrate that 'tripartite' sequences in the glacial stratigraphy must be explained in more than one way.

During the first half of this century another theory was formulated and applied generally to the various upland areas of the West Pennines where a large number of meltwater channels occur on the hillsides. These were originally 'explained' as a series of overspill channels from lakes impounded in valleys whose outlets were blocked by ice. The theory as proposed by Kendall (1902) presumed that as the ice downwasted, so the lake levels fell and new channels were cut parallel to the slope at the ice margin. This interpretation of meltwater channel formation was favoured by geologists and geomorphologists for over fifty years and for this area many 'retreat stages' in ice downwasting were identified on the hillslopes of the Rossendales, the south-west Pennines and the Bowland Forest (Jowett, 1914; Jowett and Charlesworth, 1929; Raistrick, 1933; Moseley and Walker, 1952). It became, however, increasingly difficult to use this Kendall model as more and more channels were identified throughout the West Pennines (Dean, 1953; Earp *et al.*, 1961; Taylor *et al.*, 1963; Price *et al.*, 1963), and when other researchers showed that many of the channels had originated through the work of sub-glacial meltstreams (Johnson 1963, 1965a, b, 1969b) this hypothesis was no longer regarded as tenable.

Glacial derangement of drainage too have been frequently mentioned in the regional literature (Shortt, 1880; Tonks *et al.*, 1931; Jones *et al.*, 1938; Rice, 1957; Rodgers, 1962; Johnson, 1969b). These derangements occurred at the end of the last glacial period when many Pennine valleys had become partially infilled with glacial deposits and rivers or meltstreams were forced to the valley sides where they became incised into the bedrock.

The earlier Pleistocene deposits

Within the study area there is very little evidence for any pre-Devensian glacial tills, although some weathered till deposits, 25 m thick, are preserved in a col at Pikenaze, Longdendale (SE 112007) at 350 m O.D. (Johnson and Walthall, 1979). These tills contained no foreign erratics but the sandstone content was highly variable and drawn from several outcrops not in the immediate vicinity of the col. No other pre-Devensian glacial materials have so far been identified, but at Raygill Delf (SJ 941452) in the Calder valley mammalian fauna was discovered in a limestone fissure (Miall, 1880) and subsequently recognised as being of an Ipswichian (Eemian) inter-glacial age (Earp *et al.*, 1961).

There can be little doubt, however, that the Pennine hills were covered by ice

---

Fig. 13.3 Glacial geomorphology of the south-west Pennines *1.* Reservoirs. *2.* Main water-sheds. *3.* Meltstream channels. *4.* Limits of Irish Sea/Lake District ice sheet tills. *5.* Glacial sands and gravel. *6.* Till. *7.* Till 'outliers' separate from main till depositional area.

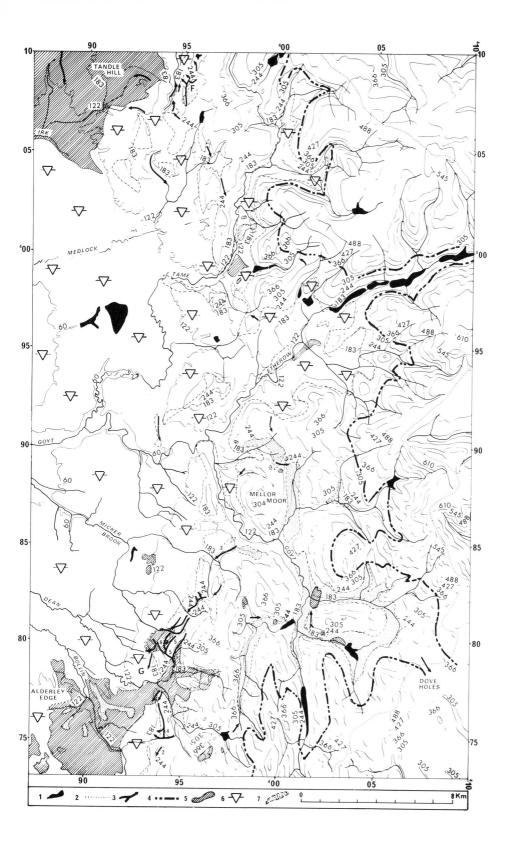

during earlier glacial episodes, for pre-Devensian deposits occur in the South Pennines and their erratic content indicates that they were derived from sources west and north of the study area. Till and individual erratics are widely distributed across the limestone plateau in the Peak District, notably in dolines on the Wye–Dove interfluve. The tills in these depressions overlie sands and clays of the Brassington Formation and therefore could be either Anglian or Wolstonian in age, but a second and more localised distribution of erratics and tills has also been identified (Dale, 1900; Dalton, 1958; Straw and Lewis, 1962; Johnson, 1969a; Burek, 1977) which marks the passage of a lobe of ice transporting Lake District and other northern erratics through the low watershed col at Dove Holes. This lobe flowed south-eastwards into the Wye and Derwent valleys where it can be dated indirectly by reference to a chronology based on cave speleothem datings. This later glacial incursion took place *c.*170 000–150 000 years BP, during a late phase of the Wolstonian glaciation (Ford *et al.*, 1983).

The Devensian glaciation

The South Pennine Uplands around Manchester

(i) *The eastern limits of ice sheet (Figs. 13.2–5)*   During this glacial episode, many glacier streams from the Lake District and the Central Pennines flowed southwards from their source areas to coalesce with other flows stemming from north Cumbria and western Scotland. As each glacier stream waxed and waned it not only eroded the local rocks over which it passed but also incorporated into its load debris reworked from both earlier glaciations and other Devensian ice-streams. The loads were carried for considerable distances, for the outer limits of this ice sheet lay south of Wolverhampton in the West Midlands and to the south-east of Lichfield in the Vale of Trent. Given such conditions, it is possible to make only very general observations about the derivation of the tills in the Manchester embayment, but in the hills to the east of the city there are further difficulties in interpreting former ice limits, as the till cover mantles only the lower slopes. Individual erratics are found on the slopes above, but their distribution is so random (Francis, 1980), and their age so uncertain, that other evidence must be used if the former ice margin is to be identified. Possible evidence includes the location of certain meltwater channels, the presence of kame mounds on some spurs and hill tops and the configuration of the relief itself. The boundary in Fig. 13.2 and 13.3 has been defined with the help of this type of evidence and shows that although Devensian ice breached the main Pennine watershed at Cliviger, Dove Holes?, Thornyleigh (SJ 9862), Rushton Marsh (SJ 9362) and Biddulph (SJ 8858), the catchment boundaries were little altered. In the Churnet valley however, some headwaters were diverted from the Trent as a result of the ice transgression (King, 1960; Johnson, 1965b).

In the Rossendales, ice transporting Scottish and Lake District erratics covered all but the highest hill summits with till or outwash sediments and it is evident

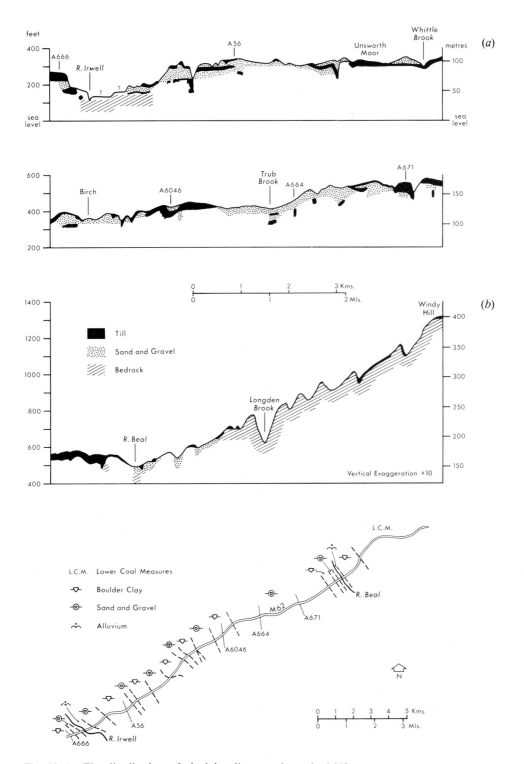

Fig. 13.4. The distribution of glacial sediments along the M62 motorway.

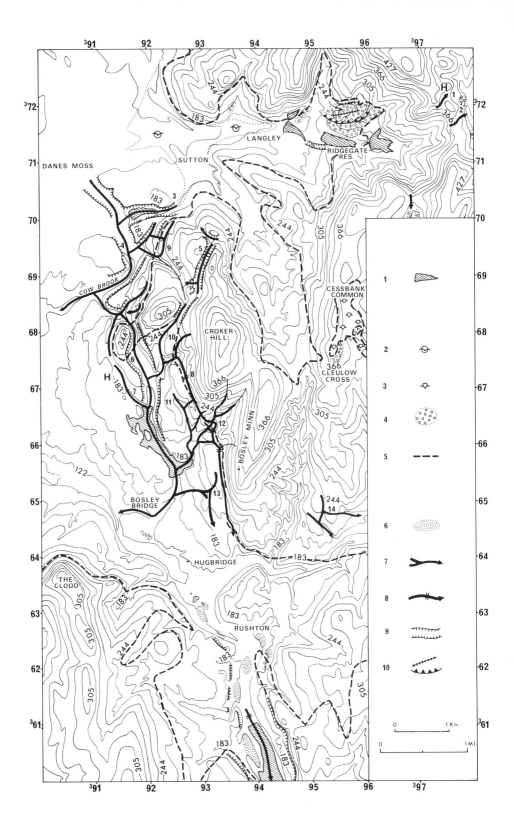

that the main ice thrust was through the low cols on the Calder–Irwell divide. Only in the north-eastern part of the Forest is there evidence for any incursion by any Ribble ice. Some of this ice reached the upper Irwell valley near Bacup, but most was diverted by the main Irish Sea ice sheet eastwards to the Yorkshire Calder valley (Jowett, 1914). On the south flank of these hills, between Bolton and Rochdale, the upper till limit decreases eastwards from 400 to 277 m O.D., and given a comparable fall of the ice surface, no ice would have reached Todmorden by way of the Walsden gorge. However, ice stood at the edge of the hills for a long time in this area, and a broad hummocky moraine belt was developed south-east of the Rossendale hills (Fig. 13.2).

In the south-west Pennines, the ice surface was at least 400 m above present sea level along the West Pennine margin, but it appears to have fallen rapidly to the east wherever ice lobes penetrated into the upland (Johnson, 1969b). In these areas most of the upper slopes of the Mersey and Derwent catchments are veneered with periglacial and not glacial regoliths. On the higher hill-summits there are deep rotted 'grus' formed in periods of severe cold climates (Wilson, 1980). Other observed periglacial phenomena include cemented screes, sandstone tors, cryoturbated soils, cambered valley sides, valley-floor bulges and almost ubiquitous solifucted and slope movement structures (Johnson, 1969b; Stevenson *et al.*, 1971; Johnson and Walthall, 1979; Wilson, 1981). It should be noted that some periglacial features have been found in the Rossendales, at Parbold (cryoturbations), on Scout Moor (SJ 820184) and in quarries at Whitworth (ice wedge pseudomorphs), all these sites being close to or inside the Devensian ice limit. These locations cannot therefore be used to determine the former Devensian ice limit as some of the features mentioned could be of a Late Glacial age.

(ii) *Glacial sediments* The difficulties of interpreting the Devensian glacial stratigraphy have been discussed extensively in the literature of this region and need no further discussion here. It is now accepted that the succession is far more complex than was once generally believed, as may be demonstrated in the succession of deposits exposed during the construction of motorways and pipelines. Figure 13.4a is a section drawn along the M62 motorway and based on borehole logs and other data. Although only shallow cores were drilled and the engineering information has been generalised a complex stratigraphical sequence is evident with several interdigitations of till and fluvio-glacial sediments. In detail there is little correspondance between the section as drawn and the pattern of drift (Fig. 13.4b) shown on the 1:63 360 geological map (Manchester sheet, 1933) and even where a 'tripartite' succession is established it may be explained in two quite different but equally viable ways: either the deposits were accumulated in morainic landforms (such as those observed along the line of the motorway near Birch) or they were

---

Fig. 13.5 Glacial meltwater channels and associated features between Macclesfield and Congleton *1.* Reservoirs. *2.* Terrace sands and gravels. *3.* Till. *4.* Kame and hummocky moraine. *5.* Upper limit of glacial tills. *6.* Ice-moulded drift topography. *7.* Smaller meltstream channels. *8.* Channels with irregular 'humped' long profiles. *9.* Major meltstream channels. *10.* Esker and kame terrace complex.

laid down as an infill within a pre-glacial valley (for example the Irwell valley on the motorway section).

At the edge of the Manchester embayment the Devensian ice penetrated deeply into the south-west Pennines to cover the lower hills. The middle and lower slopes were covered with lodgement till up to heights of about 280 m O.D. but as the ice sheet downwasted the ice margin was compressed against the hillslopes over which it could no longer ride. In such a situation supra-glacial sediments were released from within the ice and deposited over the basal tills. Hummocky moraines were formed along the edge of the hill-margin.

Within the Pennine valleys of the Irwell, Tame, Etherow, Goyt and Bollin, totally new sedimentary environments were created. Meltwater channels, kames and kame terraces (Johnson, 1963, 1965a, 1969b) now mark locations where ice blocks downwasted and where the valleys became infilled with a succession of basal tills, thick sequences of fluvio-glacial beds, and upper flow tills. The stratigraphy of these infills has led to much confusion and dispute in past interpretations. Near Broadbottom (SJ 997948) in the Etherow valley, a lodgement till at river level is overlain by thin laminated beds (rhythmites) which grade upwards into a less heavily consolidated till. This in turn is overlain by a succession of sand (or sand with gravel), twenty metres or more in thickness. Within the valley these sand beds contain sedimentary structures indicative of deposition during a high flow regime, but with considerable variation in the volumes of meltwater discharge. The sand beds are themselves capped by flow tills which were either derived from the stagnant ice lobes decaying in the valley, or were formed from tills soliflucted down from the adjacent higher hillslopes. In 1969 this valley infill was thought by the present author (Johnson, 1969b) to have formed in a sub-glacial environment, but it is now recognised that the sediments involved were deposited by meltstreams which flowed into an open ice-free valley trough aligned parallel to its margin and exposed by the ice sheet decay. In modern polar environments, such troughs are sited on the distal sides of ice-cored moraines immediately in front of the ice edge (Paul, 1983); but at Broadbottom the low hills to the west obstructed ice movement during downwasting and deflected meltwaters eastwards from the ice surface into the Etherow valley. Also, meltstreams flowing off the snow-covered hills transported debris to the same area; and, since these too had very extreme seasonal discharges, the sedimentation rate must have been very high. Unfortunately only a limited number of exposures is available in this area, but most have shown that the glacial sands and gravels have a strong cross-bedded stratification with marked lateral and vertical alterations in grading (Johnson, 1969b). Such depositional environments are typical of other Pennine valleys near Manchester, and similar stratigraphic sequences can be found in the Goyt valley near Marple, the Tame valley near Mossley, and the Irwell valley north of Bury.

South of Macclesfield there is very little evidence for a former ice-transgression on the eastern hillslopes; but, in the lee of the lower western hills infilling with glacigenic deposits took place under conditions similar to those which occurred in the Etherow valley. On the higher slopes some kame features were formed, notably at Cessbank Common and Cleulow Cross (Fig. 13.5). These are of some interest as they contain Ipswichian marine shells which were frozen with sands into the sole of the ice sheet and subsequently released at the upper margin of the ice sheet (Evans *et al.*, 1968). With further downwasting, nunatak hills emerged, so confining ice lobes to the lower valleys: hummocks of sand and gravel left in the col between the Cross and Bosley Minn mark the initial disintegration of the ice sheet in this vicinity.

In the Bollin valley, there was a prolonged stillstand of the ice front, with which several interesting landforms are associated. In the upper part of the valley a prominent sand and gravel kame with a strong ice contact face was formed, and this aggradation has diverted the river into a sandstone gorge immediately south and west of the Ridgegate Reservoir. Northwest of the kame, two esker ridges were also constructed as the ice margin downwasted, but, with the subsequent retreat of the ice, meltstreams formed a large outwash fan. The fan apex is at Langley and from here it can be traced downstream to its distal limit at the Pennine margin. Pocock (1906) argued that this fan was ice-dammed and that as a consequence the River Bollin was diverted to its present course through Macclesfield. The extensive development of hummocky moraines along the south flank of the Alderley Edge escarpment and at the foot of the Pennines clearly supports this conclusion and it would appear that once the ice was removed a large kettle hole was formed and this is now the site of the Danes Moss peat bog.

Other hummocky moraines are found in the West Pennine foothill zone, and are parts of a supra-glacial landsystem which was particularly well developed on the north and south flanks of the Alderley Edge escarpment, between Prestbury and Henbury (SJ 8974), and in an area south of Rochdale. In each of these localities the ice sheet was bounded on its flanks by local hill barriers and disgorged its load into a relatively confined space in which meltwaters were also concentrated. Thick beds of fluvio-glacial sands and gravels were laid down and then covered by flow tills. These now form the morainic ridges, but at the time of their deposition the ice margin was unstable, with much diamict sediment from the ice being incorporated into the bedding. This then became disturbed through faulting, folding or slumping at a number of sites (Johnson, 1965a). Such conditions were caused by the retreat of the ice edge as it downwasted, or by the melt out of ice from beneath the supra-glacial debris. The present topography is therefore irregular, undulating and poorly drained with trough valleys and old kettle holes separating smooth rounded ridges which at Tandle Hill (SJ 983888) stand over 75 m high.

(iii) *Meltwater channels and their distribution (Figs. 13.2, 3 and 5)*   In an earlier paper (Johnson, 1965a) the features were grouped into two categories and this division is used here. The first contains a group of widely dispersed channels eroded across watersheds and the second includes the sub-glacial drainage networks found on the lower slopes. (For purposes of description the networks are denoted by letters on the map, and some individual channels are also numbered.)

The col channels are steep-sided and although now modified by periglacial and infill deposits are quite deeply incised: for example that at North Britain (SK 010994) is some 25 m deep (Johnson, 1968). The meltstreams which cut these features either flowed directly away from the ice front or passed from one part of the ice margin to another. It would appear that, in the early stages of deglaciation, seasonally released water flowed on to the ice from the snow-covered high ground, and was at first refrozen within the ice. Later, as the ice surface became lowered, meltwaters flowed, under the force of hydrostatic pressure, beneath the ice. Passing from high to low pressures in the ice field they eroded the col channels which then emerged from beneath the ice. From studies undertaken in Labrador and elsewhere, Derbyshire (1961) has argued that sub-glacial channel erosion takes place when the ice surface is some 120–150 m above a col divide. As most col channels near Manchester occur at heights between 275 and 370 m O.D., the ice surface must have been more than 460 m O.D. at the Pennine margin when the channels were eroded.

Like the col channels, the channel network systems are not ubiquitous, but are found only where the relief favoured their development. Their location is closely linked with the spreads of morainic sand and gravel, as shown in Fig. 13.2 and 13.3, although at a higher level. The detailed configuration of one such system appears in Fig. 13.5. It shows that the meltwater stream courses evolved as part of a complex system, in which drainage from the aligned channels (parallel to the contours) penetrated into the ice by way of 'chute' gullies which are now part of the modern drainage system. As the ice surface downwasted, so the chutes extended downslope, and the older parts of the channel system were abandoned as new aligned channels were eroded at lower levels. Most of the channels were eroded in positions beneath the ice margin and display a characteristic 'hump' in their longitudinal profiles (Johnson *et al.*, 1970). Almost all the sub-parallel valleys were eroded along the strike of shale outcrops which are steeply dipping in this region. The vales are steep-sided with 25–27 degree side-walls and although some are little more than 10 m deep (H7), the largest exceeds 45 m (H6). Some channels were eroded when ice or glacial deposits formed part of the channel wall (H6, 7, 14), and where this occurred the channel system exhibits an anastomised drainage pattern (H12). Other channels were cut on the higher slopes and these acted as chute paths for meltwaters passing down on or into the ice. Such channels have a sinuous course and have developed a wide flat floor in their lower sections (H9, 10).

Figure 13.5 also shows that, although the drainage system is continuous in the mid-slope areas, separate drainage lines developed as the ice margin downwasted from the hills. In the northern sector on the lowland margin, channels were eroded across the spur of high ground at Sutton (H3) allowing meltwaters to escape into the Cow Brook channel system (H4), which at a late stage also drained the Danes Moss kettle hole. It is conceivable that where the channels enter the hummocky morainic terrain a different type of channel formation may have occurred which was similar to that observed by Marcussen (1976) in Denmark. Drainage channels directly superimposed from the ice cover onto glacial deposits would have a pattern which was influenced by the form of the original ice surface and the irregular nature of the morainic topography, but generally such a hypothesis would not apply on the Pennine slopes as the glacial deposits are very thin and the channel alignments appear to be controlled by the geology of the rocks which form the slopes.

A highly complex but separate system of channels was also developed in the upper Churnet and Trent valleys (Johnson, 1965b; see also Knowles, Chapter 12), but this network evolved where drainage could escape from the ice front into pro-glacial environments. Other complex channel systems were established on the hill-sides north of Macclesfield (Fig. 13.3). At Pott Shrigley (G), the 'aligned channels' show marked discordances in height between the intake and outlet levels, and several of the features have 'humped' longitudinal profiles (G5, 6). In some locations melt-waters were prevented from flowing downslope by thicker ice barriers, and where this took place 'one-wall' channels were eroded between the ice and the hillside: one such feature occurs at Shaw (D2). Most however evolved as parts of sub-glacial drainage networks, for anastomisation is common and the channels frequently have more than one drainage outlet (C3, 4, 5 and D3, 4). Such channels are frequently linked by downslope chutes, and often the deepest channels are not the lowest ones in the network. Very intricate channel systems have been eroded also on the south Rossendale hillslopes, east of Bury, but the largest, deepest and most extensive network is found to the west of Brinscall (Fig. 13.2 A), where the system is partly aligned along fault structures and partly developed within a zone of complex moraine relief (see Longworth, Chapter 10). In this network some of the channels were formed by meltwaters escaping from ice lobes stagnating in the Douglas valley, and the large sand and gravel mounds at Holland Lees (SD 512090) mark a former stillstand of the ice front west of Parbold.

Meltwaters have also contributed to the deepening of valleys which were not integrated into channel networks. For example, on Mellor Moor (Fig. 13.3), several small streams drain westwards to the Goyt and plunge abruptly into flat-floored, steep-sided segments within a very short distance of their spring-heads (SJ 992884). The discharge carried by these streams in the past must have been considerably greater than at present, and today some of these valleys are virtually dry. As over-deepening does not extend downslope to the lower till-covered slopes, the valleys probably served as meltwater chutes for meltstreams draining from the snowfield on the hill summit.

(iv) *The glacial alteration of drainage (Fig. 13.6)* The incursion made by Devensian and other earlier Pleistocene ice sheets resulted in a number of glacial diversions of drainage in the Mersey headwaters. For example, the Bollin at Macclesfield and the Goyt at Turf Lee (SJ 965864), were completely blocked by ice and diverted from their former valleys, the latter being a pre-Devensian diversion (Johnson, 1965a, 1968). In most cases, however, the diversions were less drastic with the rivers being diverted to the edge of old pre-glacial valleys that had become infilled with glacial tills, sands and gravels. Here, the rivers became superimposed, and eroded steep-sided, narrow gorges often over 30 m deep. A number of these valley features have large, incised meanders whose amplitude suggests that they were eroded by meltstreams with a very high discharge. Good examples of such features occur at New Mills (SK 000852), Marple (SJ 953896) and Summerseat (SD 795152).

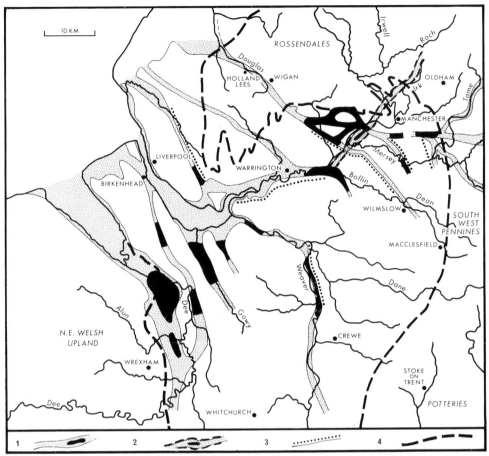

Fig. 13.6 The rockhead surface in Merseyside and Greater Manchester. *1.* 'Tunnel valleys'
and overdeepened depressions mapped by Johnson, Howell and Grayson using borehole
data available prior to 1974. *2.* Possible extension or re-alignment of tunnel valleys based
on revised and enlarged data source in 1984. *3.* Channel alignments controlled by geological
structure. *4.* Boundary separating Palaeozoic from Mesozoic rocks.

The most important changes, however, have taken place within the Manchester
embayment where a new drainage system developed as the ice downwasted. A
pattern of sub-parallel streams evolved as the ice sheet disintegrated and these
streams have very few tributaries if compared with those draining the uplands.
The stream pattern, no doubt, was influenced by the original configuration of the
till plain which generally slopes towards the west and the north and by the configuration of the disintegrating ice cover that may still have been present at the
time when the stream pattern was evolving. Today the rivers are in broad valleys
carved out of the Pleistocene sediments. In some localities the streams are incised
into the underlying Carboniferous and Permo-Triassic rocks and where this has
happened, as at Styal (SJ 8083) and Stockport, small river gorges have been eroded.
    The nature of the sub-drift or rockhead surface has been thoroughly investigated

using borehole records and other data by Gresswell (1964), Johnson (1965a), Howell (1973) and Grayson (1972). In their work Howell and Grayson examined over 2000 borehole records and the western part of the map is largely based on their findings, but with new data now available (1984) the map has been updated and modifications made. There are still several areas, however, where the data are incomplete and any attempt to reconstruct former drainage lines could be misleading for as Johnson and Musk (1974) have noted, the sub-drift surface has evolved through a long period of erosion and its form reflects the influence of fluvial, sub-glacial and glacial processes, all of which have contributed to its morphology at various times.

To the west and south of Manchester the map shows a number of sub-parallel trough depressions some of which are aligned along the strike of less resistant strata or along faults (Grayson, 1972). The borehole data are not sufficiently numerous to allow a detailed rockhead contour map to be drawn, but in general the relief appears to be subdued with broad interfluve areas separated by deep, narrow and considerably over-deepened troughs which are eroded in places to depths between 30 and 60 m below present sea level (Howell, 1973; Grayson, 1972). If the map presents a true indication of the buried relief then the role of Devensian ice as an erosive agent was limited in the embayment and largely restricted to the Dee–Mersey estuaries, the Wirral and to south-west Lancashire where the troughs are of much greater depth (– 88 m) and width. As the depressions are aligned parallel to the main direction of ice flow some selective ice-scouring of the weaker strata would have been most effective in this area (Williams, 1978).

To the north and east of Manchester, the rockhead topography is rather more dissected into narrow ridges and valleys which complement in form the hills and valleys of the adjacent uplands. In several locations these valleys have become over-deepened and in at least three locations near Middleton (SJ 8505), Audenshaw (SJ 9096), and Denton (SJ 9294), the sub-drift surface has been eroded to below present O.D.

How were these troughs and over-deepened valleys formed? It appears that at the end of the glacial episode, the troughs had been either glacially eroded or were initiated and enlarged by sub-glacial meltwaters flowing under hydrostatic pressure. Later, these troughs were infilled first, with outwash gravels and sands from the overlying ice cover, and then by outwash silts and clays, as the supply of meltwater decreased or was diverted. The valleys thus appear to be similar to the 'tunnel valleys' found in East Anglia by Woodland (1970), but they are also comparable, in their morphology, to other landforms that occur close to the former ice-sheet margin in Denmark (Marcussen, 1976). These are thought to have been formed where sub- or en-glacial drainage was superimposed directly from the ice itself onto the underlying bedrock. In certain situations, the surface configuration deflected the meltstreams into pre-Devensian valleys which were still preserved in part beneath the ice sheet, but where the streams were flowing under hydrostatic flow, or where thicker ice bodies prevented free stream movement at the ice base, meltwater erosion resulted in irregular linear depressions being formed within the sub-glacial drainage system. The linear SE–NW alignment of drainage is therefore, not only a response to structural influences, but it may also be a response to a glacial selective erosion, or alternatively, record a sequence of ice margin positions as the ice sheet retreated north-westwards, towards the Liverpool Bay sector of the Irish Sea from the Manchester embayment.

Glaciation in Ribblesdale and its surrounding uplands

(i) *Ice movement and its areal extent*   Many of the Devensian glacial landforms that occur in the Manchester area are also present in Ribblesdale, but there are some important differences which justify its separate treatment. This area, for much of its glacial history, has been invaded by Lake District, Lonsdale and the upper Ribblesdale ice; and erratics from Lake District sources have been found in Devensian tills as far east as Cliviger (Williamson, 1956). Ribble ice certainly breached the main English watershed at Cliviger and Gorple, as its tills were found in the (Yorkshire) Calder catchment to the east of Todmorden (Jowett, 1914); and the gorge at Cliviger was much modified, for it is now 120 m deep and 800 m wide, and the divide itself has been lowered 20 m and displaced 100 m north of its pre-glacial location.

At the time of its maximum extent the Ribble ice covered all low ground east of the Whalley gorge and the high hills of Pendle and Longridge. Its southern limit was on the Pennine slopes of the Rossendale and Trawden Forests but the precise limit on these slopes is uncertain. Distinctively coloured tills with characteristic Ribble ice erratic suites have been mapped up to heights of 385 m on the Bouldsworth and Trawden hills north of Cliviger (Earp *et al.*, 1961), but, if the col channels at Widdup and Steeple Stones were eroded by sub-glacial melt-waters then the ice-sheet surface must have been at least as high as 450 m O.D.

Further north, the relief features of Pendle and Longridge clearly imposed a considerable barrier to Ribble ice movement by deflecting it to the lowlands. Here, as stated earlier, the local fold structures exerted a strong control on the topography, creating a series of longitudinal vales and ridges aligned along the strike of the rocks. These features were modified by ice, for glacial striae are aligned parallel to the direction of the valleys (Tiddeman, 1872).

North of Pendle Hill there is a marked change in the glacial landform assemblage, which reflects a more dynamic ice-front condition. At the time when ice covered the Bowland and Pendle hills, the Ribble and Lonsdale ice streams were part of a Devensian ice-field with supply sources in both the Lake District and the North Pennines. The Lonsdale–Lake District ice covered much of the Bowland Forest upland and the ice or its meltwaters substantially enlarged the watershed cols at the Trough of Bowland (SD 623530), Salter Fell (SD 648588) and Cross of Greet (SD 682608). The main Ribble glacier both the Lake District and the North Pennines. The main Ribble glacier was formed by a large number of feeder glaciers (Fig. 13.7), and as it flowed southwards its ice breached the Ribble–Hodder water-shed west of Rathmell and penetrated into the Hodder valley at Slaidburn. This 'Hodder glacier' extended southwards parallel to the Ribble glacier to reach the

Fig. 13.7 Ice flows in the Central Pennines and adjacent areas.

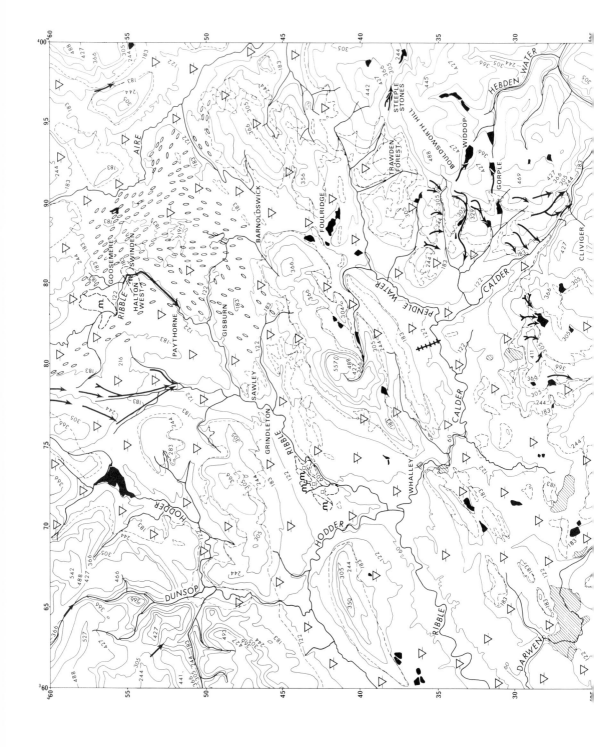

lowland south of the Bowland Forest. Here it was deflected westwards through the low col at Chipping, but was unable to escape onto the Lancashire plain since this was occupied by Irish Sea and Lake District ice. The main Ribble glacier too was checked, and at the time of its maximum growth was diverted to breach the main Pennine divide between the Aire and the Ribble. Ice also escaped southwards, passing over the Pendle, Weets Hill and Elslack uplands and through the low gap at Barnoldswick.

(ii) *Glacial sediments (Fig. 13.8)* A tripartite succession of till/sands and gravel/till has been mapped in several of these valley 'troughs', including the lower Ribble valley west of Whalley, the Whalley Gorge itself, and the valleys around Colne and Trawden (Hull *et al.*, 1875; Earp *et al.*, 1961; Price *et al.*, 1963). In these troughs glacial sediments are frequently more than 30 m thick. Other morainic sand and gravel deposits, capped by thin tills, are found on the north-western slopes of the Trawden, Bowland and Rossendale Forest hills (Jowett, 1914; Moulson, 1967). As elsewhere, they appear to be linked with the development of the meltwater channel systems located on the higher slopes. Some low moraine features also occur on the north side of the (Lancashire) Calder valley, where a degraded esker was identified by Poole at Pendle Hall (SJ 816353), (Earp *et al.*, 1961). All these features suggest that ice stagnated *in situ* on the low land south of the Pendle ridge, and that conditions during deglaciation were very similar to those that obtained in the hill areas north and east of Manchester.

In south Ribblesdale and the west Craven lowlands the topographic barriers the Ribble glacier encountered had a profound effect on the ways in which glacial sediments were deposited, and an extensive drumlin field was formed where active ice was constricted, but not halted, by the hills. The drumlin field, which is shown schematically in Fig. 13.7 forms an undulating terrain in which Raistrick (1933) identified some 245 separate drumlin and drumlinoid features. In addition to the usual lodgement till mounds, he noted that many had rock cores, and that some were formed of sand and gravel capped by till. Most of the drumlins are 100 m in length, but some exceed 175 m and are 35 m high. As such they would be classified by Rose and Letzer (1977) as mega-drumlins, and their distribution reflects a high-energy phase when ice streamed through the Barnoldswick–Foouldridge gap. Most of the smaller drumlins were formed later, when energy and sediment supply had decreased and only the lower gaps to the Aire valley and the lower Ribble vale were accessible.

The drumlins must have been formed when the ice surface was still lying above the height of the local relief, for some of the higher ridges reach 200 m O.D. But, with continued downwasting, ice velocity decreased and other landforms were developed at the retreating ice margin. One of these is located in lower Ribblesdale on the south flank of the Bowland hills between Grindleton and the Hodder–Ribble

Fig. 13.8 Glacial geomorphology of lower Ribblesdale   *1.* Reservoir. *2.* Meltstream channel. *3.* Limit of till deposits. *4.* Glacial sands and gravels. *5.* Till. *6.* Drumlin. *7.* Moraine/esker? *8.* Lake sediments. *9.* Esker.

confluence. Mapped as a moraine by Earp *et al.* (1961) it could possibly be an esker, for it is an irregular ridge 2.5 km long standing some 15 m above an adjacent lake-flat whose glacio-lacustrine sediments are 4 m thick (Fig. 13.6). Further north, a 'cross-valley' moraine occurs in the Ribble valley between Halton West and Goosemere and, at a height of 128 m O.D. it is at a lower elevation that the drumlins to the south. Upstream of this moraine there is a small lake-flat, and, at some stage in the glacial downwasting water escaped from this lake through the moraine to Swinden.

(iii) *Glacial alteration of drainage (Fig. 13.9)*   In the Ribble catchment the glacial derangements of drainage have taken place under very different conditions from those which pertained in the Manchester area for the glacier was confined by the uplands and, although diversions took place at the margin of a downwasting ice sheet, the main direction of meltwater flow was parallel to the ice itself. South of the Pendle Hill, glacial derangements were generally small and very similar to those described in the south-west Pennines, but further north drainage changes were more substantial and more complex in their origin. For example, the marked 'elbow of capture' in the Ribble valley at Hellifield was once attributed to river piracy but this hypothesis was discarded when Raistrick (1933) provided evidence for glacier transfluence across the Ribble–Aire divide. Watershed breaching by ice, however, is only part of the explanation and other factors must also be considered.

From Swinden, downstream to Paythorne, the Ribble now occupies a former meltwater channel which was eroded by Devensian ice meltstreams escaping southwards through the cross-moraine between Halton West and Goosemere. This diversion was probably one of the last major drainage changes to take place but the rockhead contours (Fig. 13.9) show that a valley had existed between Hellifield and Sawley prior to the Devensian glacial episode and that the rockhead surface is much lower at Swinden than it is on the former Aire–Ribble divide. If the Ribble ever was an Aire headwater then the 'capture' must have occurred much earlier in the Pleistocene Epoch.

Further downstream, between Paythorne (SD 830513) and Sawley (SD 778467), the present river course is deeply entrenched into Carboniferous rocks and this gorge is the work of meltwaters released when ice downwasted on the north side of the Pendle ridge. These streams used the Stockton Beck and Ribble valleys to escape westwards at a time when most of the valley was either still ice-covered or had become infilled with glacial debris. A new channel were eroded to the south of the former valley (revealed by the rockhead contours), and this subsequently became the main trunk artery for the streams draining the upper Ribble catchment. East of Sawley, the former channel remains infilled with glacial sediments, but bedrock exposures in stream channels north of the present gorge indicate that the former channel floor was somewhat above present river level in this area.

Fig. 13.9   The rockhead surface in Ribblesdale.   *1*. Reservoirs. *2*. Sub-drift contours. *3*. Former course of the Ribble. *4*. Drift-free areas.

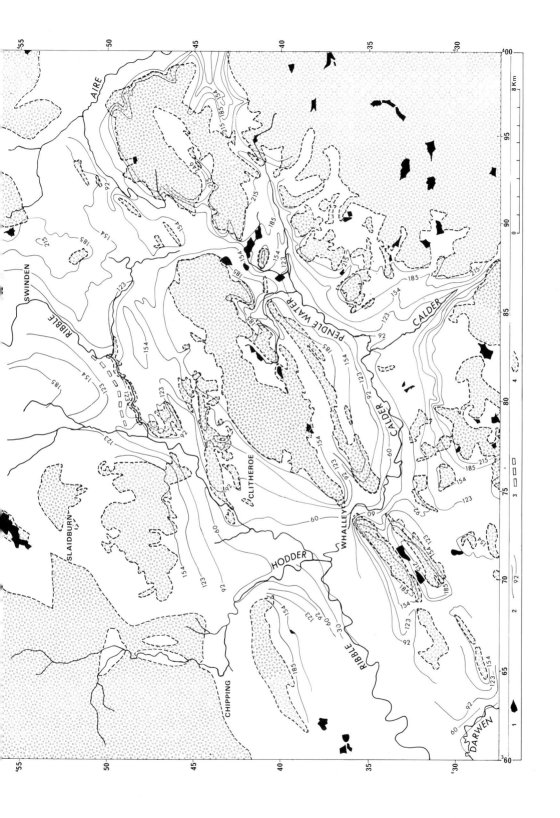

Between Paythorne and Gisburn, and Sawley and Clitheroe, the Ribble has re-excavated its former valley, but at three places (SJ 838577, 830508 and 735425) it has been diverted and incised into bedrock. Some exhumation of the former valley has also taken place west of Clitheroe, but at Sale Wheel (SJ 676360) the river is again superimposed, and the old channel at this point lies at 45 m O.D., i.e. *c.*5 m above present river level (Dean, 1950).

The rockhead surface configuration (Fig. 13.9) also indicates that the lower Ribblesdale embayment is an old feature with a vale floor which at one time was over 5 km wide at Clitheroe. This broad vale extended up the tributary Calder and Darwen valleys and through the fault-controlled gaps at Whalley and Hoghton (SD 630275). Such vales were eroded at times when the rivers were able to migrate freely across the valley floors, and before the Carboniferous limestone reef knolls in the vale become exposed. They thus contrast markedly with the buried relief at the edge of the Bowlands where the sub-drift surface was mapped by Worthington (1972) using geophysical methods. Here, there are a number of steep-sided, narrow valleys up to 20 m deep and infilled with glacial sediments: these predictably have a till/sands/till succession which in places is up to 40 m thick. The complex of hummocks and channel infills have diverted the Wyre, Brock and Calder rivers from predominantly north–south strike vales at the hill margin to their current east–west alignments (see also Longworth, Chapter 10).

Conclusion

In North-west England three distinct glacial landsystems are readily identifiable. In the Lake District the glacial features are all parts of a landsystem in which valley glaciers were the main geomorphic agents. In the lowlands of Lancashire and Cheshire the till plain is essentially part of a sub-glacial landsystem, where the landforms and sediments were created at the base of the Devensian ice sheet. Further south, at Ellesmere this system is replaced by one in which moraines are the dominant landform, and where the sediments are the result of till flowage, englacial meltout, meltwater sorting and stream deposition. Such a simple arrangement of landsystems is readily recognised if a traverse is made along the direction of maximum ice-sheet growth, but where a traverse is made laterally across the ice sheet to the Pennines the landsystems become telescoped and difficult to separate. To overcome this difficulty a new landsystem is proposed which is shown schematically in Fig. 13.10. The system is for glaciated upland areas marginal to lowland ice sheets. It contains elements found in other landsystems, but is sufficiently integrated to be profitably used to interpret landforms and sediment associations in the Welsh Borderland, the Isle of Man, west Cumbria, and the Irish Sea coastal (upland) areas of Wales and south-east Ireland.

The model contains four distinct, but not entirely separate, sectors. First, there is an outer sector in which ice formerly penetrated through low ground between the hills or passed over the lower summits to cover all but the highest relief. Next to it lie the upland areas which were snow-covered, ravaged by frost processes

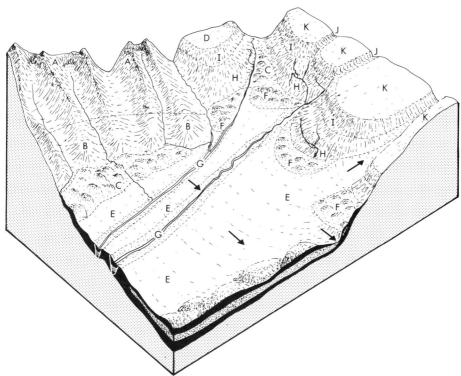

Fig. 13.10 A glacial landsystem model for the West Pennine margin. *A.* Lake District. *B.* Cumbrian Lower Fells. *C.* Drumlins. *D.* East Cumbrian Plateaus. *E.* Till plains (sub-glacial ice-sheet landforms). *F.* Moraines and supra-glacial ice-sheet landforms. *G.* Main river valleys of post-glacial formation. *H.* Lower Pennine slopes; drift covered with channel networks. *I.* Drift-free Upper Pennine slopes. *J.* Ice-breached watersheds. *K.* Pennine Upland surfaces.

but were not themselves high enough to generate their own ice-fields. The second sector is where ice was held against the hillslopes by the mass of the ice sheet to the west, but could not surmount them during deglaciation. Next, there are the till plain and the drumlin fields which lie peripheral to the hill margin, and, lastly a transition zone in which features of the till plain are replaced increasingly by landforms associated with upland valley glaciers and their sediment associations.

The outermost sector is located where the upper slopes of the hills are charac-terised by kame mounds, col channels, erratics and lodgement tills. It is part of the 'broad marginal zone of erosion which swept forward to the maximum extent of the (Devensian) ice sheet' (Boulton *et al.*, 1977), but on the hillslopes it was very constricted by the high relief. Lodgement tills were deposited on the lower and middle slopes of the valleys within the uplands, once the ice became stabilised at its margins, but where supra-glacial ice abutted against the higher slopes, then kame sand/gravel mounds and occasional erratics were deposited. At these high levels, meltwaters were able to escape to pro-glacial areas east of the main English water divide.

With downwasting and general retreat, the ice margin became greatly attenuated in this sector. Ice lobes occupied the valley floors and lower slopes, but with the emergence of the hills these became increasingly isolated, and eventually many stagnated *in situ*. Large volumes of sediment were deposited, to be re-sorted by meltstreams. These infills are now dissected, but in the past the deposits obstructed river drainage in the valleys and caused its derangement.

The foot slopes along the outer edge of the Pennines constitute the second sector. During the period of active ice growth, the slopes were overlain by thick ice, but, as it downwasted, meltwater was released to erode the complex network of channels which are so characteristic of this sector. Such channels undoubtedly carried sediments to the lower hillslopes, but the hummocky moraines now located in this sector were formed for the most part by the meltout of the ice-load direct from the ice surface itself. According to Boulton *et al.* (1977), supra- and en-glacial debris reaches the ice margin whenever interior basal ice containing enriched ice debris loads is sheared forward over basal ice layers frozen to the underlying ground surface. In Cheshire and Shropshire these conditions resulted in a supra-glacial sediment association, and created a zone of lateral and terminal moraines which form the main landform element in this sector.

Supra-glacial/pro-glacial sediments and landforms were no longer formed, once the ice had backwasted from the upland hill-margin, but some vestige of high relief is contained in the third sector where drumlins were constructed by the movement of active ice over its own till plain. These were developed under thick ice, but at times of decreased ice-velocity. Outside the drumlin fields there is an almost featureless surface formed by lodgement tills, lacustrine clays and sand/gravel outwash. This depositional plain extends from the Lake District to the Ellesmere–Whitchurch moraine, broken only by the mid-Cheshire hills, the low sandstone escarpments of the Wirral and south-west Lancashire and the undulating hummocky terrain at Kirkham.

The fourth-sector landforms occur in the Ribble valley north of Hellifield and in the southern fells of the Lake District. These are, for the most part, glacially eroded features, but the valleys also contain transverse 'cross-valley' ridges, drumlins, drained lake-flats and meltwater channels. The valley floors and lower slopes are mantled with thin tills and other deposits, so this sector marks a transition between the till-plains and the glaciated uplands.

Pennine karst areas and their Quaternary history

Introduction

In general terms the Pennine range of hills may be described as a broad anticline of Carboniferous rocks with its crest eroded to varying depths such that the oldest rocks now outcrop at the core of the range. Karst landforms have developed on the Carboniferous Limestones which are exposed in the south and north Pennines, the two areas being separated by an area of Millstone Grit moorland stretching from Castleton to Skipton (Fig. 14.1). The southern outcrop is a compact area almost 40 km long (north–south) and up to 20 km wide which is surrounded by impermeable lithologies and forms part of the Peak District of Derbyshire and Staffordshire. The northern outcrop is more complex with numerous inliers and outliers of older and younger rocks and three distinctive karst regions may be identified: northwest Yorkshire (which includes parts of Lancashire and Cumbria), Morecambe Bay, and the northern Dales. Only one of these regions, northwest Yorkshire, is considered in the present chapter as the Morecambe Bay karst has already been discussed (Chapter 9), while the northern Dales are cut into a number of thin limestones and are therefore poorly karstified. The limestones which outcrop to the south of the northwest Yorkshire region are also largely un-karstified.

The Peak District and northwest Yorkshire karsts are developed on almost horizontal limestones of similar geological age but differences in lithology, thickness, and position in relation to impermeable rocks, together with the contrasting tectonic and Quaternary histories of the two areas, have resulted in the development of distinctive karst regions. Both have been the subject of con-siderable research and excellent syntheses have been published (Waltham, 1974; Ford 1977a) together with detailed descriptions of the many caves (Brook *et al.*, 1972–6; Ford and Gill, 1979), the latter being regularly updated as new caves are discovered and existing systems extended. As might be expected much of this work is essentially area-specific and there have been relatively few comparative studies, the main exceptions being Warwick (1958) and Sweeting (1972a) who outlined the main characteristics of all the British karst regions and Pitty (1972) who contrasted the calcium carbonate content of karst waters from Derbyshire and Yorkshire. The aim of the present chapter is to provide a more detailed comparison by focuss-ing on the individual elements of the 'Geomorphological Equation' – materials, processes, landforms – and on their variation through time, rather than on regional description *per se*.

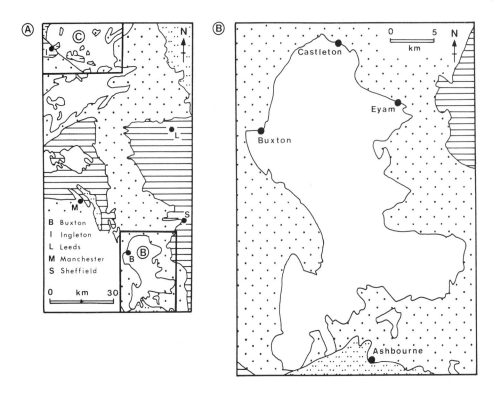

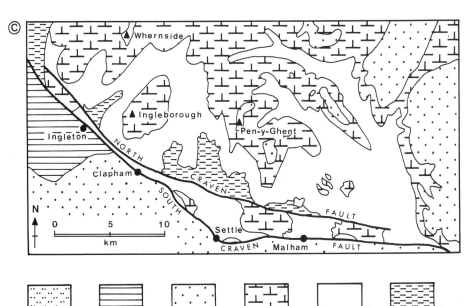

Fig. 14.1 Pennine karst areas: *A*. general location; *B*. Peak District; *C*. Northwest Yorkshire. Key: 1. Permo-Trias; 2. Coal Measures; 3. Millstone Grit; 4. Yoredale Series; 5. Carboniferous Limestone (Peak District)–Great Scar Limestone (Northwest Yorkshire); 6. Pre-Carboniferous.

Materials

One of the prime reasons for the contrasting karst landforms in the Peak District and northwest Yorkshire is that the limestones differ in lithology and thickness and in their relationships to surrounding non-carbonate rocks. Superficial materials are also important both as controls on landform development and as indicators of early stages in the evolution of the present landscape. All aspects of materials in the Peak District are reviewed in Ford (1977a) and shorter reviews of materials in northwest Yorkshire have been published in Waltham (1974) and by Halliwell (1979).

In the Peak District karst the Carboniferous Limestone is made up of three main lithological types: bioclastic limestones formed in a lagoonal environment which prevailed over most of the area; massive, fine-grained reef limestones which formed at the fringes of, and less commonly within, the lagoons; and thinly bedded, fine-grained calcisiltites and calcilutites, often associated with siliceous cherts, which were formed in basinal or gulf environments. Bedding planes are rare in the reef limestones but common in other lithologies where they form important controls on water movement and cave development. Three main types of bedding discontinuity may be recognised: pyrite-rich shale bands which mark a short period of terrestrial sedimentation; changes in limestone lithology as a result of a relatively rapid change in the environment of deposition; and thin green clay bands ('wayboards') which originated as showers of volcanic dust. Lateral movement of water is also encouraged by impermeable basaltic lavas, tuffs and dolerites ('toadstones') which are locally interbedded with the limestones. The overall thickness of the limestone sequence is generally thought to be in the range 700–800 m although the underlying strata are nowhere exposed. The form of the pre-Carboniferous surface is probably highly irregular as the only two boreholes to penetrate to basement did so at depths of 274 m and 1803 m, the former having been drilled from river level with at least 150 m of limestone above its top. However, the limestones are surrounded by impermeable lithologies (Triassic sandstones in the south and Millstone Grit elsewhere) forming a '*karst barré*' (ponded karst) and there is no evidence to suggest that the basement surface exerts any control on the direction of water movement or that it forms a lower limit to water circulation. Allogenic streams which rise on the surrounding strata form important point-sources of aggressive recharge to the limestones and most of the major cave systems are located around the edges of the outcrop. Nearer the centre recharge is entirely autogenic although the situation is locally complicated by toadstones and by superficial deposits of both Tertiary and Quaternary ages. The Tertiary deposits are Mio-Pliocene fluvial sediments which provide evidence of early stages in the evolution of the Peak District karst as they are preserved in solution collapse pockets which form part of a fossil karst surface (Walsh *et al.*, 1972; Ford, 1972, 1977b). Quaternary deposits comprise remnants of a formerly more extensive glacial till, or possibly tills, thought to be of Wolstonian and/or Anglian age (Burek, 1977) and a more extensive veneer of loam and chert-gravel composed largely of a silty loess which is often mixed with insoluble limestone residue (Pigott, 1962; Burek and Cubitt, 1979).

In contrast to the Peak District the Carboniferous Limestone of northwest Yorkshire has two divisions: the Great Scar Limestone which is the principal karst rock, and the Yoredale Series. The Great Scar Limestone consists mainly of fine grained bioclastic calcilutites with about 2 per cent insoluble residues and 98 per cent matrix of which about half is fossil material and the remainder micrite and sparite in proportions which vary stratigraphically though not rhythmically. Each bed reacts differently to weathering and erosion but in general the sparry limestones are more massively bedded, have greater mechanical strength and are less soluble than the biomicrites. This gives rise to a characteristic 'step and stair' relief in which steep sparite cliffs (scars) are separated by more gently sloping benches. As the area was not subject to any Carboniferous volcanic activity the toadstones and wayboards of Derbyshire are absent but the calcareous succession is broken by about twenty shale bands which have had an equally important influence on water movement and cave development (Waltham, 1970). They are concentrated in three main stratigraphic zones and range in thickness from paper-thin partings to 2 m beds, although most are in the range 1–50 cm (Waltham, 1971). Pre-Carboniferous strata have been exposed by erosion in several valleys within the karst and they also outcrop widely west of the Dent fault. Although undulating the basement surface appears to be less irregular than in the Peak District with a relative relief of approximately 100 m. However, the thickness of the Great Scar Limestone is only 100 to about 200 m, much less than in Derbyshire, so that the impermeable basement strata constitute a significant lower limit to water circulation although the detail of their hydrological and geomorphological significance is still uncertain (Halliwell, 1979; Waltham, 1977a). As the valleys which cut through the limestones to basement contain rivers which drain south across the Craven faults to a lowland area the limestones are to a large extent freely draining unlike their counterparts in the Peak District. This has had a major influence on the type of karst relief which has developed, northwest Yorkshire being a classic holokarst whereas the Peak District as a whole is best described as a fluviokarst. The Great Scar Limestone is overlain by the Yoredale Series, a repetitive series of limestone, shale and sandstone cyclothems with a total thickness of about 300 m. Although the Yoredale limestones are lithologically fairly similar to the Great Scar they are usually more thinly bedded and less pure. In addition their surface exposure in northwest Yorkshire is relatively limited and hence they are generally not significant as karst rocks although there are several important caves in the Yoredale limestones east of the River Wharfe. Overall, the main physical influence of the Yoredale strata is to encourage stream development which provides point-sources of recharge to the Great Scar Limestones and hence preferential points for the development of underground drainage and cave systems. In common with the Peak District point-sources are frequently located close to the periphery of the karst area but there are also a large number of sinks and several extensive cave systems located around the Yoredale outliers (Ingleborough, Pen-y-ghent and Gragareth). Finally, superficial deposits of solifluvial and/or glacial origin cover much of the limestone, particularly at higher altitudes, and may facilitate surface drainage within the karst area.

Processes

Karst systems are distinguished from others by the predominance of one erosion process, the aqueous solution of limestone, which is usually accompanied by a greater or lesser degree of carbonate deposition as speleothems and/or subaerial tufa. Slope processes are generally confined to slow mass movement of superficial deposits and rockfall from free-faces neither of which contributes greatly to land-form evolution. Where drainage has remained on the surface mechanical erosion and fluvial sediment transport processes operate and corrasion may also contribute to the downcutting of cave passages. Few quantitative data on the operation of these processes are available but it seems unlikely that they have contributed in any major way to the moulding of karst landscapes in northwest Yorkshire and the Peak District. Hence, only the limestone solution/deposition process and the glacial/periglacial processes which operated during the Quaternary cold periods are considered in this section.

Solution processes

The importance of solution in the development of the Pennine karsts has been recognised for over 100 years but there are surprisingly few published data on water hardness (Table 14.1) or solutional erosion rates (Table 14.2). Moreover, many of the hardness data are derived from a small number of samples (sometimes only one) collected at widely spaced time intervals and unrelated to discharge. The use of this type of data in erosion rate computations inevitably increases the error limits and this problem has been exacerbated, first by most authors' use of the Corbel (1957) formula which has been shown to give inaccurate results under most normal conditions (Gunn, in press), and secondly by a failure to take account of solute inputs in precipitation and from allogenic sources. Hence, there is an urgent need for reliable estimates of net solutional erosion and its temporal and spatial variability based on hydrochemical budgeting techniques (Gunn, in press). On the basis of data presently available it would appear that solute concentrations (expressed as total hardness) are generally higher in the Peak District (Table 14.1) although this is partly offset by a lower rainfall and hence runoff so that overall erosion rates in the two karsts are roughly comparable. Pitty (1972) has suggested that about 52 per cent of the difference in hardness values can be accounted for by flow-through times (Peak District 97 days, Yorkshire 59), 37–40 per cent by dilution (Yorkshire karsts receive greater allogenic inputs from the Millstone Grit), and the remaining 8–11 per cent by soil fertility (soils in the Yorkshire karsts are more acid). Flow-through times were obtained by a statistical technique developed by Pitty (1966) and widely used in the Pennine karsts in which water hardness is related to antecedent climatic variables (Pitty, 1968b, 1968c, 1971a, 1974; Pitty *et al.*, 1977; Ternan, 1972, 1974). However, the long flow-through times obtained by these workers may be partly a result of the sampling interval (monthly or fortnightly) as Gunn (1974) derived much shorter times from an analysis of data collected at daily and weekly intervals from a Peak District cave. In a more recent

Table 14.1   Total hardness data for risings and percolation water in Pennine karsts
(mg/1 $CaCO_3 + MgCO_3$).

	Mean	Range	Comment
Peak District			
Sweeting, 1964		230–250	
Pitty, 1968a	206	137–255	11 risings sampled approx. monthly for 2 years
Edmunds, 1971	285	178–436	30 risings mostly sampled only once.
Pitty, 1972	180		35 surface streams, sampled approx. monthly, mostly for 2 years.
Christopher and Wilcock, 1981	259	182–319	34 risings sampled at least twice. Sites sampled by Edmunds excluded.
Northwest Yorkshire			
Sweeting, 1964		140–180	
Sweeting, 1965	120–140 winter, 180 summer		
Sweeting, 1972b	180 'average' 60–80 under flood conditions		
Richardson, 1968		118–238	7 springs ⎫ Alum Pot area,
Richardson, 1974	227		3 seepages ⎬ sampled only once
	84		5 resurgences ⎭
Pitty, 1972	124		36 surface streams sampled monthly for 1 year.
Pitty, 1974	185	156–227	9 seepage sites in Ingleborough Cavern, the cave stream and a nearby spring sampled monthly for 1 year.
Ternan, 1974	148	101–195	Mean of means, mean max. and mean min. for 5 risings on Darnbrook Fell sampled fortnightly for 1 year.
Pity *et al.*, 1977		99–104 142–188	Main stream ⎫ White Scar Cave 13 seepages ⎬ samples from 19 sites ⎭ over 11 months.
Halliwell, 1979	89	81–99	9 large risings
Halliwell, 1980	143 >200	100–200	27 permanent-flow smaller risings 12 risings mainly along Craven Fault zone.
Pentecost, 1981	151	129–191	7 springs near Malham Tarn

study, Halliwell (1979) used a short sampling interval and a wider range of statistical techniques and obtained consistent results across the northwest Yorkshire karst. Three lagged correlations were identified: (1) a strong negative at one day prior to sampling interpreted as a dilution pulse mainly from allogenic waters; (2) a delay factor of 30–40 days interpreted as the mean flow-through time for diffuse vadose seepage water originating from areas of thin soils or bare limestone pavement and

(3) a weaker delay factor of 75–85 days thought to reflect the influence of non-limestone rock, peat and thick soil cover. These results confirm the importance of allogenic recharge (dilution) in the Yorkshire karst but they suggest that Pitty (1972) may have overemphasised both the average length and the overall significance of flow-through times in controlling total hardness.

Table 14.2 Solutional erosion rates for Pennine karsts (m^3/km^2/a) (a = annum).

		Comment
Peak District		
Pill, 1948	~44	Castleton rising. Present author's calculation from quoted figure of 840 tons/year.
Corbel, 1957	80	Area as a whole.
Dearden, 1963	77	
Douglas, 1964	26–79	R. Derwent above Matlock Bath.
Pigott, 1965	50–100	
Sweeting, 1965	51–85	
Ford, 1966	61	Castleton area.
Pitty, 1968a	75–83	Area as a whole. Estimates for 11 sites are 41–193.
Christopher, 1981	55±12	Area as a whole. Estimates for 7 drainage basins are 22–99.
Northwest Yorkshire		
Hughes, 1886 Goodchild, 1890	~50–80	Based on weathering of monuments and tombstones.
Sweeting, 1964, 1965, 1966	83	40 surface lowering; 43 in main body of bedrock.
Gascoyne, 1981	22–80	Based on rate of cave passage entrenchment.
Gascoyne et al., 1983a	20–50	

Ternan (1972) suggested that the coefficient of variation (C.V.) of dissolved calcium carbonate in risings in the Yorkshire karst provides an index both of the rate of groundwater circulation and of the mode of recharge. Although there are problems of interpretation this technique has been extended by Halliwell (1979, 1980) who used the variability of temperature, calcium, magnesium and bicarbonate alkalinity together with geographical location to identify five groups of sites in the Ingleborough area: (1) surface waters sampled prior to sinking; (2) large risings with significant allogenic recharge; (3) permanent flow smaller risings; (4) high hardness and high C.V. risings; and (5) high hardness low C.V. risings.

The most detailed study of karst solution processes in the Pennines is that of Christopher and Wilcock (1981) who examined the geochemical controls on water composition over the whole Derbyshire karst. Cluster analysis was used to identify four chemically distinct water types: (1) grit/shale surface waters; (2) general limestone waters including those with evidence of contact with lava; (3) dolomitic

waters and (4) mineral/thermal waters. Dominant geochemical controls were identified as biogenic carbon dioxide in conjunction with calcium carbonate and the
effect of shale contact. Contact with dolomite or toadstones and the equilibrium
conditions under which limestone is dissolved (open or closed system) are less
dominant controls. Christopher (1980) also found that the temporal variability
of solute concentrations at conduit-fed risings is largely a function of rainfall. If
equally detailed research could be undertaken in the Yorkshire karst then a clearer
understanding of the contrasts in solution processes in the two areas should emerge.
However, at the present time it would appear that the most important factors are
(1) mode of recharge, the Peak District receiving more autogenic inputs; (2) biogenic
carbon dioxide, which is probably higher in the Peak District; and (3) extent of
contact with non-carbonate rocks. The influence of variations in limestone lithology
on hardness variations at the drainage basin and regional scale is uncertain but
research in Yorkshire has identified local differences in bed solubility. In particular
Sweeting and Sweeting (1969) found that in Littondale the heterogeneous biomicrite
beds are twice as soluble in a given time as the more homogeneous sparite beds,
while Glover (1974) has shown that the low solubility of the micritic Porcellaneous
Band has had an important influence on cave development.

Although the caves of the Pennines contain a wide variety of speleothems
(stalactites, stalagmites, helictites, flowstones) there has been no research on
underground carbonate deposition rates and processes. However, some consideration of this topic is likely in the future as it is now possible to obtain growth rates
for stalagmites using uranium-series dating. Deposition of carbonate in Malham
Tarn and of tufa in Gordale Beck have been discussed by Pitty (1971b,c); Pentecost
(1981) has carried out research on the origins of the extensive subaerial carbonate
deposits of tufa and travertine in the Malham district; and Thorpe *et al.* (1980)
have examined carbon-14 levels in tufa deposits at Gordale and in the Peak District.
Although tufa deposition is often associated with and assisted by certain mosses
and algae it is essentially a chemical process caused by loss of carbon dioxide to
the atmosphere when underground waters supersaturated with calcite reach the
surface. The Malham district contains the most extensive deposits of exposed tufa
in the British Isles and Gordale Beck is probably the best British example of a
tufa-depositing stream but there are few deposits elsewhere in the northwest
Yorkshire karst. This is probably due to the greater influence of allogenic waters
and conduit flow outside the Malham area as these waters frequently rise in an
aggressive (undersaturated) state. In the Peak District karst there are more tufa
deposits than in northwest Yorkshire and the sites are more widely dispersed with
concentrations around Matlock, Youlgreave and Bakewell (Fig. 14.2). As at
Malham they are all fed by long-residence diffuse flow springs although two of
the deposits near Matlock are associated with warm springs. The carbon-14 levels
of actively forming tufa at Gordale and at Matlock are similar and are directly
related to those of the present stream waters suggesting that depositional processes
in the two areas are essentially the same.

Finally, it is necessary to consider the long-term temporal variability of solution
processes and erosion rates. In northwest Yorkshire the average rate of postglacial

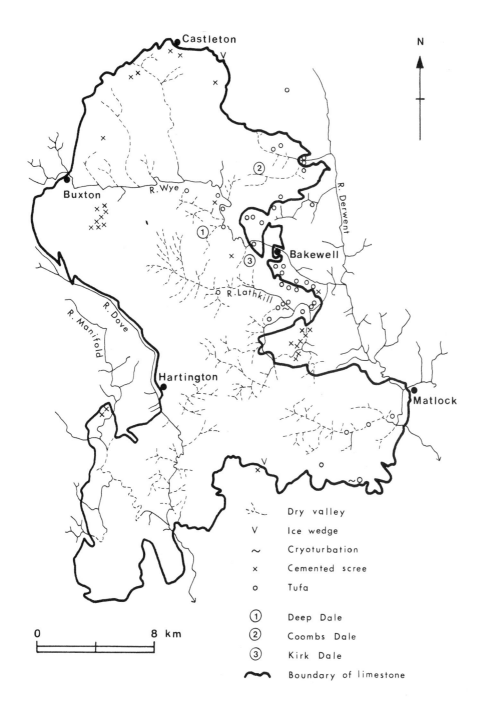

Fig. 14.2 Dry valley systems and periglacial deposits in the Peak District karst (after Burek, 1977).

surface lowering has been estimated as 30–40 mm/ka (1 ka = 1000 years) on the basis of 40–50 cm high pedestals which have formed where erratic blocks have protected the underlying limestone pavement from solution. As this is remarkably similar to Sweeting's (1966) estimate of 0.04 mm/a (a = annum) for present-day surface lowering it may be inferred that solution processes are currently operating at a rate which is close to the average for the Holocene although spatial variability is to be expected, with some localities experiencing markedly higher erosion rates. How representative the Holocene erosion rates are of those in earlier interglacial periods is a matter for speculation and there is also some uncertainty over the operation of solute processes during cold periods. King (1976, p. 75) states that 'Solutional activity and new sink initiation took place mainly in cold phases, by the action of meltwater'. However, as Warwick (1971) has pointed out solutional erosion is likely to be at a minimum during cold periods owing to lowered biological activity and production of carbon dioxide. In addition groundwater percolation would be severely restricted beneath glacier ice and would cease altogether under periglacial permafrost conditions. This is reflected in a paucity of dated speleothems from Pennine caves during the Pleistocene cold phase (Table 14.3). Hence, it may be concluded that solutional activity has been largely restricted to warmer and wetter periods before, between and after the major cold stages during which solutional erosion either ceased or was confined to modification of surface landforms (Waltham, 1977a).

Table 14.3 Possible correlation of periods of speleothem deposition and non-deposition in Pennine ka (Ford *et al.*, 1983; Gascoyne *et al.*, 1983b) with the British Quaternary sequence (Mitchell *et al.*, 19 and marine isotope stages (Hays *et al.*, 1976).

Period[1]	Duration (ka BP)		Possible British Quaternary stage	Marine isotope stage and duration (ka BP)	
	Ford *et al.*	Gascoyne *et al.*			
1 D	0–17	0–13	late Devensian–Flandrian	1	0–11
2 A	17–45	13–34	main Devensian glaciation	2	11–30
3 D	45–75	34–80	mid-Devensian interstadial	3,4	30–71
4 A	75–90	80–85	early Devensian cold	5a ↓ 5e	71–129
5 D	90–145	85–140	Ipswichian interglacial		
6 A	145–170	140–165	Wolstonian glaciation	6	129–189
7 D	170–225	165–350[2]	warm (? Hoxnian)	7	189–249
8 A[2]	225–350		cold (? Anglian)	8–10	249–355
9 D	>350	>350	warm (interglacial)	11	355–426
			cold (extensive glaciation)	12	426–457

[1] D = Speleothem deposition, A = Absence of speleothem deposition.
[2] Gascoyne *et al.* record speleothem growth throughout this period; see text for discussion.

Glacial and periglacial processes

The number and extent of ice advances during the Quaternary and the glacial/inter-glacial chronologies of northwest England are the subject of considerable debate and continuing investigation as evidenced in other chapters in this volume. A major problem is that few early surface deposits survive scouring during more recent events. However, caves are the longest-lived elements in our landscapes and they provide ideal environments for the preservation of evidence for past geomorphic events on the earth's surface. To date the main research focus has been radiometric dating of speleothems by the ^{230}Th/^{234}U method. In the Peak District Ford *et al.* (1983) found that speleothem deposition occurred in five distinct periods separated by four periods of non-deposition. As similar periods had been obtained by Atkinson *et al.* (1978) in northwest Yorkshire a tentative correlation with the British Quaternary sequence was put forward (Table 14.3). Subsequently, Gascoyne *et al.* (1983a,b) have published the results of a more comprehensive dating programme in northwest Yorkshire which narrow the first three phases of non-deposition (periods 2, 4 and 6 in Table 14.3), and, more significantly, show continuous speleothem growth through period 8 (225–350 ka) which had previously been identified as a phase of non-deposition and correlated with the Anglian glaciation. This poses a major problem as there is substantial evidence for an early glaciation in the Pennine karsts. One possible solution is that the period 225–350 ka was not a full glacial although speleothem deposition was reduced by periglacial conditions. The major glacial advance currently labelled as 'Anglian' could then be correlated with marine isotope stage 12 (~ 430–460 ka BP) as evidence from marine cores suggests that more ice accumulated during this stage than in any of the succeeding glacial episodes. Resolution of this problem and provision of a more exact chronology for the middle Pleistocene, with possible extension into early Pleistocene times will require further dates on uranium-rich speleothems supplemented both by other speleothem-dating techniques such as thermoluminescence and electron spin resonance which have a greater time range, and also by palaeomagnetic dating and quantitative analysis of clastic cave sediments (Shaw, 1984).

The precise limits of the various ice advances are uncertain but it is generally agreed that northwest Yorkshire was ice-covered during each advance and that ice extended into the Peak District on at least one and probably two occasions although the area lay outside the limit of Devensian ice. Even during the earlier advances the main stream of ice is thought to have been outside the Peak District with only slow flow down the Derwent and Wye valleys and as a result the glacial removal of limestone from this area is not thought to have been great (Burek, 1977). By way of contrast there was greater down-valley flow of ice in the Yorkshire Dales (Clayton, 1966, 1981) and valley deepening has been attributed chiefly to glacial scour, with suggested values of 20–50 m per glaciation (Brook, 1974) and 60 m (Sweeting, 1974) or 100 m (Clayton, 1981) for the sum of all glaciations. It has also been suggested on the basis of surface karst features that glacial scour was responsible for considerable stripping back of the Yoredale strata in northwest Yorkshire and that the cover around Ingleborough was stripped back by 250–600 m during the last glaciation (Sweeting, 1974). However, Warwick (1971) was not convinced that glacial erosion was so all-powerful in northwest Yorkshire and this opinion has been supported by recent estimates based on speleothem dating which suggest that the erosional effects of ice may have been overestimated and that the maximum average rates of lowering for Kingsdale and Chapel-le-Dale are between 6 and 24 m per glacial/interglacial cycle (Gascoyne *et al.*, 1983a). Thus it is possible that the general appearance of the northwest Yorkshire karst has given a false impression of the degree of ice moulding which has taken place. However, the limestone pavements which are a major feature of the Yorkshire karst are a product of glacial removal of superficial debris and of the effects of ice scour on beds of different strength and to that extent they may be regarded as glacio-karstic landforms (Williams, 1966). Their absence from the Peak District karst is directly attributable to its location outside the Devensian ice limits and to the accumulation of superficial deposits under periglacial conditions (Pigott, 1962, 1965; Burek, 1977).

In addition to the direct action of ice, meltwater may have played an important part in the enlargement of cave passages and the deepening of some valleys (King, 1976; Clayton, 1981). Deposits of glacial till and of fluvioglacial sands and gravels are also of importance in both of the Pennine karsts. In northwest Yorkshire the deposits include kame and kettle moraine south of Malham Tarn and extensive drumlin fields at Ribblehead and the Aire Gap (Clayton, 1981). However, the primary importance of the deposits associated with ice sheets is that they have totally filled some minor surface depressions, vertical shafts (potholes), cave entrances, and cave passages and have partially filled many others (Warwick, 1971). Similarly in the Peak District most cave mouths are thought to have been choked by till and fluvioglacial sediments and although some were cleared out to a large extent by runoff during the Ipwichian interglacial and the succeeding Devensian periglacial activity others remained blocked until cleared in relatively recent times by lead miners or cavers (e.g. Moorfurlong Mine near Bradwell (Worley and Beck, 1976)). During those parts of the Quaternary cold stages when ice was not covering the Pennine karsts periglacial conditions would have prevailed. In northwest Yorkshire this aspect has received relatively little attention although Waltham (1976) has suggested that the Yordas gorge and dry valley and Malham Cove are of periglacial origin. The periglacial phenomena of the Peak District karst have been the subject of more detailed research and Burek (1977) has suggested a threefold division:

1. The formation of new deposits of loess and cemented scree. Loess deposits are widespread, covering most of the limestone plateau and, like the glacial deposits discussed above, they have also been washed into many of the cave systems. Scree is formed as a result of frost action and both fossil and active screes are widespread in the Peak District (Fig. 14.2). Water circulating through voids in the fossil screes has deposited calcareous material which has cemented the loose clasts together and formed a hard, compact rock. Burek (1977) has suggested that the cemented screes are indicative of a periglacial climate with discontinuous permafrost followed

by a climatic amelioration and hence they are thought to be of Late Devensian age.

2. Modification of existing structures in superficial deposits. No patterned ground has been observed in the superficial deposits of the Peak District karst but Burek (1977) has recognised ice-wedge and cryoturbation structures (Fig. 14.2).

3. Landform modifications, including dry valleys (considered later as fluviokarst landforms), tors, and mass movement features. Tors may form in a variety of environments and the Millstone Grit tors of the Pennines have been the subject of considerable debate (Linton, 1955, 1964; King, 1958; Palmer and Radley, 1961). Dolomite tors in the Peak District karst have been studied by Ford (1963) and by Burek (1977) who suggested a complex origin involving both deep chemical weathering during interglacial conditions and also extreme cryogenic activity during a succeeding glacial or glacials. Most of the superficial deposits in the Peak District karst have been subject to some mass movement either by solifluction or by soil creep and landslips have occurred where limestone masses overlie inclined tuff or lava horizons (Burek, 1977). However, the Hobb House landslide in Monsal Dale has no apparent tuff or lava horizons and may instead be a result of former under-cutting of the outside of a meander on the River Wye (T. D. Ford, pers. comm.). Mass movement processes have also operated on superficial deposits in northwest Yorkshire.

Landforms

In general terms the northwest Yorkshire karst landscape may be described as a benchland with its upper surface at 400–500 m above sea level which surrounds several residual peaks of impermeable caprock and which is dissected by glacially modified valleys drained by surface streams. Differential erosion of limestone beds has resulted in parallel bands of scars separated by benches, and glacial stripping of surface soil and vegetation during the Devensian has exposed substantial areas of bare limestone. Streams which rise on the residual hills sink soon after crossing onto the limestone and flow through extensive cave systems before emerging in the valleys.

By way of contrast the karst landscape of the Peak District is essentially a soil-covered, gently undulating limestone plateau surface which ranges in altitude from 275 to 450 m and is devoid of residual hills. The plateau surface is dissected by a complex network of sinuous valleys, most of which are permanently dry under present climatic conditions and only two of which carry perennial surface streams (the Dove and Wye). The large land area devoid of surface drainage and the com-paratively high effective precipitation (~ 800 mm) suggest that well developed under-ground drainage systems should exist but to date only a few are known to any extent and most of these lie in small areas close to the shale/limestone margin.

The two areas may also be contrasted in terms of their individual exokarst land-forms (karren, small and large closed depressions and fluviokarst) and in their endokarst landforms (caves and cave sediments).

Karren

Karren are small-scale superficial solution features measuring from a few millimetres to several metres which may occur either as isolated forms or in large karren fields. Although they may form either subaerially or beneath a partial or complete soil cover, karren are only visually apparent in areas where bare limestone is exposed. Hence, they are relatively rare in the Peak District and those which have been described have suffered considerable degradation (Warwick, 1975). This is in marked contrast to the northwest Yorkshire karst where large areas of exposed bedrock are divided into a geometrical pattern of blocks (clints) by the intersections of solutionally widened fissures (grikes), thus forming a limestone pavement (Williams, 1966). The pavements are one of the most distinctive features of the limestone surface and have been the subject of study from the nineteenth century (Goodchild, 1890) to the present day (Vincent and Rose, in press). It is now generally believed that they are the outcome of glacial stripping of bedrock although the detail of their surface morphology is due largely to postglacial chemical and biological processes, including grazing by animals and human actions (Goldie, 1973, and in press). Structure and limestone lithology have also contributed to the great morphological variety of pavements which makes them particularly suitable for morphometric study (Goldie, 1973; Paterson and Chambers, 1982). The role of lithology has been particularly emphasised by Sweeting (1966, 1974, 1979) whose observations suggest that pavements on the more massive and sparry limestones generally have larger clints and deeper, narrower grikes than those developed on the less strong biomicrite beds which are more prone to physical weathering and scree production. Most clints are scoured by rundkarren (runnels) which were probably formed by acid peaty waters draining through a vegetation cover which has now been largely removed. Removal may have been partly a result of climatic change but human actions seem a more likely cause, particularly since shrubby vegetation is regenerating in an area which has been fenced in to restrict grazing.

Closed depressions

Closed depressions of various types and sizes are a characteristic feature of karst regions and a distinction is generally made between small depressions ranging from 5 to 1000 m in diameter and 2 to 100 m in depth and larger forms (poljes) which usually exceed 1 km in length. There are no poljes in the Pennine karsts but smaller depressions of a variety of types are common in both areas. Apart from the fossil Tertiary collapse pockets discussed above (p. 265) there is no published information on the Peak District depressions although an unpublished morphometric study by Al Sabti (1977) has shown that over 90 per cent of the depressions occur on the relatively impermeable and highly fissured shelf and basin limestones with only 4.9 per cent on adjacent reef limestones which are of the same age but have a much higher porosity. In another unpublished account Beck (1980) notes that analysis of the closed depressions is complicated by mining activity, particularly in the north and east of the area, and he suggests that the highest concentrations of depressions lay along mineral veins prior to mining, so that only the isolated ones now remain undisturbed. Subjective observation by the present author suggests that both simple solution dolines and alluvial dolines (shakeholes) are present and in this respect the Derbyshire karst is similar to that of northwest Yorkshire. However, in the northwest Yorkshire karst collapse dolines are very much more frequent than in the Peak District and there are also several large, complex uvalas which are not found in the Peak District. Shakeholes are cone-shaped depressions often about 3 m deep and 8–11 m in diameter which form where superficial materials have fallen into solutionally enlarged joints in the limestone. Their size and depth depend upon four main factors: (1) the width and depth of joints in the limestone; (2) the thickness of superficial deposits; (3) the amount of surface drainage entering the hole; and (4) the general shape of the land (Sweeting, 1972b). In northwest Yorkshire they occur on the Great Scar Limestone and on limestones within the Yoredale Series and most are of recent (Holocene) age. Shakeholes in the Peak District which are formed in loess are probably also of Holocene age, but those containing glacial till may well be older. Clayton (1966, 1981) has provided comprehensive accounts of the origins of shakeholes in northwest Yorkshire. In general collapse dolines have abrupt and cliff-like walls at the time of formation, but weathering back of the doline sides towards equilibrium solution slopes occurs from the time of collapse so that old collapse dolines may be difficult to distinguish from solution dolines on morphometric grounds. Nevertheless several dolines which are clearly of collapse origin occur in northwest Yorkshire: Gavel Pot on Leck Fell (60 m long, 30 m wide and 40 m deep) and Hull Pot on Pen-y-ghent (100 m long, 10 m wide and 20 m deep) being particularly good examples. Sweeting (1972b) has suggested that recent falls of the limestone along shaley or ferruginous mudstone bands have contributed to their enlargement. The largest closed depressions in the Pennine karsts, and probably also in Britain are the saucer-shaped features 300–900 m in diameter and up to 100 m deep which are found to the northeast and southwest of Malham Tarn (Figure 14.1C). Sweeting (1972b, 1974) has suggested that they are uvalas (compound depressions) which originated at boundaries between the top beds of the Great Scar Limestone and Yoredale shales which have since been removed, but Clayton (1966, 1981) regards them as solution dolines at a stage of early maturity and probably of preglacial origin. Both Sweeting (1974) and Clayton (1981) have further speculated that depression development may have been favoured by the warmer climate and deeper regolith in the area during late Tertiary times. Speleothem age data from Victoria Cave provide some support for the ancient karst hypothesis, but Gascoyne *et al.* (1983a) caution against over-reliance on results from a single site.

Fluviokarst

Fluviokarst landforms are produced by flowing water at the earth's surface and as a result some authors have suggested they are not true karst forms (Sweeting, 1972b). They may be subdivided into karst valleys which have active surface rivers and dry valleys which are the most numerous valley form in karst areas. Karst valleys may be further divided into allogenic valleys formed by rivers which have their origins outside the karst area and pass completely through it, blind valleys which form where a stream goes underground within a karst area, and pocket valleys (reculées) which occur in association with large risings at the foot of a limestone massif. All three types of karst valley are found in both of the Pennine karsts but their relative numbers differ markedly. Most of the Yorkshire dales are allogenic valleys and although flow is intermittent in some (e.g. upper Chapel-le-Dale) the only major sinking river is the Nidd to the east of the karst area. By way of contrast there are only three allogenic valleys in the Peak District, the Manifold, the Dove and the Wye and of these only the latter two carry perennial streams. The reasons for this contrast lie in the shape of the limestone outcrops (Fig. 14.1B,C) and the greater time span since glacial disruption of drainage in the Peak District. These factors also account for the greater numbers of blind valleys in the Peak District and for the presence of numerous half-blind valleys (where swallow holes absorb flood overflow below the normal point of disappearance) in northwest Yorkshire. Pocket valleys are less spectacular in areas where the limestones form lower ground at the 'downstream' side of the outcrop and hence they are less in evidence in the Peak District than in northwest Yorkshire where Malham Cove forms a spectacular example about 75 m in height. Dry valleys are also found in both Pennine karsts, but they are most numerous in the Peak District which is traversed by the most complex dry valley network in Britain (Fig. 14.2). It seems likely that many of the Peak District valleys originated on an impermeable cover and were superimposed onto the limestone (Warwick, 1964; Ford and Burek, 1976). Desiccation may then have resulted either from a lowering of the regional water table as a consequence of downcutting by major streams (Warwick, 1964) or from a gradual karstification of drainage as a consequence of solutional erosion and the geohydrological properties of limestones (Smith, 1975). In either case the valleys would have been reactivated during periglacial periods when the ground was frozen and the frozen ground could also have formed the impermeable cover required by the superimposition theory (Burek, 1977). Similar considerations probably apply in northwest Yorkshire where there are fewer dry valleys and it is considered unlikely that any of the valleys owe their origin to glacial ice, though some may have been formed largely by meltwater (e.g. Trow Gill and the Watlowes valley below Malham Tarn (Sweeting, 1974)). King (1976) has suggested that Trow Gill and some of the other dry valleys in Yorkshire are a result of cavern collapse and similar ideas were once prevalent for the deeper dry valleys in the Peak District, such as the Winnats near Castleton (Fig. 14.2). This seems highly unlikely as cave collapse is generally only a feature of landscapes in which erosion has reached an advanced stage and also because the form of the dry valleys resembles that of 'normal' fluvial valleys rather than cave passages.

Caves and cave sediments

More than 200 km of cave passage have been explored and surveyed in northwest Yorkshire, including 80 km in the Three Counties Systems (Waltham and Brook,

1980), and about 30 km in the Peak District. Their geomorphology is the subject of thirteen chapters in Waltham (1974) and nine in Ford (1977a). More recently substantive contributions have been made by Beck (1980), Christopher and Beck (1977), Ford *et al.* (1983), Gascoyne *et al.* (1983a), Waltham (1977a, 1977b, 1981), and Waltham *et al.*, (1981). In both areas four primary controls on cave development may be identified:

1. Lithology: the variable solubilities of micrites and sparites determine whether cave development will occur above or within a particular bed, and chert or other insoluble residues may be locally important.

2. Horizontal discontinuities: in northwest Yorkshire the shale horizons in the Great Scar Limestone may act as barriers to downward penetration of ground-water and in the Peak District the shale bands, toadstones and clay wayboards have a similar function. In both cases weathering of contained pyrite may release sulphuric acid and enhance limestone solution. The nature and distribution of bedding planes is also important and there are marked contrasts between the regular bedding in northwest Yorkshire and the highly variable bedding in the Peak District.

3. Structure and mineralisation: in northwest Yorkshire both vadose and phreatic cave passages are often aligned along faults or prominent joints but in the Peak District joint and fault control over passage development is more limited as most joints are not strongly developed and few extend more than the height of a single bed. However, connections between cave levels are almost exclusively formed on strong joints or faults and Ford and Worley (1977) have suggested that deep phreatic flow through a series of vein and fault cavities may provide an explanation both for the initiation and early development of many Peak District caves and also for the alternation of passages in certain systems from vadose to phreatic and back to vadose again.

4. Regional landform evolution and climatic change during the Pleistocene and in particular variations in the local base level and in the volume of allogenic inputs of water and clastic sediments.

As no two cave passages have had exactly the same factors controlling their development there is a great morphological variety in Pennine caves and most systems are complex features with contrasting components. For example, the Swinsto Hole to Keld Head cave system in Kingsdale is a complete hydrological system, with three main components: a modern vadose zone, 130 m deep, through which a stream descends by following dipping bedding planes and shale beds and older (pre-Devensian) passages; a modern phreatic zone, 20 m deep, through which the stream loops by descending joint lines and gently rising up dipping bedding planes; and a middle section consisting of rejuvenated and largely abandoned phreatic passages which date back to the Hoxnian or earlier interglacials (Waltham *et al.*, 1981). Although the modern vadose and phreatic zones in this particular system are distinct many Pennine caves exhibit greater complexity with undulating profiles in which the low sections are still phreatic, while the higher sections are presently vadose. Despite this complexity, morphological studies of cave passage distribution and of their altitudes, clastic sediments, and speleothems has shown that similar situations existed in the Bradwell, Eyam–Stoney Middleton, Castleton and Lathkill Dale areas of the Peak District (Beck, 1980). Hence, Ford *et al.* (1983) identified ten developmental stages for the north Derbyshire karst and suggested correlations with the British Quaternary sequence. In summary it is suggested that speleogenesis began in the early Pleistocene with the development ofphreatic tube networks. Periods of rejuvenation and vadose incision together with the development of new tube networks at successively lower levels are correlated with the Cromerian, Hoxnian and Ipswichian interglacials. The early glaciations (Anglian and Wolstonian) are regarded as stillstands, but the Devensian is subdivided into periods of sand/silt/clay deposition separated by a period of speleothem deposition and limited downcutting during the mid-Devension interstadial. Finally, the Holocene has been a period of vadose downcutting through sediment fills and the development of new phreatic tube networks.

Early Pleistocene development of phreatic tube networks also took place in north-west Yorkshire and all workers are agreed that large phreatic tubes had developed and been drained before 350 ka BP. Atkinson *et al.* (1978) suggested that they developed under a stable water table of regional extent and that this was terminated by glacial entrenchment of about 70 m. However, Gascoyne *et al.* (1983a) do not support the idea of a deep glacial cut but instead argue in favour of erosion and drainage having continued in lesser steps through an unknown number of glacial–interglacial cycles. The general sequence since draining of the tubes is similar to that of the Peak District: enlargement of cave passages and development of new sink-points during the warmer and wetter interglacial phases; valley incision during glacial phases both by ice and by surface streams (including meltwater) flowing over permafrost, leading to rejuvenation of phreatic passages; deposition of clastic sediments by meltwater streams during the later stages of cold phases, and post-glacial vadose and phreatic passage development, together with incision and removal of clastic sediment fills. However, it should be emphasised that this is very much a generalised pattern and that there is no clear regional sequence in northwest Yorkshire. Indeed, it is usually difficult to categorise the genesis of a complete cave system and even local correlations are difficult without radiometric or biostratigraphic control (Waltham *et al.*, 1981; Gascoyne *et al.*, 1983a).

Conclusions

Examination of materials, processes and landforms and of their variations through time has revealed more contrasts than similarities between the two main Pennine karst areas. Unfortunately it has not proved possible to consider the equally marked contrasts in their hydrogeology, and for this reference should be made to the summaries in Waltham (1974) and Ford (1977a) and to more recent work by Christopher (1980), Christopher *et al.* (1981), Halliwell (1979, 1980) and Smith and Atkinson (1977). Finally it must be noted that although the distinctive character of the Pennine karsts is primarily due to their respective Quaternary histories, human activities have also had a pronounced, and growing impact. In the Peak District lead mining began in Roman times and reached a peak in the eighteenth and nineteenth centuries (Ford and Rieuwerts, 1975). As a result collapsed shaft hollows, overgrown waste hillocks and modified closed depressions are common on the limestone plateau, making morphometric studies extremely difficult. In sinking shafts and driving adits it was not uncommon for miners to break into natural cave systems and most then suffered considerable modification (e.g. Water Icicle Close, near Monyash). However, the miners' major impact came through the driving of drainage levels (soughs) which have substantially altered the area's hydrogeology (Oakman, 1979). Lead mining did not take place in northwest Yorkshire, but this area has suffered its own particular form of human impact: the removal of clints from limestone pavements for decorative purposes (Goldie, in press). However, the most dramatic human impact on the Pennine karsts has come from quarrying of limestone for aggregates, cement making and the chemical industry. In an early attempt to quantify this impact Dearden (1963) estimated that seventy times more limestone was being removed from the Peak District by quarrying than by natural processes. Figures provided by the Peak Park Joint Planning Board indicate a doubling of production between 1963 and 1971. Moreover, 'this high rate of removal is concentrated at certain spots, whereas natural removal is more general, largely hidden underground, and at a rate so slow as to be unnoticeable in its effects on scenery' (Dearden, 1963). Thus quarrying has substantially modified some landforms, particularly valleys; completely destroyed others, a prime example being Victory Quarry fissure near Buxton, which had yielded the Peak District's only early Pleistocene fauna; and created new landforms such as worked-out quarries with steep, sometimes benched, sides and extensive flat floors. At the time of writing there are several controversial proposals for quarry expansion in the Pennine karsts, one of which could involve the removal of a complete hillside and the cave system it contains. It is therefore apparent that studies of the Quaternary history of Pennine karsts should be complemented by research on the quarrying industry and its environmental impact, and one such investigation has recently commenced in north Derbyshire.

15 *Stephen J. Gale*

The Late- and Post-glacial environmental history of the southern Cumbrian massif and its surrounding lowlands

Introduction

Probably the most intensive studies of Late- and Post-glacial environmental change in the world have been made in the mountain massif of central Cumbria and, to the south, its surrounding fringe of coastal lowlands. Part of the attraction of the area for this type of work is that it provides an almost unique combination of adjacent and contrasting environments for investigation, from mountain to valley and from coast to interior. Equally importantly in a national context, the region includes, at Windermere, the type site for the sequence of Late-glacial environmental change in Britain; the region also represents the frontier of man's colonisation of the country during the Late-glacial and early Post-glacial.

In the following review, we consider the effects of successive environmental changes on the landscape and environment inherited from the Quaternary glaciations. The period studied extends from the end of the last full glaciation until the time of the elm decline and the arrival of Neolithic cultures in the area at ~ 5.0 ka BP. After that time, man began to have a more marked impact on the environment, and natural conditions, particularly with regard to vegetation, can no longer be considered to have existed.

All dates are presented in terms of uncorrected radiocarbon years before present (BP) and the location of all the sites mentioned in the text is shown on Fig. 15.1.

Late-glacial environmental history

The end-glacial phase (prior to ~ 14.5 ka BP)

At the maximum of the last glaciation, ice from the southern part of the central Cumbrian massif extended as far west as the Cumbrian lowlands, where it came into contact with Irish Sea ice (Marr, 1916; Huddart *et al.*, 1977), as far east as the Lune valley, where it came into contact with ice from the Yorkshire Dales (Tiddeman, 1872; Marr, 1916; Moseley and Walker, 1952) and at least as far south as Low Furness where the ice again came into contact with Irish Sea ice (Grieve and Hammersley, 1971). A number of workers, in particular Gresswell (1951, 1962), who interpreted drift deposits in the valleys of southern Cumbria as terminal moraines, regarded the deglacial phase of this glaciation as having been punctuated by episodes of glacial readvance. Better evidence for such events was provided by Huddart (1971a,b; Huddart *et al.*, 1977) who used stratigraphic, sedimento-

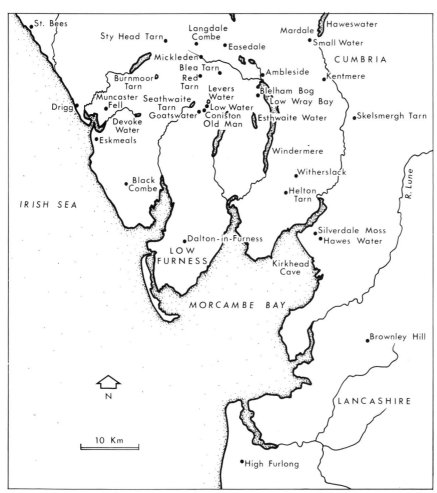

Fig. 15.1 The southern Cumbrian massif and its surrounding lowlands showing the sites mentioned in the text.

logical and geomorphic methods in the southwest Cumbrian lowlands to demonstrate that one and probably two readvances of Irish Sea ice had occurred. Grieve and Hammersley (1971) also provided sedimentological evidence for one of these readvances in Low Furness. Huddart *et al.* (1977) argued that these readvances must have preceded the date of 12 810±180 BP obtained from a kettle hole on the terminal moraine at St. Bees; although Huddart and Tooley (1972) have debated the validity of this date.

Other workers, however, have questioned the reality of these readvances, particularly since the major lakes of the area are known to have been ice-free since before ~14.5 ka BP (see below). These workers have suggested that climate ameliorated rapidly at the end of the last glacial and that deglaciation was largely achieved by *in situ* melting and thinning. Thus, great masses of ice may have survived for a while out of equilibrium with the prevailing conditions and probably with little increment from winter snowfall. Pennington (1978) has therefore implied that the terminal moraines identified in the valleys of southern Cumbria by Gresswell (1951, 1962) may be interpreted simply as hummocky drift formed by downwasting *in situ*. As a consequence of this *in situ* decay, it is suggested that areas of high ground, such as Muncaster Fell (Smith, 1969), became ice-free before the valleys and lowlands.

Meltwaters flowing beneath, through and over the ice occasionally cut channels where they impinged on the subglacial surface (Gresswell, 1962; Smith, 1967, 1969). In some areas, meltwaters also laid down spreads of sands and gravels. Such deposits have been interpreted by some workers as proglacial sediments laid down in front of readvancing ice fronts. This appears to be the case in the southwest Cumbrian lowlands where the sands and gravels indicate an upsection increase in flow regime, and are overlain by readvance tills (Huddart, 1971b). Elsewhere, however, the spreads of clays, sands and gravels appear to have been laid down in front of a retreating ice sheet, as in Morecambe Bay itself, where a single, complex sequence of glaciofluvial and glaciolacustrine deposits overlies a till (Knight, 1977).

Vincent and Lee (1981) and Vincent (1982) have suggested that the complex proglacial depositional environment formed in Morecambe Bay provided a source for the deposits of loess which overlie till on the limestone uplands surrounding the bay. This is corroborated by a study of the palynomorph assemblages in sediments reworked from the loess, which shows that the loess is likely to have been derived from sediments laid down by Irish Sea ice (Gale and Hunt, 1985). Vincent (1982) described the loess as 'Late Devensian' in age. Although no evidence was offered in support of this view, it seems extremely unlikely that widespread loessic deposits could have survived from an earlier glaciation. Unfortunately, the approximate date of 11.0–10.5 ka BP assigned to thick beds of reworked loess in Kirkhead Cave provides only a minimum date for the original deposition of the loess (Gale and Hunt, 1985).

Underlying and to some extent associated with the loess deposits around Morecambe Bay are deposits of cemented scree (Gale, 1981b; Vincent, 1982). Since the screes also overlie striated limestone surfaces, their formation may be best attributed to an end-glacial phase of gelifraction immediately after the retreat of the ice from the area. Other fossil periglacial features, such as screes, block fields, stone stripes, and gelifluction deposits and terraces, found particularly in the central Cumbrian uplands (Tufnell, 1969), may also date from this phase. However, these features could also have been formed during the second Late-glacial stadial. This certainly seems to have been the case for those screes found within basins occupied by ice at that time, as well as for the involution structures recorded by Johnson (1975) in gelifluction deposits at Dalton-in-Furness. Numerous examples are also given below of gelifluction deposits of this age found within datable stratigraphic contexts.

In the valley lakes of central Cumbria, such as Windermere (Pennington, 1943, 1947), Esthwaite Water (Franks and Pennington, 1961) and Kentmere (Walker, 1955), a basal diamicton, almost certainly laid down by ice of the last glacial, is overlain by laminated beds. At Windermere and Kentmere, these beds exhibit coarse–fine couplets which have been interpreted as the result of seasonal meltwater deposition in standing water bodies; in Windermere, the upward transition from wide to narrow couplets is also regarded as indicating the retreat of active ice from the lake basin (Pennington, 1947, 1978). That the deposition of the laminated beds took place under proglacial conditions is indicated by the almost total absence of organic material within the deposits (Pennington, 1943, 1947; Franks and Pennington, 1961; Coope, 1977b) and by the high calcium, sodium and potassium content of the deposits, almost equivalent to that of the unweathered parent Borrowdale Volcanic Series rocks (Pennington, 1964; Mackereth, 1966; Holmes, 1968). These elements are likely to have been derived from freshly-exposed rock debris, and their presence suggests that both the lake waters and the surrounding sediments would probably have had a comparatively high base status at this time.

The process of valley infilling accompanying this phase of deposition is likely to have had important geomorphic results causing, in particular, the build-up of the flat deltaic deposits found at the heads of most of the valley lakes of central Cumbria.

Away from the central Cumbrian massif, as at Skelsmergh Tarn (Walker, 1955), Witherslack and Helton Tarn (Smith, 1958), Hawes Water and Silverdale Moss (Oldfield, 1960b), the basal diamicton of the last glacial ice are overlain by deposits which may be interpreted as the result of gelifluction and local seasonal melt, often into standing water bodies. This may indicate that by this time these basins were beyond the reach of meltwater deposition from the major ice bodies.

Late-glacial stadial (∼ 14.5–13.0 ka BP)

By before ∼14.5 ka BP, ice had disappeared from areas as far north as the head of Windermere (Pennington and Bonny, 1970; Pennington, 1977), and even before that date, silty clays containing organic materials were being laid down over the glacial diamictons at Blelham Bog (42 m O.D.) (Pennington, 1970). Pollen analysis at Blelham Bog suggests a vegetation at this time dominated by *Rumex* (dock and sorrel), Gramineae (grasses) and Cyperaceae (sedges), with some *Juniperus* (juniper), *Salix* (willow) and *Betula* (birch). In this context, the *Juniperus* pollen may be interpreted as the product of prostrate shrubs dependent on a winter snow cover for survival (Pennington, 1970, 1975a). The annual pollen influx during this period never exceeded 200 grains cm^{-2}, a rate characteristic of contemporary tundra (Pennington, 1975a); this is consistent with a pollen spectrum suggestive of treeless tundra vegetation and with the presence of a typically northern Cladoceran (water-flea) fauna within these deposits (Harmsworth, 1968). The dominant diatoms within the sediments suggest that the waters of the pre-Blelham Bog lake were alkaline and probably oligotrophic, whilst the presence of *Melosira teres* and *Achnanthes suchlandtii* supports the case for cold climatic conditions (Evans, 1970). These tundra-like conditions seem to have been maintained at Blelham Bog until ∼13.0 ka BP (Pennington, 1975a).

Similar environmental conditions are recorded in cores taken from Low Wray

Bay in Windermere (41 m O.D.). Here the laminated meltwater deposits are over-lain by a thin clay unit deposited prior to ~14.5 ka BP, which, although apparently barren of pollen (Pennington, 1977, 1978) and Coleoptera (beetles) (Coope, 1977b), contains larvae of Trichoptera (caddis-flies) and Chironomidae (midges) (Coope and Pennington, 1977), as well as the stems of mosses (Pennington, 1947). Above the clay, a silt unit laid down between ~ 14.5 and ~ 13.0 ka BP contains an increasing proportion of organic material at higher levels in the unit. At the start of the deposition of this unit, between ~ 14.5 and ~ 14.0 ka BP, the pollen content is suggestive of a vegetation cover of *Salix herbacea* (shrub willow) and Cyperaceae, and an annual pollen influx of $\leqslant 100$ grains cm^{-2} (Pennington, 1977), both characteristic of treeless tundra. However, the associated Coleopteran assemblage, although very sparse, contains no alpine species and provides no evidence of cold conditions (Coope, 1977b; Coope and Pennington, 1977). Furthermore, during the subsequent millennium, the area appears to have been colonised by a suite of Coleoptera which would be at home at Windermere today, including a number of species absent or rare in the northerly parts of Europe (Coope, 1977b). This fauna is indicative of a sudden and intense climatic amelioration, and contrasts with the palaeobotanical record for this period at both Blelham Bog and Windermere itself, where pollen indicates a *Rumex-* and Gramineae-dominated assemblage with some *Betula* (tree birch) present (Coope and Pennington, 1977; Pennington, 1977). Although the presence of tree birches suggests that climatic amelioration had taken place (as suggested elsewhere in northern England by Beckett, 1981; Hunt *et al.*, 1984), there may be several possible reasons why even the vegetation of the lowlands was kept at a comparatively pioneer stage at a time when mean July temperatures must have reached 14 °C by 13.0 ka BP (Coope, 1977b); winters may have been very cold, conditions of relative drought may have existed, spring thaws and floods might have been severe, soils might have been highly unstable, or temperate vegetation species might have had insufficient time to migrate from their glacial refugia many hundreds of kilometres to the south (Huntley and Birks, 1983).

Cores taken from Blea Tarn (187 m O.D.) provide some evidence of environ-mental conditions at this time in the mountains of central Cumbria. The basal deposits at this site resemble those of Blelham Bog in their low pollen concen-tration, contributed mainly by Gramineae, Cyperaceae and *Salix*. They differ from both Blelham Bog and Windermere, however, in their low values of *Rumex* (Pennington, 1973). This may be indicative of the existence of harsher conditions in the region at altitude. On the other hand, *Rumex* is present in assemblages dating from this time at Burnmoor Tarn (254 m O.D.) and Sty Head Tarn (436 m O.D.), although at the latter site, the rate of pollen deposition appears to have been so low that secondary and far-travelled pollen dominate the assemblage (Pennington, 1977). Both at Blea Tarn and Burnmoor Tarn, the carbon content of the sediments rose steadily through this period, suggesting the accumulation of humus in developing soils after the end of glaciation, and little, if any, disruption of these soils by solifluction (Pennington and Lishman, 1971). At Blea Tarn this period was also marked by the start of a continuous decline in the sodium content of the sediments; according to Mackereth's (1966) hypothesis this represents the onset of a regime in which leaching predominated over erosion in the drainage basin (Tutin, 1969).

The environmental history of numerous sites near the coast of south Cumbria has been investigated, but in the absence of any chronostratigraphic control, it is difficult to interpret this work easily. However, Johnson, Tallis and Pearson (1972) have investigated Late-glacial deposits at Dalton-in-Furness. During this period, the pollen record indicates a vegetation dominated by Gramineae and Cyperaceae, with some *Artemisia* (mugwort or wormwood), *Rumex*, *Betula* and *Salix*. The alpine rock plant *Saxifragia* (saxifrage) also grew in the area at this time. Of some interest are the high values of *Potamogeton* (pondweed) pollen, probably indicative of the local presence of open water conditions. Similar palaeobotanical conditions appear to have existed even further to the south, as at High Furlong on the lowland plains of north Lancashire (Hallam *et al.*, 1973).

Late-glacial interstadial (~ 13.0–11.0 ka BP)

The vegetational record from low-altitude sites in central Cumbria (Blelham Bog and Windermere) is characterised by a rapid increase in the rate of deposition of *Juniperus* pollen after ~ 13.0 ka BP (Pennington, 1975a, 1977). This was interpreted by Pennington and Bonny (1970) as indicating an immediate response to climatic improvement by *Juniperus* already in the area, but hitherto stunted and sparsely flowering. Such an improvement is also indicated by the order of magnitude increase in pollen influx, both at Blelham Bog and at altitude at Blea Tarn, to ~ 10^3 grains cm^{-2} a^{-1} throughout the period 13.0–11.0 ka BP (Pennington, 1973). At the same time, the evidence of Coleoptera from Windermere and Cladocera from Blelham Bog is indicative of an increase in mean July temperatures from ~ 14 °C at the start of the period to ~ 15 °C by 12.5 ka BP (Harmsworth, 1968; Coope, 1977b).

At mid-altitudes in central Cumbria (Blea Tarn), the vegetational response seems to have been similar to that at lower altitudes. Here, however, there is evidence of a temporary climatic recession at ~ 12.7 ka BP (by interpolation from the Blelham Bog sequence), when conditions at this altitude appear to have become unfavourable to *Juniperus*, with a return to a vegetation characterised by *Betula nana* (dwarf birch) and *Lycopodium selago* (fir clubmoss). However, there is no evidence in the stratigraphy for any renewed frost disturbance of the soils of the area, and the vegetational response may therefore have been due either to an increase in snowfall, with its associated adverse effects on the flowering of *Juniperus* (Pennington, 1970), or to the onset of conditions of relative drought or cooler summers. At higher altitudes still, the dominance of herbaceous and disturbed ground species, such as *Rumex* and *Artemisia*, was greater during the early part of the interstadial, and these species totally dominated the assemblage at the altitude of Sty Head Tarn (Pennington, 1977).

By ~ 12.5 ka BP, the occurrence of *Betula* at low-altitude sites had reached its Late-glacial maximum (Pennington, 1975a, 1977). From the evidence of macrofossil remains found at Esthwaite Water (Franks and Pennington, 1961) and Windermere (Pennington, 1947), it is clear that most of the birch was *B. pubescens* (white birch), with *B. pendula* (silver birch) and *B.nana* either rare or absent. The tree birches must have steadily dispersed towards central Cumbria from distant sites in the period since the end of full glacial conditions (Huntley and Birks, 1983). *Betula* appears to have competed with and suppressed *Juniperus*, probably as the result of tree birches shading out juniper (Pennington and Bonny, 1970). Interestingly, the characteristic Late-glacial assemblage of *Helianthemum* (rock-rose), *Hippophaë* (sea buckthorn) and *Artemisia* also reached its maximum at Windermere during the period 12.5–12.0 ka BP. This assemblage must represent a community of unshaded habitats. Some factor may therefore have continued to inhibit the growth of tree birches locally on particular types of terrain, to give a mosaic of woodlands and open areas with unstable soils (Pennington, 1977). A similar conclusion was drawn by Walker (1955) from his work at Kentmere (168 m O.D.) and Skelsmergh Tarn (114 m O.D.). Alternatively, however, the mixture of pollen spectra could be a function of the wide range of habitats of different altitudes within the catchment area of the basin.

The maximum temperatures of the interstadial were reached during this period. They must have been sufficient to result in the disappearance of all ice and permanent snow from the central Cumbrian mountains. Even in Haweswater, fed by the great cirques of Mardale where post-interstadial ice was slowest to melt, the organic muds deposited at this time show no evidence of sediment input from ice-melt (Pennington, 1970, 1978).

The trend of temperature increase was interrupted briefly during the period ~12.0–11.8 ka BP when the deposition rate of relatively thermophilous plants such as *Betula*, *Juniperus*, *Filipendula* (meadowsweet) and *Myriophyllum alterniflorum* (alternate-flowered water milfoil) fell at the low-altitude sites of Blelham Bog (Pennington, 1975a), Windermere (Pennington, 1977) and Esthwaite Water (Franks and Pennington, 1961); at the latter site the herbaceous maximum coincides with an episode of greater mineral deposition. At the same time, the Coleopteran record from Windermere indicates a decline in the presence of southerly species and the first incoming of more northerly species, including *Helophorus glacialis*, a Boreo-alpine species often associated with snow patches (Coope, 1977b). This implies a sharp fall in mean July temperatures to 12 °C (Coope, 1977b), which appears to have been sufficient to have caused soil movement in the Windermere catchment. Thus, the organic content of the sediments laid down in Blelham Bog fell slightly during this period, whilst at Low Wray Bay in Windermere a thin bed of silt containing rock fragments was laid down during this phase (Coope and Pennington, 1977). Similarly, geochemical data from Low Wray Bay show a return to the chemical composition of the deposits of before 13 ka BP (Pennington, 1977). By contrast, at altitude this climatic deterioration clearly had a greater impact. At Sty Head Tarn, for instance, this period is characterised by the dominance of a *Rumex–Lycopodium selago* assemblage (Pennington, 1977).

For the remainder of the interstadial, the pollen evidence at Blelham Bog and Windermere is indicative of an interplay between *Betula* and *Juniperus* within a fluctuating environment (Pennington, 1975a), suggesting a breakdown of the ecological equilibrium established before 12.0 ka BP. Pennington and Bonny (1970) considered that an irregular breakup of birchwoods, with a consequent increased flowering of *Juniperus*, would have produced this result. At the same time, there was an increase in the frequency of Gramineae, Cyperaceae and *Empetrum* (crowberry) pollen at Windermere (Pennington, 1977). This supports the evidence of the Coleoptera from Windermere (Coope, 1977b) and the Cladocera from Esthwaite Water (Goulden, 1964) and Blelham Bog (Harmsworth, 1968) that no climatic recovery took place after ~ 11.8 ka BP and that the prevailing cool temperate conditions must have been very much colder than during the earlier part of the interstadial.

By contrast, at mid- and high altitudes at Blea Tarn and Sty Head Tarn, *Betula* dominated the assemblage (Pennington, 1973, 1977). Furthermore, by the end of the interstadial, these sites, and others such as Devoke Water (233 m O.D.) and Seathwaite Tarn (375 m O.D.) were experiencing severe winter erosion. The evidence for this comes from the increased mineral and sodium content of the lake sediments, from the increase in frequency of pollen of disturbed ground species and, at Sty Head Tarn, from the presence of a pollen assemblage dominated by Cyperaceae and Caryophyllaceae (Pennington, 1970, 1973, 1977). At Blelham Bog, this episode probably corresponds to the rise in the concentration of *Rumex*, Compositae and Caryophyllaceae (Pennington, 1975a), and at Windermere to the local subzone of Cyperaceae and *Selaginella* (Pennington, 1977). Nevertheless, conditions were probably not so severe at these lower altitudes, since there is no evidence of any increase in the deposition rate of disturbed ground species. In fact, the lack of any evidence of soil disturbance at most sites in the area, together with the evidence of chemical analysis of the lake sediments, suggests that the entire period may be viewed as one characterised by soil stability and a continuous vegetation cover. A fall in the iron and manganese content of the sediments towards the end of the period can be interpreted as showing the presence of humic acids and reducing conditions, leading in many areas, such as Blea Tarn, to podsolisation and the spread of *Empetrum* heaths (Pennington, 1970, following the interpretation of Mackereth, 1966). Despite this, the diatoms found within lake sediments at Windermere (Pennington, 1943), Kentmere (Round, 1957), Esthwaite Water (Round, 1961) and Blelham Bog (Evans, 1970) all indicate that the water bodies continued to be base-rich throughout the Late-glacial.

The environment of the coastal lowlands appears to have been rather different from that of central Cumbria during this period. At Dalton-in-Furness, the opening of the interstadial was heralded by the expansion of *Betula* and *Salix herbacea* into the area, and by the first appearance of *Juniperus*, *Empetrum*, Ericaceae (heaths) and *Hippophaë* (Johnson *et al.*, 1972). At Hawes Water, by contrast, the deposits attributed to the interstadial show evidence of a rapid colonisation by *Salix herbacea* and then *Juniperus*. This was followed by a rapid expansion alongside the shrubs of *Betula*, which eventually became dominant throughout the area (Oldfield, 1960b). This expansion of arboreal species in the coastal regions may well have taken place earlier than in central Cumbria, where ice may have lingered in the valleys until later in the interstadial. Interestingly, both at these sites and at Helton Tarn and Witherslack (Smith, 1958), *Juniperus* is far less prominent than in central Cumbria, and tree birches appear to have dominated the vegetation. This may have been because tree birches were able to expand relatively early in the period in the coastal lowlands, whilst further north suitable habitats were limited and *Juniperus* was able to flourish on the hillslopes above the decaying ice bodies.

There is some evidence of a brief environmental deterioration in the coastal lowlands during the interstadial, which may be comparable to that experienced in the period ~12.0–11.8 ka BP further north. During this time, the proportion of *Betula* pollen in the assemblages may have declined and the proportion of herbaceous pollen may have increased (Smith, 1958; Oldfield, 1960b; Johnson *et al.*, 1972). However, the evidence is by no means incontrovertible, and it is clear that the magnitude of the change, if it occurred, was far less than that experienced further north in central Cumbria.

Finally, it would seem that *Betula* may have reached its maximum in the coastal lowlands later rather than earlier in the interstadial (Smith, 1958; Oldfield, 1960b; Johnson *et al.*, 1972). At Dalton-in-Furness, this event is associated with a reduction in vegetation diversity, and this may indicate rather stable, although poor, environmental conditions in contrast to the fluctuating conditions in evidence to the north. This is supported by evidence from further south at High Furlong. Here the *Betula* maximum has been carbon-14 dated to 12 200±160 BP and 11 665±140 BP (Hallam *et al.*, 1973). In this area, *Betula* seems to have increased at the expense of *Juniperus*.

Of equal importance was the discovery within the interstadial deposits at High Furlong of the skeleton of an elk (*Alces alces*) in association with barbed points. The implications of this find are of some significance, for it demonstrates that post-last glacial recolonisation of Britain by man had reached this far north by the middle of the interstadial. Moreover, since the antlers of the elk were just about to be shed, it also indicates the presence of winter hunting groups on the north Lancashire lowlands at this time. Whether this may be interpreted as indicating a seasonal lowland–upland migration pattern by Later Upper Palaeolithic groups, as appears to have been the case for Mesolithic groups can, however, be no more than speculation.

Late-glacial stadial (11.0–10.0 ka BP)

An abundance of lithostratigraphic and biostratigraphic evidence indicates that at ~11.0 ka BP, the region experienced a major deterioration in climate, resulting in the onset of glacial and severe periglacial conditions, and the reappearance of ice in the area. Cirques formed during previous glacial phases were reoccupied by ice (Evans and Clough, 1977), as may be seen from the absence of interstadial sediments in those cirques found at higher altitudes in the mountains of central Cumbria. Perennial snowbeds also existed in the area down to altitudes of ~200–300 m O.D. (Sissons, 1980). Using the evidence of deposits and landforms developed during this time, Manley (1959) attempted to reconstruct the equilibrium snowline in the area during the period. It ranged from 365 to 425 m O.D. in Easedale to 670 m O.D. on Black Combe in the south of the area.

A more detailed analysis by Sissons (1980) showed that at the maximum extent of the glaciers, 55 km^2 of the area was covered by ice, and that firn-lines extended from below 400 m O.D. in the central Cumbrian mountains to above 650 m O.D. on Coniston Old Man. He suggested that the central part of the east–west mountain axis received the most snowfall during this period, and that the distribution and nature of the ice indicated snowfall associated with winds from a southerly direction and hence with warm or occluded fronts moving across the area from the west or southwest. At the time of the glacial maximum, the mean July temperature at sea level was calculated as slightly below 8 °C, a value similar to that obtained by Manley (1959), and by Coope (1977b) from his studies of Coleopteran assemblages.

In those basins below the snowline, but with active ice in their catchments, the organic deposition of the preceding interstadial gave way to the deposition of laminated clastic beds, as at Seathwaite Tarn (Pennington, 1964). In Windermere and Hawes Water, the interstadial sediments are conformably overlain by 400–450 paired couplets, interpreted as the annual input of sediment from melting ice in the catchment (Pennington, 1947, 1978).

In those basins without ice in their catchments, as at Devoke Water, Esthwaite Water, Blelham Bog and Blea Tarn, the deposits of this phase generally consist of unstratified fine-grained materials, often containing subangular clasts (Franks and Pennington, 1961; Pennington, 1964, 1975a). These may be interpreted as the result of gelifluction, an interpretation supported by the high calcium, sodium, potassium and magnesium content of these deposits at Blea Tarn and Esthwaite Water, suggesting unweathered drift sludged by periglacial processes into the lake basin (Mackereth, 1966, Pennington, 1970).

These conclusions are borne out by the palynological evidence. Absolute rates of pollen deposition in the area fell markedly: to approximately 400 grains cm^{-2} a^{-2} at Blelham Bog, to approximately 1000 grains cm^{-2} a^{-1} at Blea Tarn, and to less than 100 grains cm^{-2} a^{-1} at Low Wray Bay in Windermere (Pennington, 1973). This fall was particularly marked in the case of arboreal and shrub pollen, and Pennington and Bonny (1977) have suggested that the deposition rates of *Betula* and *Juniperus* pollen during this period are too low to indicate a local presence of these genera. Only Filicales (ferns), *Artemisia* and Caryophyllaceae experienced an increase in pollen influx during this period (Walker, 1955; Franks and Pennington, 1961; Pennington, 1975a, 1977). *Artemisia* appears to have been dominant at this time at all altitudes below the snowline, and this has been interpreted as indicating that the soils of the area were sufficiently disturbed by frost to break up the vegetation cover and to expose the bare soil (Pennington, 1970). It is noteworthy, however, that even at the altitude of Blea Tarn, the contribution of both *Salix* and *S. herbacea* to the pollen influx remained low during this period, suggesting the absence of snowbed vegetation at that site (Pennington, 1973). On the other hand, Walker (1955) has recorded high relative frequencies of *Salix* pollen at Skelsmergh Tarn and Kentmere, suggesting that snowbed vegetation may have existed there during this period.

By ~10.7 ka BP, absolute rates of pollen deposition of all species except Filicales, *Artemisia* and Caryophyllaceae were beginning to increase again (Pennington, 1975a). The resultant herb-dominated pollen assemblage zones, characterised by Gramineae, Cyperaceae and *Rumex*, may be interpreted as representative of pioneer vegetation colonising and stabilising soils as climatic amelioration brought glacial and periglacial conditions to an end. Confirmation of such climatic amelioration is obtained from the fact that by this time ice had disappeared from the more shallow or peripheral cirques of the area, such as Goatswater, Low Water, Red Tarn and Levers Water, enabling them to accumulate organic sediments which may be assigned to the prevailing Gramineae–*Rumex* pollen assemblage zone or earlier (Pennington, 1978). The retreating ice left in its wake extensive morainic deposits. The consistent pattern of end and retreat moraines of this period is indicative of active retreat for 100–150 m, followed by decay *in situ*. It is difficult to explain this as representing anything other than a synchronous, climatically controlled event acting across the whole area (Sissons, 1980).

This trend of environmental amelioration was confirmed by ~ 10.5 ka BP. Assuming that the laminated sediment couplets in the large lakes of the area were annually deposited, it is likely that by this stage, deposition by melting ice in the region had ceased. Moreover, by this time, full organic deposition had recommenced, and the rate of deposition of taxa such as *Juniperus*, which might be taken as reflecting increasing warmth, had increased at Blelham Bog and Blea Tarn (Pennington, 1973, 1975a). It is possible that this was an immediate response to climatic improvement by *Juniperus* already in the area, but previously stunted and sparsely flowering.

As was the case in central Cumbria, environmental deterioration at the start of the stadial in the coastal lowlands of south Cumbria and north Lancashire is indicated by the increase in the incidence of open-environment pollen such as Cyperaceae, Gramineae, Filicales, *Artemisia*, *Rumex*, *Thalictrum* (rue) and *Salix* (Smith, 1958; Oldfield, 1960b, Johnson *et al.*, 1972; Hallam *et al.*, 1973). The relative frequency of *Betula* also declined at the start of the period, and by the middle of the stadial very low absolute rates of pollen influx were being recorded at Helton Tarn (Smith, 1958), along with barren layers of soliflucted material at Dalton-in-Furness (Johnson *et al.*, 1972). By contrast with central Cumbria, however, the late-stadial recovery of *Betula* was earlier than and greater than that of *Juniperus*. This sequence appears to have replicated that which occurred in the area immediately prior to the warming of the Late-glacial interstadial, and similar reasons may presumably be given in explanation.

Deposits of this age in Kirkhead Cave contain pollen characteristic of open, treeless habitats: *Plantago lanceolata* (ribwort plantain), Cyperaceae, Ericaceae, Coryloid (hazel or sweet gale), *Phragmites*-type (reed) and Gramineae. However, towards the end of the stadial, the relatively high pollen and spore productivity of herbaceous species and Filicales, which appear to behave in the cave context as a thermophile, suggests conditions of amelioration. This is borne out by the appearance in the deposits of the land snail *Oxychilus cellarius*. This species is a thermophile with an affinity for rock rubble and cave habitats, and is indicative of a warmer environment in the area than that suggested by the pollen. The most likely explanation of this apparent anomaly may be simply in terms of the time lag between climatic change and vegetational response in the area (Gale *et al.*, 1984; Gale and Hunt, 1985).

By the middle to late part of the stadial, Kirkhead Cave was in use as a Later Upper Palaeolithic occupation site (Gale *et al.*, 1984). It is unclear whether man had existed in the area throughout the entire stadial, but given the harsh environmental conditions he would have had to have faced this is perhaps unlikely. It is more probable that with the rapidly rising temperatures and vegetation recolonisation of the second half of the stadial, man followed the game which migrated into the area from further south. The remains of *Megaloceros* sp. and ?ox found close to flint artifacts in the cave testify to the importance of large mammals in the local Later Upper Palaeolithic economy (Gale and Hunt, 1985).

Post-glacial environmental history

Early Post-glacial (~ 10.0–8.5 ka BP)

From the evidence of Coleopteran assemblages in the English Midlands (Osborne, 1976) and from foraminiferal analysis of North Atlantic deep-sea sediments (Ruddiman *et al.*, 1977) it is clear that within a matter of a few centuries at the start of the Post-glacial, temperatures had risen to close to those of the present day. Certain thermophilous fauna, such as the land snail *Oxychilus cellarius*, reached the area almost immediately (Gale *et al.*, 1984), but the floral response to the higher temperatures lagged by perhaps several millennia as trees migrated from glacial refugia further south (Huntley and Birks, 1983) and as soil development occurred.

At the start of the period there was a vigorous shrub vegetation of *Juniperus* and *Betula* at almost all sites in the area, with *Salix* important in the wetter places. This was eventually dominated by *Betula*, and by the end of the period, the relative frequency of *Salix* and *Juniperus* had been reduced by the spread of birch woodland, which quickly became dominant on all available land up to altitudes of over 500 m O.D. (Pennington, 1964). By the end of the period, all sites had experienced a marked increase in the relative frequency of deposition of arboreal pollen, although the woodland does not appear to have been totally closed by this stage (Oldfield, 1960b; Pennington, 1964; Walker, 1965; Dickinson, 1973).

The stabilisation of the land surface by vegetation and the development of soils are both indicated by the decline in the sodium and potassium contents of the sediments being laid down in the lakes. During the period ~ 9–5 ka BP, the lakes seem to have experienced a minimum of chemical input from the soils of their catchments, presumably indicating a phase of maximum soil and vegetational stability (Mackereth, 1966).

During the early part of the Post-glacial, the diatoms in the lakes indicate generally alkaline and eutrophic conditions. The lakes were obviously just rich enough to support eutrophic forms, possibly due to the leaching of materials from the surrounding deposits. However, starting in this period and continuing into the middle Post-glacial, there occurred an increase in the frequency of planktonic forms. This appears to have been due to a change in the composition of the lake waters, possibly resulting from an increase in the presence of natural chelating elements and nitrogen. Both of these factors were probably at least partly the result of the establishment of woodland in the area (Round, 1957; Haworth, 1969; Evans, 1970).

By extrapolation from evidence from elsewhere in the country, it might have been expected that the Post-glacial rise in temperatures and development of woodland would have had a remarkable effect on the fauna of the area. Elsewhere, *Equus ferus* (horse), *Rangifer tarandus* (reindeer) and *Megaloceros giganteus* (giant deer) were replaced by *Cervus elaphus* (red deer), *Capreolus capreolus* (roe deer), *Alces alces*, *Sus scrofa* (wild boar) and *Bos primigenius* (aurochs). Unfortunately, south Cumbria has provided little evidence of Late- and Post-glacial faunal remains in an interpretable stratigraphic context, the only exceptions being the record of small fauna at Kirkhead Cave (Gale and Hunt, 1985) and the molluscan record from Hawes Water (Sparks, 1962).

Similarly, with the exception of charcoal fragments found throughout early and middle Post-glacial deposits in Kirkhead Cave (Gale and Hunt, 1985), no certain evidence of Earlier Mesolithic man has yet been found in the area. Nevertheless, since Earlier Mesolithic sites have been found at altitudes of up to 450 m in the Pennines (Radley and Mellars, 1964; Switsur and Jacobi, 1975), as well as possibly in Scotland (Mercer, 1970; Coles, 1971), it seems unlikely that south Cumbria remained uncolonised during this period. More probably, settlement may have reflected the coastal pattern characteristic of Later Mesolithic sites in the area, with the consequence that most sites dating from this time would have been submerged beneath eustatically-rising Post-glacial sea levels (Huddart *et al.*, 1977) and now remain hidden offshore. Jacobi (1973), in fact, has suggested that the extended lowland plains of Morecambe Bay and Lancashire that existed at this time would have provided suitable wintering grounds for groups whose summer hunting extended to the western edge of the Pennines.

Middle Post-glacial (~ 8.5–5.0 ka BP)

The Post-glacial migration of plant species into the region from distant glacial refugia continued during the period ~ 8.5–5.0 ka BP (Huntley and Birks, 1983). Unfortunately, environmental events in the region are rather poorly dated during this period and the sequence of events is based largely on biostratigraphy and in particular on pollen zonation. The entire pollen subzone which opens this period, VIa, has been carbon-14 dated to 8450±340 BP at Small Water (Pennington, 1970). At the start of this subzone, the early *Betula*-dominated vegetation appears to have been displaced by the sudden increase of *Corylus* (hazel, the major producer of

Coryloid pollen in the Post-glacial), which rose to dominate the entire pollen assemblage at many sites in the area. Iversen (1960) has argued that the presence of *Corylus* may prevent the regeneration of *Betula*. However, Oldfield (1960b) has suggested that in the limestone areas to the south of the region, *Corylus* may have simply been able to colonise areas of bare limestone which had been scoured during the glacial maximum and from which *Betula* was absent. *Corylus* does not appear to have been so dominant at altitude in the central part of Cumbria (for example, at Goatswater at 500 m O.D. (Pennington, 1964)), but even here Walker (1965) has suggested that *Corylus* may have colonised the steeper and more thinly soil-covered slopes.

At the end of the subzone, the area experienced the beginning of the spread of *Ulmus* (elm) and *Quercus* (oak) into the region, both genera appearing up to altitudes of at least 510 m O.D. at Red Tarn (Pennington, 1964). From this time until ~ 5.0 ka BP, all sites in the area at altitudes below 750 m O.D. exhibited herbaceous pollen frequencies of less than 5 per cent, which Pennington (1970) has interpreted as indicating the presence of closed woodland.

In the south of the area, the succeeding pollen subzone, VIb, was characterised by a relative fall in the frequency of *Betula* and *Corylus*, and by the expansion of *Quercus* to dominance within the arboreal assemblage. Oldfield (1960b) has suggested that the most favourable soils, previously occupied by hazel scrub, may have come to support oak woodland, whilst *Ulmus* may have succeeded *Betula* on all but the poorest soils.

By contrast, at low altitudes in central Cumbria, as at Blelham Bog, *Quercus* did not become dominant until the close of pollen subzone VIc (Pennington, 1965a). At higher altitudes, such as Mickleden (130 m O.D.), Devoke Water, Langdale Combe (425 m O.D.) and Red Tarn, *Quercus* only experienced a relatively minor expansion, and *Betula* and *Corylus* remained the dominant components of the assemblage (Pennington, 1964; Walker, 1965).

The upper part of the succeeding pollen subzone, VIc, was carbon-14 dated to 7560±160 BP at Burnmoor Tarn (Pennington, 1970). This subzone was marked by the appearance of *Alnus* (alder) in the region. At certain sites, *Alnus* appears to have replaced *Betula* in terms of relative frequency, as at Devoke Water (Pennington, 1964) and Mickleden (Walker, 1965). Elsewhere, however, perhaps where rocks and screes were important, *Alnus* may not have been able to compete effectively with *Betula*.

This subzone seems to have experienced the highest relative frequency of *Pinus* (pine) pollen deposition at those sites where *Pinus* was present. However, *Pinus* values appear to vary remarkably from site to site, and it must be concluded that the distribution of *Pinus* was edaphically controlled all through the early and middle Post-glacial periods (Pennington, 1970).

Iodine in Post-glacial sediments at Burnmoor Tarn and Blea Tarn has been interpreted as having been derived in part from rainfall. The fall in iodine values during late pollen zone VI was therefore interpreted by Pennington and Lishman (1971) as indicative of dry conditions at that time. During the subsequent pollen subzone, VIIa, however, higher iodine values were regarded as indicative of rather wetter conditions, a view subscribed to by Lamb (1977b), who considered this period to be the wettest phase of the Post-glacial. It was under these conditions, therefore, that the general expansion of the moisture-loving *Alnus* took place across the region at ~ 7 ka BP.

Further chemical studies of the sediments of the major lakes demonstrated that this period, ~ 7–5 ka BP, was one of rapid leaching and minimal erosion, with the result that the base-status of the soils declined markedly (Mackereth, 1966). One consequence of this can be seen in the decrease of alkaliphilous diatoms and the increase of acidophilous diatoms in Kentmere (Round, 1957), Esthwaite Water (Round, 1961), Blea Tarn (Haworth, 1969) and Blelham Bog (Evans, 1970). The peak frequency of the diatom *Melosira distans*, an indicator of extreme, acidic, peaty water, also occurred in Esthwaite Water at this time (Round, 1961).

Under these conditions, a long period of approximately two millennia was entered upon during which relatively little change took place in the pollen rain of the region. In the south of the region, mixed oak forest with *Corylus*, *Betula* and some *Tilia* (lime) was dominant (Smith, 1959; Oldfield, 1960b; Oldfield and Statham, 1963; Dickinson, 1973; Pennington 1979). The high frequency of *Corylus* at these sites may have reflected in part the presence of base-rich soils on the local limestone bedrock. Elsewhere in the south of the region, this assemblage must have occupied hillslopes and valleys up to altitudes of at least 425 m O.D., as at Brownley Hill (Moseley and Walker, 1952).

In central Cumbria, the expansion of *Alnus* was not always so marked, for instance at Blelham Bog and Red Tarn (Pennington, 1964). Nevertheless, *Alnus* must have predominated in valley mires at altitudes up to at least 370 m O.D. (Pennington, 1970). On the better-drained soils, *Quercus–Ulmus–Corylus* forests extended up to altitudes of at least 425 m O.D. (Walker, 1965; Pennington, 1970), with an upper fringe of *Betula–Pinus* woodland on the central Cumbrian hills (Pennington, 1975b).

One particularly noteworthy aspect of the vegetational history of this period is the spread throughout the region of *Tilia cordata* (small-leaved lime), perhaps the most thermophilous of the native tree species. Although south Cumbria forms the northern limit of its distribution in Britain, *T. cordata* became a significant component of the vegetational assemblage at this time (Birks, 1982). Pigott and Huntley (1980) have discussed the spread of *T. cordata* through the area from its earliest appearance at ~ 7.4–7.0 ka BP until it reached its maximum northward extension in central Cumbria at some time after 5.0 ka BP.

Although Kirkhead Cave was used by man throughout the entire early and middle Post-glacial (Gale and Hunt, 1985), it is only within pollen zone VIIa that there is any other clear evidence of man in the region. Despite the identification of more than 35 Mesolithic sites in the area, mainly located on the coastal lowlands around Morecambe Bay and the Irish Sea (Barnes and Hobbs, 1951; Nickson and MacDonald, 1956; Cherry, 1963, 1965, 1967, 1969, 1977a, 1977b; Pennington, 1965b; Fell, 1971; Cherry and Cherry, 1973; Bonsall, 1976, 1981, 1982, 1983; Bonsall and Mellars, 1977), only two major groups of sites have so far been dated. At Drigg, a hearth, with Mesolithic artifacts nearby, was carbon-14 dated to

6085±90 BP and 5730±55 BP (Cherry, 1977b), whilst at Eskmeals, a hearth associated with a rich flint industry, probable stake holes and a possible Mesolithic storage pit was carbon-14 dated to 6752±156 BP (Bonsall, 1976, 1981). Since Huddart *et al.* (1977) have suggested that the mean high water mark of spring tides reached perhaps 5 m O.D. at this time, then the location of these occupation sites at around 7 m O.D. may indicate their situation on a contemporary shoreline.

With the exception of a single flint flake recorded at Ambleside (Fell, 1971), definite Mesolithic artifacts are absent from the uplands of the region. Since Later Mesolithic use of upland areas is well documented in other parts of northern England (see, for example, Jacobi *et al.*, 1976), this absence in south Cumbria might be interpreted as the result either of insufficient fieldwork or of poor preservation of sites in the uplands. However, Bonsall (1981) has suggested that a seasonal migration into the uplands, as practised elsewhere in Britain (see, for example, Clark, 1972), would have meant man leaving the coast at precisely the time of year when it had most to offer in terms of potential food resources. He has therefore concluded that year-round occupation of the coastal lowlands would have both been possible from a subsistence point of view and probable in the absence of any clear evidence of seasonal occupation of the coastal sites.

Conclusions

A remarkably detailed picture of the Late-glacial environmental history of south Cumbria and its surrounding lowlands is available. Not only has this enabled the broad pattern of biological and geomorphic events in the area to be reconstructed, but it has also demonstrated the clear influence of such factors as altitude and proximity to the coast on the rate and sequence of operation of these events. Rather less information exists concerning Post-glacial environmental change in the area, but a good picture of Post-glacial environmental history may nevertheless be reconstructed. The broad pattern of environmental change over the entire period was remarkably simple: a general, though by no means constant, environmental amelioration from the end of the last full glacial until the middle of the Post-glacial, superimposed upon which was a single phase of environmental deterioration at ~11.0–10.5 ka BP. In detail, however, the environmental record is one of great complexity, consisting of rather rapidly fluctuating environmental conditions, each event not necessarily being recorded by every environmental indicator, but only by those for which particular environmental thresholds were crossed at that time. It is also clear that the various environmental indicators reacted at vastly differing rates, the difference between faunal and floral indicators perhaps being the most marked. The most responsive indicators clearly demonstrated that many environmental changes operated rather rapidly. Thus, the faunal indicators show that climatic changes sufficient to bring about glacial conditions were able to operate over a matter of a few centuries.

Whilst all these changes were taking place, man began to colonise the area as part of his recolonisation of the country after the last full glaciation. Until recently, it had been considered that Palaeolithic and Mesolithic populations were too small either to have exerted any appreciable influence on the environment or to have required that such changes took place. However, recent work has suggested that Mesolithic communities in particular may have altered the environment over wide areas. No clear evidence yet exists that this was the case in south Cumbria, but the density of Later Mesolithic occupation along the southwest Cumbrian coast and the recently discovered evidence of vegetational alteration at ~ 6.0 ka BP which might be attributed to the effects of man (see, for example, Birks, 1982) suggest that anthropogenic impact may have been as marked in this area as it was elsewhere in Britain.

The distribution and origins of the lowland mosslands

Introduction: some definitions and objectives

A glance at the Ordnance Survey Quaternary Map of the United Kingdom (1977) reveals that, after the great arcuate expanse of Fenland peat, encircling the Wash, the second most extensive lowland peat deposits in Britain are to be found in south-west and central Lancashire where they are commonly referred to as *mosslands*. The term came into popular use in the late eighteenth century when, in 1786, the land utilisation maps of William Yates were first published and his terminology adopted by John Holt in 1795 with the publication of *A General View of the Agriculture of the County of Lancaster*. As the name suggests, the mosslands were characterised by a predominance of species of the bog-moss, *Sphagnum*, growing in nutrient-poor conditions and decaying to form slightly concave, raised peat bogs. In most areas, reedswamp and fen-carr peats were first formed in more nutrient-rich conditions from the partly humified remains of the common reed (*Phragmites australis*), species of sedge (*Carex*) and the herbs and shrubs typical of alder–willow fen carr in the process of hydroseral succession (Walker, 1970). Gradually, as the nutrient level dropped and conditions became acidic, the transition to raised bog via the increasing competitive dominance of *Sphagnum* occurred. Many of the mosslands developed to the raised bog stage, but because of peat extraction and drainage several mossland areas today comprise only the lower horizons of reedswamp peat. Hence, recent soil surveys have recognised the distinction between raised bog and reedswamp peats, labelling the former as the Turbary Moor Series from the fact that many of the mosses were cut for the common right of turbary, and the Altcar Series, after the village of Great Altcar in south-west Lancashire (Hall and Folland, 1967).

The review of Walker (1970) on post-glacial hydroseres takes much of its information from palynological and stratigraphical studies carried out on the peat deposits of the lowlands of North-west England, including the earlier classic studies of Walker (1955), Godwin (1959) and Oldfield (1960a,b) and a wealth of later studies which have answered many queries on the post-glacial history of the landscape and vegetation of the British Isles. Thus, Pennington (1970) remarks that the Lake District region 'has been investigated by pollen-analysts more fully than any area of comparable size in Britain'. Similarly, the researches of Birks (1964, 1965a,b) and Tallis and Birks (1965) have shed much light on the origin and development of the mosses in and around the Greater Manchester conurbation, while the works of Tooley (1978b) have enabled similar interpretations for the mosslands of central and south-west Lancashire. From these regionally-specific groups of research information comes a topographical and geographical definition for the present discussion on the lowland mosslands of North-west England – all those peat deposits found at altitudes of less than 200 m O.D. in the counties of Cheshire, Lancashire, Greater Manchester, Merseyside and the South Lakeland district of Cumbria. The chapter has three objectives, namely, an inventory and classification of the mossland types; a review of research which has thrown light onto the origin of the mosslands; and finally, an analysis of possibilities for future pure and applied research in the field.

A classification of the peatlands

The peatlands are classified primarily according to their suggested origins, secondly according to their geographical location, and thirdly in terms of their topographical relationships to river catchments and drift geology. Hence, three major origin types, ten geographical groups and twenty-five topographical series are recognised and their distributions shown in Fig. 16.1 and 16.2. The maps have been prepared from data derived from 1:10 000 Ordanance Survey, geological and soil survey sheets and from a literature survey supplemented by personal fieldwork. The basic typology follows that of Walker (1970) in recognising peatlands developed in coastal and estuarine sites, large inland basins and small inland basins. Certain refinements and more specific definitions are provided in the following classification and there are occasional alterations of category, the reasons for which are discussed later, and also several problems of categorisation.

Type A. Coastal and estuarine peatlands developed over a layer of marine clay and silt, at altitudes between 3 and 15 m O.D.

Group 1. Estuarine and valley mosslands of South Lakeland
1. Duddon Estuary (Fig. 16.1 : 1) – coastal raised bogs at the mouth of the River Duddon developed between estuarine salt marshes and low fells formed from boulder clays and morainic drift overlying igneous and sedimentary strata of Silurian age. Shaw Moss (SD 1985) to the west of the estuary and Herd House Moss (SD 2284), White Moss (SD 2285), Heathwaite Moss (SD 2387) and Black Moss (SD 2288) in the broad valley of Kirkby Pool are the principal areas; marginal agriculture and rough pasture at 5–10 m O.D.

2. East Leven Estuary (Fig. 16.2A : 2) – approximately five square kilometres of raised bog on the eastern side of the Leven Estuary developed between low Carboniferous limestone exposures at Old Park Wood (SD 3378) and Roundsea Wood (SD 3382) and the Ellerside escarpment formed ot rocks of Silurian age, south of the River Leven. Ireland Moss (SD 3384) and Rusland Moss (SD 3388) are additional sites in the narrow valley of Rusland Water; cutover raised bog, silviculture and marginal agriculture; 3–10 m O.D.

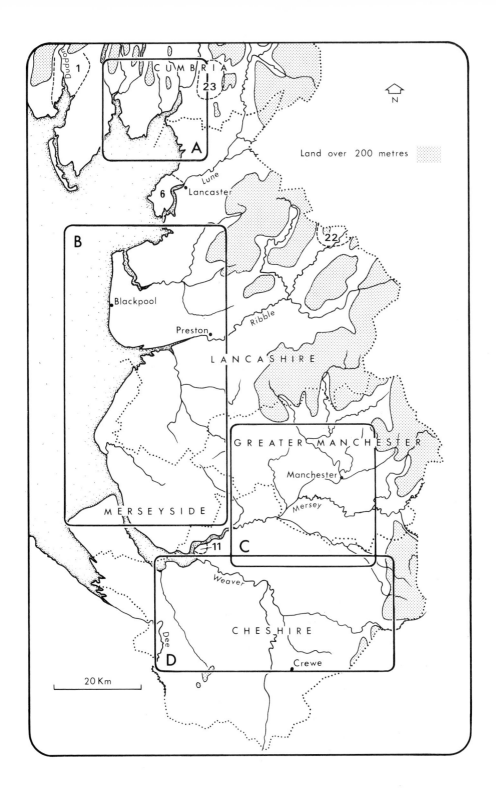

Land over 200 metres

N

CUMBRIA
1
23
A

Lune
Lancaster
6

22
B
Blackpool
Preston
Ribble
LANCASHIRE

GREATER MANCHESTER
Manchester
Mersey
MERSEYSIDE
11
C

Weaver
Dee
CHESHIRE
Crewe
D

20 Km

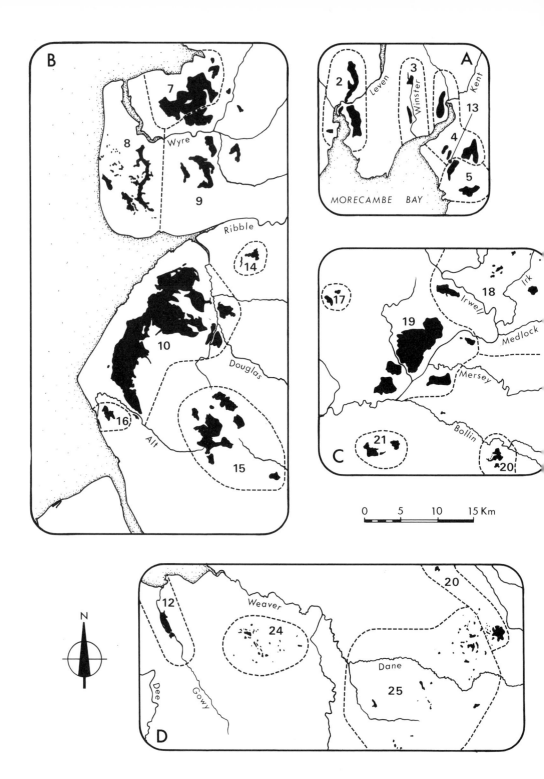

3. Winster Valley (Fig. 16.2A : 3) – a series of small raised bog and reedswamp peats in the lower Winster Valley from Wilson House (SD 4281) in the south to Swallow Mire (SD 4287) in the north; marginal agriculture, 7–12 m O.D.

4. Lyth Valley/Kent Estuary (Fig. 16.2A : 4) – a large raised mossland, Foulshaw Moss (SD 4682), formed between the Kent Estuary and low Carboniferous limestone fells to the west of Milnthorpe Sands; Arnside (SD 4678) and Silverdale Mosses (SD 4777) south of the estuary developed in alluvium-filled basins between Carboniferous limestone hills, and White Moss (SD 5076) and Hale Moss (SD 5077) in the upper reaches of the catchment of Leighton Beck; cutover raised bog, silviculture and marginal agriculture; 3–10 m O.D.

5. Myers Dike Catchment (Fig. 16.2A : 5) – two areas of raised moss and associated tarns in the drainage area of Myers Dike formed between Carboniferous limestone outcrops and coastal salt marshes; the larger of the two, Leighton Moss (SD 4875) is 200 ha in extent and classed as a National Wildfowl Refuge by the Royal Society for the Protection of Birds.

Group 2. Mosslands of Wyre and Fylde
6. Western Lune Estuary (Fig. 16.1 : 6) – approximately 4 square kilometres of raised mossland around Heysham Moss and Heaton-with-Oxcliffe (SD 4361) formed in a shallow basin at 5 m O.D. between the coastal boulder clays at Higher Heysham (30 m O.D.) and the low (15 m) glacial till ridge along the inner Lune Estuary Between Heaton and Overton; agricultural.

7. Over Wyre: Pilling Water Catchment (Fig. 16.2B : 7) – extensive areas of mossland in the catchment drained by Ridgy Pool Water, Pilling Water and the River Cocker at 5–10 m O.D. bounded on three sides by a chain of low (15 m) glacial till hills along the northern edge of the River Wyre floodplain; predominantly agricultural, except at Winmarleigh Moss (SD 444477) where a 60 ha Site of Special Scientific Interest represents the last remaining extensive area of active raised bog on the Lancashire coastal plain.

8. Fylde Lowlands (Fig. 16.2B : 8) – a ribbon development of fen-carr and reedswamp peat developed along the Main Dyke running from the River Wyre at Little Thornton (SD 357410) inland to Weeton-with-Preese (SD 3735) and the Main Drain/Wrea Brook catchment entering the Ribble Estuary east of Lytham (SD 3827); raised moss peatlands at Lytham Moss (SD 3430) and Great Marton Moss (SD 3431), partly warped, and scattered peat deposits around Marton Mere (SD 3435); agricultural/horticultural (3–12 m O.D.).

9. Lower Wyre Catchment (Fig. 16.2B : 9) – a series of raised moss and reedswamp deposits comprising Rawcliffe Moss (SD 4442) and around Garstang (SD 4945) in the northern catchment, and in the catchments of Thistleton Brook (SD 4137) and Woodplumpton Brook at Inskip Moss (SD 4539) and Myerscough (SD 4938); agricultural (5–10 m O.D.).

Group 3. Mosslands of south-west Lancashire in the Douglas–Alt hinterland
10. Mosslands of the Douglas–Alt hinterland (Fig. 16.2B : 10) – an extensive area of about eighty square kilometres between Tarleton (SD 4421) on the River Douglas in the north-east and Great Altcar (SD 3206) in the south-west consisting of a string of drained raised bogs, notably Tarleton Moss (SD 4321), Burscough Moss (SD 4414), Halsall Moss (SD 3511) and Downholland Moss (SD 3308). Also included in this series are Croston and Mawdsley Mosses (SD 4816, 4815) in the confluence of the Rivers Douglas and Yarrow, and Hoscar Moss (SD 4612).

Group 4. Mosslands of the lower Mersey Estuary
11. Halton Moss (Fig. 16.1 : 11, SJ 5684) – a small mossland developed in the catchment of Keckwith Brook between the River Mersey and the boulder clays overlying New Red Sandstones of Daresbury Hill.

12. Gowy Valley (Fig. 16.2D : 12) – reedswamp peatlands developed between and around the River Gowy and Mill Brook in Wimbolds Trafford (SJ 4372).

Type B. Peatlands of large inland basins generally more than 100 ha in size, developed over boulder clay, glacial sand and gravels and blown sand deposits at altitudes ranging from 20 to 200 m O.D.

Group 5 North Lancashire mosslands
13. Hawes Water (Fig. 16.2A : 13) – an inland lake basin with a sequence of deposits from freshwater muds and clay to fen and raised bog peats (SD 4776).

Group 6. Mosslands of south-central Lancashire
14. Farrington and Hoole Mosses (Fig. 16.2B : 14, SD 4923) – an irregular area of raised moss at 18–30 m O.D. developed over boulder clay deposits between the Douglas and Lostock Rivers.

15. Mosslands of the River Alt and Sankey Brook Catchments (Fig. 16.2B : 15) – a group of large basin mosses developed mainly over inland blown sands of the Shirdley Hill Series at altitudes ranging between 30 and 50 m O.D. Bickerstaffe Moss (SD 4302) and Simonswood Moss (SJ 4499) are reclaimed for agriculture while Holiday Moss, Rainford (SD 4901) and Parr Moss, St. Helens (SJ 5494) are tipped with industrial waste.

16. Moss Lake, Liverpool (Fig. 16.2B : 16) – raised bog over inland lake deposits (SJ 3591).

17. Red Moss, Horwich (Fig. 16.2C : 17, SD 6310) – a small area of mossland formed over boulder clays in a basin on the watershed between the headwater tributaries of the Rivers Yarrow and Croal; extensively cutover for peat and now used for the tipping of refuse; a classic palynological site.

Group 7. Mosslands of the upper Mersey Catchment
18. Mosslands of the Irwell–Irk–Medlock hinterland (Fig. 16.2C : 18) – a group

of ten to twelve raised bogs developed in basins overlying the boulder clay deposits between the Rivers Irwell, Irk and Medlock at altitudes between 90 and 120 m O.D. The three largest are Ashton Moss (SJ 9299) reclaimed for horticulture, White Moss (SD 8804), a golf course, and Linnyshaw Moss (SD 7005) reclaimed for agriculture.

19. Mosslands of the central Mersey Valley (Fig. 16.2C : 19) – large basin mosslands developed over boulder clays and Late-glacial flood gravels at 18–20 m O.D. The largest, Chat Moss (SJ 7096) is some 25 square kilometres in extent and separated from Holcroft Moss (SJ 6893) and Risley Moss (SJ 6691) by the Glaze Brook Channel. Trafford Moss (SJ 9678) lies between the Manchester Ship Canal and the River Mersey and is virtually covered by an industrial estate, while Carrington Moss (SJ 7491) occurs between the River Mersey and Sinderland Brook and has been reclaimed for agriculture after peat extraction.

20. Mosslands of the upper Bollin Catchment (Fig. 16.2C,D : 20) – two medium-sized raised bogs occupying shallow, saucer-shaped depressions in boulder clay and surrounded by higher land underlain by glacial sands and gravels; Lindow Moss (SJ 8281) is at approximately 70 m O.D. and Danes Moss (SJ 9071) is at 160 m O.D.

21. Mosslands of north-central Cheshire (Fig. 16.2C : 21) – Sink Moss (SJ 6783) and Gale Moss (SJ 6581) are formed in shallow depressions over boulder clays at 60 m O.D. on the watershed of Bradley Brook and Gale Brook draining northwards towards the Mersey, and Ardley Brook draining southwards to the Weaver catchment.

Group 8. Raised mosses of the upper Ribble Valley
(Fig. 16.1 : 22) – a number of medium-sized mosses (20 ha) in the upper Ribble catchment at altitudes of 150–200 m O.D. in Lancashire and Yorkshire; White Moss (SD 792546) on the watershed between the River Ribble and Tosside Beck is a good example of an active mire surrounded by drier moorland.

Type C. Mires formed in small basin sites and kettle-holes, generally less than 10 ha in extent and underlain by a complex of glacial till deposits.

Group 9. Basin mires of the south-east Lakeland drumlin fields
(Fig. 16.1 : 23) – small basin mires formed of fen-carr and reedswamp peats usually less than 10 ha, developed in shallow basins or along the lower slopes of the south-east Lakeland drumlin field between the River Kent and River Lune, as many as seventy per 100 square kilometres, yet only accounting for 2 per cent surface area cover. Two altitudinal series are recognisable – a *foothills series* of larger mires at altitudes of 100 to 200 m O.D., commonly associated with tarns, as for example, Eskrigg Tarn (SD 576888) and Terrybank Tarn (SD 593825), and a *lowland series* at altitudes of less than 50 m O.D. in the catchment of the River Beala, such as those between Higging Head and Chapel Hill (SD 5244810) and at Moss Side (SD 537823).

Group 10. Mires and meres of north Cheshire
(Fig. 16.2D : 24, 25) – small mosses developed in kettle-holes in the drift, often associated with underlying saliferous beds of Triassic age, seldom more than 3 ha in area. Two geographical series are recognisable – *Delamere Forest series* (2D : 24) – a group of over thirty mires, mostly more than one kilometre from the edge of saliferous outcrops; *Dane Catchment series* (2D : 25) – forty-odd mires, mainly in association with underlying saliferous beds of Upper and Lower Keuper strata.

Origins of the main mossland types

The general classification of Walker (1970) divides mire sites according to their origins in either coastal and estuarine sites, in large inland basins or in small inland basins. In all these situations the usual sequence of hydroseral development is: submerged macrophytes, floating-leaved macrophytes, reedswamp, fen, fen-carr, bog. This sequence produces a stratigraphical sequence commencing with basal clays and silts and continuing through nekron muds (gyttjas) containing coarse plant detritus to fen peat, fen-carr peat and finally raised-bog peat. The rates of succession vary according to a number of characteristics of the mire site, but especially the original water depth and the rate of shallowing by inwashed materials. In shallow estuarine lakes the transition from open water through to raised bog is short while in deeper inland lake basins the transitional time period may be more than four or five times longer. For example, at the coastal sites of Foulshaw Moss (Smith, 1959) and Ellerside Moss (Oldfield and Statham, 1963) the direct transition from open water to bog in the former and the gradual transition from open water to fen to bog in the latter took a mere 500 years, contrasting with the large inland basin of Moss Lake (Godwin, 1959) where the open water to bog transition took 2500 years and with the small inland basin of Skelsmergh Tarn (Walker, 1955) in which the open water to fen stage took 2000 years.

The date for the commencement of hydroseral development obviously varies according to the stability of the environment around the lake basin. A series of palynological and radiocarbon dating studies have enabled the categorisation of a number of mossland types according to the approximate date of the inception of peat formation and Table 16.1 summarises the types in relation to the Blytt–Sernander and Godwin Numerical Zonation schemes for the Late Weichselian and Flandrian periods with the citation of classic examples for the region.

The formation of present-day coastal and estuarine peatlands clearly depended upon the stability of the coastline of the north-west, especially the number and frequency of marine transgressions. The detailed work of Tooley (1978b) has summarised sea-level changes during the Flandrian period and from his analysis and data of earlier authors it is possible to define five major periods of origin when areas of open water, brackish at first, but then swelled by river flooding and converted to freshwater lakes, became isolated behind coastal deposits. In theory, and with further research it should be possible to relate all the peatlands

Table 16.1 A summary of the major mire types and their approximate dates of origin.

Years BP	Blytt–Sernander Zonation	Godwin numerical Zonation		Coastal and estuarine mosslands	Large inland basin mosslands	Small inland basin mosslands
0	Sub-Atlantic	VIII	Flandrian			
1000						
2000						
3000	Sub-Boreal	VIIb				
4000						
5000	Atlantic	VIIa				
6000						
7000	Boreal	VI / V				
8000						
9000	Pre-Boreal	IV				
10 000	Upper Dryas & Allerød	II & III	Late Weichselian			
11 000						
12 000	Lower Dryas	I				
13 000						

Mossland sites:

Coastal and estuarine mosslands — T5 T4 T3 T2 T1:
- Fylde & West Lune (T5)
- Pilling Water & Douglas–Alt (T4)
- Lakeland estuaries (T3)
- Lakeland valleys (T2)
- Silverdale Moss (T1)

Large inland basin mosslands — T1 T2:
- Hawes Water, Moss Lake, Chat Moss (A), Red Moss (T1)
- Chat Moss (B), Holcroft, Lindow, Danes Moss etc. (T2)

Small inland basin mosslands — T1 T2 T3:
- Skelsmergh Tarn, Bagmere & Flaxmere (T1)
- Abbots Moss, Hatchmere, etc. (T2)
- Oakmere (T3)

to the ten marine transgressions recognised by Tooley (1978b) – the Lytham I to X series – but at present it seems practicable to recognise just five periods of origin. At Silverdale Moss peat formation commenced in Flandrian IV prior to the Lytham I transgression and was interrupted by the Lytham IV transgression (6710–6160 years BP) to recommence after 5730 BP. This site is atypical of the coastal peatlands and early growth represents development in a large inland basin similar to that of Hawes Water. A second type, located in the middle sections of many of the South Lakeland valleys, originated in Flandrian IV after Lytham III (7600–7200 BP) and is typified by Ireland and Rusland Mosses in the Leven Valley. The mosslands at the river mouths and estuaries of South Lakeland constitute a third type which originated in Flandrian VIIa after Lytham IV (6710–6150 BP). Examples include Kirby Pool in the Duddon Estuary, Ellerside Moss in the Leven Estuary, Foulshaw Moss in the Kent Estuary and Storrs Moss in the Myers Dike Catchment. The vast expanses of mossland in the Pilling Water Catchment and the Douglas–Alt hinterland began their modern phase of development after the Lytham VI (5570–4900 BP) transgression during Flandrian VIIb and comprise a fourth type, while the more localised deposits of the Fylde Lowlands and the Western Lune Estuary probably originated after Lytham IX (1795–1370 BP) and in consequence, are only thin peats which were easily reclaimed for agriculture.

Large inland basin peatlands are commonly formed in valleys which have been glacially dammed or overdeepened, in extensive depressions in drift landscapes and in solution or subsidence hollows. The basins are often, shallow and saucer-shaped, with gentle gradients at their borders and the raised bogs formed within are greater than 100 hectares in extent. The deepest deposits are freshwater muds, clays and gyttjas generally at depths of 8–12 m and there is no evidence of phases of marine deposition. Many of the classic palynological studies in the region have been undertaken on the deposits of these large basins as, for example, at Hawes Water (Oldfield, 1960), Moss Lake (Godwin, 1959), Red Moss (Hibbert *et al.*, 1971) and Chat Moss (Birks, 1965a), to characterise a full sequence of deposits from Late Weichselian I to Flandrian VIIa, VIIb or VIII. In these localities, the basin deposits of Late Weichselian age varied in thickness from 40 cm at Moss Lake to 185 cm at Hawes Water, to be succeeded by Flandrian fen and raised-bog peats of maximum thickness of 760 cm and 695 cm at Hawes Water and Chat Moss respectively – two sites in which the surface peats had not been removed for turbary.

All these four sites are medium-sized basins between 10 ha and 20 ha in area and yet, the peat deposits in their vicinity cover a much greater surface area. This phenomenon has been explained in the research by Birks (1964, 1965b) by stratigraphical analyses which demonstrate the extension of peat growth from the central reservoir into subsidiary hollows or by the initiation of new, discrete peat deposits during Flandrian VIIa. Thus, the greater proportion of Chat Moss, all of Holcroft Moss and Carrington Moss began to form in the period 7100–5000 BP. Similarly, Lindow Moss and Danes Moss in the catchment of the River Bollin have basal peats of Flandrian VIIa overlying boulder clay and stratified sands and gravels (Tallis, 1973b), and it seems that similar origins may also obtain for the remainder of the peat deposits of large inland basins in the region. There are some suggestions that the basins in north Cheshire may have formed at the start of Flandrian VIIa by the subsidence of strata rich in saliferous beds which were rendered unstable by climatic worsening and the natural dissolution of rock salt.

Small basin mires, usually less than 3 ha in area, formed in the hollows of drumlin fields and in kettle-holes in superficial drift deposits, are found in several areas of North-west England. In South Lakeland, notably to the south and east of Kendal, the small basin mires are mainly developed within an extensive drumlin field at altitudes ranging from 30 to 200 m. The drumlins show a pronounced NE–SW orientation (mean direction 203°) and the stony, grey-brown, silty clayloam till was most probably derived from the Silurian greywackes, slates and flagstones of the Howgill Fells (Furness and King, 1972). The wasting of detached ice blocks in the hollows between the larger drumlins produced a number of freshwater bodies which became gradually terrestrialised through the process of hydroseral development to form reedswamp and fen-carr peats. Many of the deposits are in the depth range of 0.6 m to 2 m in the low-lying open basins to the south-west of the drumlin field, but at higher altitudes (100–200 m) in the north-east, several basins have apparently been overdeepened by the persistence of ice blocks, or flooded after eighteenth century peat cutting and drainage, and in many, open water persists. In such situations deposits are usually in excess of 4 m while at Skelsmergh Tarn (SD 5396) 6 m of basal clays, muds and succeeding peats reveal a detailed sequence from Late Weichselian I through to Flandrian VIIb (Walker, 1955).

An understanding of the evolution and development of the mires and meres of small basins in north Cheshire has greatly benefited from the works of Birks (1965a), Tallis and Birks (1965) and Tallis (1973a). The latter author has demonstrated that although stratified sands and gravels, deposited after ice-melt and stagnation, cover only 33 per cent of a drift-covered surface dominated by unstratified boulder clays, 66 per cent of the moss sites overlie the former deposits. To the eastern areas of the Dane Catchment (Fig. 16.2D : 25), many of the mire sites occupy kettle-holes formed by the embedding and melting of detached ice blocks near the decaying ice margins of the Pennines. A similar origin can be postulated for the mires of the Delamere Forest region (Fig. 16.2D : 24), with ice blocks detaching from the decaying glacier margins of the Mid-Cheshire Ridge, becoming buried in the deltaic sands of a glacial lake and subsequently melting to form kettle-hollows. Such hollows might be expected to contain deposits of Late-glacial age but as yet only two sites, Bagmere (Birks, 1965a) and Flaxmere (Tallis, 1973b), have yielded clear evidence. In the greater number of stratigraphical studies, basal deposits of Flandrian IV have been recorded, though it may be possible that Late-glacial deposits underlie horizons impenetrable by normal borers. Other basins have Flandrian VI basal deposits and a smaller number, Flandrian VIIb. Clearly this diversity of dates for the commencement of peat formation demands an alternative hypothesis to the kettle-hollow theory. Several authors have thus suggested that the natural and more modern, industrially-induced subsidence of underlying saliferous beds of Triassic age have produced steep-sided or stepped, linear or isodiametric hollows at various times of major climatic shift in the Flandrian period.

Tallis (1973b) has calculated that in an area underlain by 22 per cent of saliferous beds, 35 per cent of the mire sites overlie or fall within one kilometre of the outcrop while the remaining 65 per cent overlie other rock types. It thus seems that a good proportion of the mires of north Cheshire may have originated in hollows formed by the solution of underlying rock salt deposits and the subsequent collapse of overlying strata.

The mosslands today: their conservation and research potential

In the Tudor period few incursions had been made into the mosslands of the north-west and in some parishes they accounted for as much as 75 per cent of the total land surface. Rodgers (1955) has recorded that 74 per cent (5700 ha) of the Ormskirk group of townships was waste mossland, the Leigh group had 70 per cent (5200 ha) of similar terrain, while the figure for the Pilling group was 55 per cent (2285 ha). This latter area was formerly a centre of the salt manufacturing industry and the earliest controlled peat cutting in the region was that in the thirteenth century of Pilling and Cockerham Mosses for fuel for this industry (Taylor, 1975). Most parishes were allocated areas of mossland for the common right of turbary in the parliamentary enclosure acts of the mid-eighteenth century and peat was the universal fuel of the Fylde and West Derby districts. Some of the earliest mossland improvement followed enclosure at Halsall, near Ormskirk, in the period 1750–60 when approximately 400 ha were turned into productive agricultural land after drainage, peat cutting, burning and seeding with two crops of rye. This method was not completely successful, especially on the deeper peats. At Rainford Moss, for example, the continual outbreak of fires and soil impoverishment led to the abandonment of the reclamation scheme in 1787 after only seven years. In other localities, drainage and flooding posed problems and much of the open water areas now known as Martin Mere was developed in 1786 with the flooding of a good proportion of the 1450 ha that had been reclaimed. At Chat Moss by 1849 approximately 800 ha had been successfully reclaimed by cutting metre-deep main drains and shallower cross-drains, burning, ploughing and paring and by adding marl or a mixture of lime and salt to break up the surface (Fletcher, 1982).

From sequences of Ordnance Survey maps, aerial photographs and field surveys Riggall (1980) has described the general process of mossland reclamation and changing land use from 1845 onwards. His data for the South Lakeland and Lancashire lowlands, supplemented and modified by personal fieldwork are presented in Table 16.2. The data sets for the two regions are contrasting in both the original and existing total areas of mossland and the major periods of reclamation. In Lancashire, it may be seen that the main phase of agricultural reclamation was from 1845 to 1880 in response to the industrial development and population growth in the region, whereas the greatest effects on the South Lakeland mosses were due to peat cutting and drainage in the period 1900 to 1940, followed by afforestation between 1950 and 1970. Today, there are just over 150 ha of

Table 16.2 The progressive reclamation of the lowland mosslands, 1845–1978 (area in ha / %).

Period	1845–60	1888–97	1945–51	1970–5	1978
South Lakeland					
Moss	1783/100	1328/75	494/28	175/10	156/9
Peat cuttings	–	–	270/12	263/15	204/11
Drained moss	–	7/0.5	435/24	61/2	43/2
Woodland	–	87/5	177/10	796/45	810/46
Agriculture	–	361/19	465/25	483/27	565/3
Urban	–		5/0.5	5/0.5	5/0.5
Lancashire lowlands					
Moss	4229/100	1005/24	247/6	106/2	11/0.5
Peat cuttings	–	40/1	382/9	296/7	207/5
Drained moss	–	60/1	178/4	201/5	228/5
Woodland	–	100/2	180/4	213/5	311/7
Agriculture	–	3010/72	3130/74	3241/7	3297/78
Urban	–	14/0.5	112/3	172/4	175/4
Chat Moss complex					
Moss	1276/94	486/36	100/8	46/3	3/0
Peat cuttings	75/6	40/3	340/25	246/18	205/15
Drained moss	–	–	50/4	71/5	79/6
Woodland	–	26/2	48/4	106/8	181/13
Agriculture	–	799/59	773/58	812/60	813/60
Urban	–	–	40/3	70/5	70/5

Based on the data of J. Riggall, Nature Conservancy Council, modified and supplemented by personal fieldwork.

natural, actively growing raised bog in South Lakeland and a meagre 10 ha in the Lancashire lowlands. Consequently such prestigious sites as Roundsea Mosses (SD 3382) in the Leven Estuary, Winmarleigh Moss (SD 4447) in the Pilling Catchment, White Moss (SD 7954) in the upper Ribble valley and Abbots Moss (SJ 6069) in north Cheshire are designated as Sites of Special Scientific Interest (Ratcliffe, 1977). Other sites which have been partly reclaimed by drainage and subsequently flooded, such as Leighton (Storrs) Moss (SD 4875) and Martin Mere (SD 4214) are of scientific interest primarily for their ornithological value, while partly drained and cutover mosses cn the urban fringes of Greater Manchester and Merseyside have been conserved as local nature reserves, for example, Risley Moss (SJ 6691) and Sutton (Parr) Mosses (SJ 5494). At all these sites there is an urgent need for conservation, and detailed hydrological and applied geomorphological studies are needed to ensure their continued existence.

Peat cutting and drainage operations are still being undertaken in the Chat Moss complex of mosses. In 1983, the North West Water Authority commenced a major drainage scheme in the Moss Brook Catchment, a tributary of Glaze Brook which drains much of the Chat Moss area. Three special drainage areas, in which existing channels are being resectioned and regraded and new channels excavated have been designated for the Astley part of the moss. These major operations provide opportunities for applied geomorphological research on methods of mossland drainage and reclamation and the monitoring of resultant land use changes.

In terms of pure geomorphological and palaeo-ecological research potential, the mosslands of the north-west still offer several opportunities for investigations into their origins and development. The following topics seem worthy of attention and hopefully, some answers will be forthcoming in the next decade or so:

the correlation of coastal and estuarine peatlands with the Lytham I to X marine transgressions;
the origins of mosslands of large inland basins, especially those at mid-altitudes (150–250 m O.D.) in the foothills of the Pennines;
the origins of the north Cheshire mires in relation to subsidence of underlying saliferous strata;
the origins of the mires in the south-east Lakeland drumlin field.

Erosion of blanket peat in the southern Pennines: new light on an old problem

Introduction

More than one million hectares (10 000 km^2) of the land surface of Great Britain is peat-covered (Taylor, 1983). Much of this peat cover is in the uplands, where it lies across the major watersheds in areas of subdued relief, thus modifying the patterns of stream flow and erosion in the headwaters of the rivers originating there. Such blanket peat is a major geomorphological influence.

A series of early ecological investigations in the uplands (for example: Moss, 1904, 1913; Pethybridge and Praeger, 1904–5; Lewis, 1908; Pearsall, 1941; Osvald, 1949) demonstrated clearly the frequency of erosion in these blanket peats, and the interest aroused was cogently expressed by Verona Conway in a classic paper in 1954:

Few people can see severe peat erosion for the first time without a sense of astonishment, and the curious-minded will wonder why there is any peat still left when it is evidently exposed to such destructive forces. This question leads to others, such as the age of the peat beds, the date when erosion set in, the cause of erosion, the rate of it, and the nature of the surface which will be left when all the peat has vanished. . . . Those with an interest in problems of land-use will ask whether peat erosion can and should be checked or whether it should be hastened or modified in some way to give a surface with at least some small degree of productivity.

In the decade following the publication of Conway's paper, a series of investigations was carried out on the morphological features exhibited by eroding peats (Johnson, 1957a; Bower, 1960a,b; Radley, 1962), and on the possible causes of their erosion (Bower, 1962; Radley, 1962; Barnes, 1963; Johnson and Dunham, 1963; Tallis, 1964). These studies emphasised the need for further intensive work on the dynamics of peat build-up and degradation in the uplands, and over the ensuing two decades generated a three-pronged approach to the problems of peat erosion: pollen–stratigraphic studies of peat profiles, to try and deduce the time of onset of peat erosion (Tallis, 1965; Bostock, 1980); detailed monitoring of rates of peat erosion (Crisp, 1966; Imeson, 1971; Tallis, 1973a, 1981a); and studies of the hydrological properties of blanket peat and of the action of environmental factors on it (Chapman, 1965; Crisp, 1966; Ingram, 1967; Tallis, 1973a; Crisp and Robson, 1979; Burt and Gardiner, 1981). The Pennines have figured prominently in this research programme, because 'there are probably greater expanses of deep and heavily eroded peat than can be found in any other mountain region of the British Isles' (Conway, 1954). In the last seven years, increasing concern over the re-vegetation of bare peat, both that of long standing and that resulting recently from accidental moorland fires and footpath erosion (Phillips *et al.*, 1981), has further contributed to the search for answers to the problems posed by upland peats. To many of the questions put by Conway we can now give partial answers.

The extent of erosion

Anderson and Tallis (1981) attempt to evaluate the extent of peat erosion on the moorlands of the Peak National Park (which includes all the major areas of erosion in the southern Pennines). They present the results of a survey of eight moorland parishes (comprising 54 per cent of the total moorland area within the Park, but excluding, most notably, the whole of the Kinder plateau). The survey was based partly on field mapping and partly on examination of aerial photographs. Peat areas were categorised as:

1. Uneroded – with few or no drainage gullies visible on the aerial photographs;

2. Type I dissection (Bower, 1960a) – closely gullied, deep peat areas on flat or nearly flat ground, with an intricate network of branched drainage gullies and bare peat pools encircling discrete dry hummocks;

3. Type II dissection (Bower, 1960a) – peat on sloping ground with a system of sparsely branched drainage gullies aligned nearly parallel to each other.

The representation of these three categories in the eight parishes was:

Uneroded peat	39.4 km^2 (26% of total peat area)
Type I dissection	19.6 km^2 (13%)
Type II dissection	92.3 km^2 (61%)

In addition, areas of bare or partly bare peat were mapped and measured. These totalled *c.*17 km^2 (11% of the total peat area). Much of the bare or partly bare peat was associated with Type I dissection, but substantial discrete areas of bare peat are known to have resulted from accidental moorland fires within the last 40 years (Tallis, 1981b).

Anderson and Tallis (1981) found that only 19 per cent of the blanket bog margin (abutting on to non-peat-forming vegetation) in these eight parishes was uneroded; 43 per cent was classified as 'slightly eroded' and 38 per cent as 'severely eroded'.

Eroding peat is not uniformly distributed on the Peak District moorlands. It is rather commoner in the west than in the east (Table 17.1), and much commoner at higher altitudes; Type I dissection, in particular, is only widespread above 550 m altitude (Fig. 17.1). The two groups of parishes listed in Table 17.1, lying to the west and east of the main north–south water-divide of the southern Pennines, have similar areas of high ground (above 457 m altitude) and similar mean annual rainfalls, so it is unlikely that the prevalence of eroding peat in the westerly parishes is directly associated with a more severe climate. Topography could be important, as the westerly parishes have considerable areas of high flat ground flanked by steep slopes, while in the easterly parishes there are greater expanses of gently rolling moorland. There is a clear difference in the extent of heather moor in the two

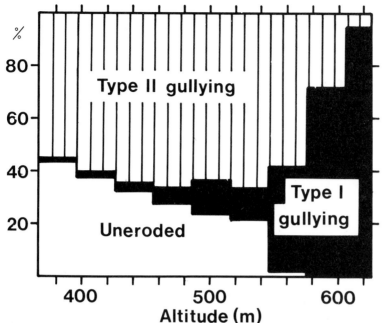

Fig. 17.1 The proportion of peat-covered ground at different altitudes in eight moorland parishes of the Peak District that is uneroded or affected by Type I or Type II gullying (from Anderson and Tallis, 1981). The data are plotted for 100-ft (30.5-m) altitudinal intervals.

groups of parishes (four times as much in the easterly group), and this could be interpreted as indicating that land management influences the prevalence of eroding peat. Conversely, however, the prevalence of eroding ground could itself determine the extent of heather moor.

Table 17.1 Topographic and peat characteristics of two groups of south Pennine parishes (data from Anderson and Tallis, 1981).
Westerly group: Charlesworth, Holmfirth, Saddleworth, Tintwistle.
Easterly group: Bradfield, Dunford, Hope Woodlands, Langsett.

	Westerly group	*Easterly group*
Total area (km²)	126.8	152.9
Land above 457 m altitude (km²)	66.9	64.0
Steeply sloping ground (> 10°) (km²)	31.1	21.8
Mean annual rainfall (mm)	1436	1384
Heather moorland (km²)	13.4	55.2
Uneroded peat (km²)	14.5 (22%)	24.9 (29%)
Type I dissection (km²)	10.2 (16%)	9.4 (11%)
Type II dissection (km²)	41.1 (62%)	51.3 (60%)
Bare or partly bare peat (km²)	12.3 (19%)	4.5 (5%)

The overall picture, nevertheless, is in conformity with Margaret Bower's pioneer survey (1960b) of the distribution of eroded peat in the Pennines as a whole. She found that the areal extent of erosion increased with altitude; that the most advanced stages of erosion occurred on the highest moors; that areas of Type I dissection were restricted to deep peat on flat ground; and that the proportion of ground affected by Type I dissection was greatest on the highest moors. Any 'explanation' of the severity of peat erosion in the southern Pennines must accordingly take into account not only the special character of the southern Pennine situation, but also its conformity in general terms with the whole Pennine area.

The causes of peat erosion

Type I and Type II dissection systems in the southern Pennines occupy separate and distinct topographic situations, which in other upland areas of Britain are characterised by pool-and-hummock complexes and by bog-slope communities, respectively. The latter show little microtopographic differentiation of the bog surface, but pool-and-hummock complexes (or hummock–hollow complexes) comprise a well-defined mosaic of drier hummocks and wetter *Sphagnum*-filled hollows or pools (Tallis, 1969). Standing water is usually evident in a pool-and-hummock area, even in dry weather, but is seldom apparent on the bog slope, where water drains away rather quickly over the bog surface or through the more permeable superficial peat layers, sometimes along well-marked 'water tracks' (Ingram, 1967), towards drainage streams issuing from the peat margins. There is fairly general agreement that Type I dissection systems represent eroded pool-and-hummock complexes, where excess water has been drawn off by a general lowering of the water table, and that on the bog slope Type II gullies form by gradual incision of the surface run-off waters into the peat mass. Peat erosion thus represents a major change in the water relations of the peat mass; and as both Type I and Type II dissection systems are generally (but not invariably – e.g. Kinder) present within the same blanket peat catchment, and are linked to head-waters of the same main streams flowing from the catchment, it is logical to assume that these changed water relations derive from geomorphological processes affecting the whole catchment. Several different processes have been suggested.

One suggestion is based on evidence derived from a comparison of successive editions of the 1:63 360 Ordnance Survey maps of the Peak District and later aerial photographs (Moss, 1913; Johnson, 1957a; Mayfield and Pearson, 1972; Ovenden and Gregory, 1980). The maps and photographs show an apparent extension head-wards of certain streams draining the peat blanket by 1.2 km between 1830 and the 1870s, by a further 0.4 km by 1912, and by 0.2–0.4 km between 1912 and 1953. Such a rapid headward extension of the main drainage streams (probably along channels that were in existence before blanket peat spread over the catchment) could have led eventually to a drawing off of water from a pool-and-hummock complex on the watershed, and a gradual lowering of the water table throughout the blanket bog. The lowered water tables then resulted in increased run-off of water over the bog surface, and the gradual incision of drainage gullies into the peat of the bog slope.

The basic evidence on which this scheme is founded is suspect, however, because of uncertainties about the accuracy of the surveying for the early Ordnance Survey maps in moorland country, and the different criteria used for mapping stream courses (Burt and Gardiner, 1982). Moreover, even if such an extension of the peat streams headwards had occurred, it is still uncertain whether an entirely natural process is to be invoked or whether a change in the erosive capability of the drainage network is required. The lack of a mineral load in water flowing over a peat mass may give peat streams a low erosive capability (Radley, 1962), particularly where the vegetation cover is intact, and accordingly water erosion of peat (whether by actual solution or by the transport of low-density peat particles) may only be significant when bare peat is exposed (Radley, 1962; Johnson and Dunham, 1963).

Possible mechanisms for exposing bare peat are varied, and there is no agreement as to their relative importance. Bower (1962), studying the incidence of peat erosion in the Pennines as a whole, found clear relationships between the severity of erosion and altitude, and suggested that climate and topography were the major factors responsible for erosion. Later workers have increasingly favoured a biotic 'explanation' of peat erosion, seeing regular burning (Radley, 1962), heavy sheep grazing (Shimwell, 1974; Evans, 1977), artificial drainage, and particularly air pollution (Tallis, 1964, 1965; Lee, 1981), as potent mechanisms for disrupting the bog vegetation. Moreover, the high intensity of these biotic factors, and specially air pollution, in the southern Pennines affords a logically satisfying reason for the unusually high incidence of peat erosion there.

Various workers have noted the occurrence of subterranean drainage channels either within or below the peat mass in upland areas, and have suggested that Type II gullies might originate by the caving-in of the roofs (Pearsall, 1950). Johnson and Dunham (1963) proposed that the peat blanket develops an internal drainage system analogous to that in limestone regions, but it is difficult to see how this could arise without far greater vertical movements of water taking place in a peat mass than are known to occur. The significance of these underground 'pipes' and 'channels' remains obscure, but it is at least possible that some of them are secondary features associated with drainage, brought into being by water seeping from vertical or near-vertical exposed peat faces. Such lateral water seepage may occur at the peat base, or along layers of more permeable *Sphagnum* remains within the peat mass.

The high water content (over 90%) and low cohesion of an undrained peat mass make it an inherently unstable system, and instability must normally become more pronounced as peat continues to accumulate. The well-documented occurrences of bog-flows and bog-bursts in the past (e.g. Colhoun *et al.*, 1965), when a slurry of semi-liquid peat and massive peat blocks has gushed forth from localised points around the bog margin (e.g. Crisp *et al.*, 1964) is telling evidence of this instability. Conway (1954) believed that instability in southern Pennine blanket peats was accentuated by climatic change around 600 BC, leading to the rapid build-up of uncompacted *Sphagnum* peat above more humified peat, and was then relieved by bog-bursts around the margins of the peat mass. Johnson (1957a) also suggested that bog-bursts were a feature of the 'post-mature' stage of peat accumulation, when erosion of the unstable water-charged peat mass became inevitable, and he claimed that bursts were widespread in the southern Pennines. Later workers make no such claims, and attach relatively little importance to bog-bursting as a general mechanism of blanket peat erosion.

These various theories of peat erosion – what we may call the fluvial (stream extension), biotic (burning, grazing and pollution), karstic (subterranean drainage) and catastrophic (bog-bursting) hypotheses – are not necessarily mutually exclusive. Weighing up their relative merits is not easy, but clearly in order to do so it is vital to know the time of onset of peat erosion in specific localities, and the rate at which erosion is currently proceeding. Evidence on these points is just becoming available.

Rates of peat erosion

Bare peat surfaces are clearly unstable at the present day. Abundant evidence of this instability can be seen in the field. Finely-divided peat, washed downslope, accumulates against objects in its path, such as walls and wire fences; even clumps of bilberry may be partly buried by this peat wash; and residual patches of snow, persisting after snow-melting in spring, quickly acquire a superficial layer of peat particles. In areas of scour, established individual plants come to be raised up on peat pedestals several centimetres high, as the surrounding bare peat surface is gradually lowered. After heavy rain and during snow-melt, turbid water rushes down the drainage gullies, carrying away peat in suspension and solution. In dry weather the superficial layers of peat shrink and crack, and may blow away as dust, while the cracks extend downwards. In frosty weather these superficial layers take on a spongy 'puffed-up' texture, as water freezes and expands in the pore spaces and then drains away on thawing. Partial thawing after a heavy frost produces a highly unstable slurry of water-saturated surface peat overlying still-frozen peat lower down. Young seedlings colonising bare peat are worked upwards by frost-heaving in winter, and the roots may be pushed completely clear of the peat by spring – death generally ensues.

However, bare peat is not irrevocably unstable. Recent experiments at a number of sites in the Peak District show that a plant cover can be gradually re-established even on deep peat, but that it is probably necessary to apply fertiliser (N, P, K and lime), to augment the natural seed supply, and to fence to keep out sheet (Tallis and Yalden, 1983). Situations can be observed in the field where recolonisation is occurring naturally – for example, on the floors of non-functional drainage ditches (Mayfield and Pearson, 1972) and blocked-up gullies. Even the steeply-sloping bare peat sides of gullies at lower altitudes may show negligible peat losses, at least in the short term, and gradually acquire patches of colonising moss: for example, Wet Moss, Rossendale (SD 763185) at 385 m altitude, where no overall erosion was measured over a 12-month period (Tallis, 1981a).

Table 17.2 Rates of peat erosion: yearly changes in level of bare peat surfaces (surface-lowering) at several sites in the southern Pennines. Each value (mm yr^{-1}) is the mean of a number of closely-spaced point records. Data of Tallis (1981a) and Tallis and Yalden (1983).

Topographic situation	Site	Surface lowering (mm)	No. of points	Period of measurement
(1) *Gently-sloping expanses of bare peat*				
Exposed peat margin	Holme Moss	34	60	2 yr
Sheltered peat margin	Snake Pass	5	19	1 yr
Exposed peat mound	Holme Moss	33	20	2 yr
Exposed peat mound	Cabin Clough	18	20	2 yr
Sheltered hagg	Doctor's Gate	10	20	2 yr
(2) *Steeply-sloping gully sides*				
	Snake Pass	8	97	1 yr
(3) *Floors of drainage gullies*				
	Snake Pass & Holme Moss	6	30	1 yr
(4) *Seepage lines at heads of drainage gullies*				
	Snake Pass & Holme Moss	8	17	1 yr

Grid references and altitudes of sites:
Holme Moss: SE 0903; 530 m
Snake Pass: SK 0992; 490 m
Cabin Clough: SK 074927; 425 m

Yet until recently there have been almost no quantitative assessments of the rate of peat erosion (Tallis, 1981a). The measurements in Table 17.2 refer to surface-lowering of bare peat surfaces in the southern Pennines, calculated as the mean of the individual changes in level at a number of closely-spaced points over a specified period of time using either 'erosion straws' (Tallis, 1981a) – not a wholly reliable method – or 'erosion transects' (Tallis and Yalden, 1983).

The values for rate of erosion range from 5 to 34 mm yr^{-1}, with the highest values recorded from gently-sloping expanses of bare peat in exposed situations. The values for the floors of drainage gullies (situations where the peat is compacted, and constantly wet for much of the year) and for seepage lines give an approximation to losses of peat by solution in flowing water.

There measurements can be used to derive an approximate time-scale for some of the erosion features visible in the southern Pennines today. An annual surface lowering of 3 cm at the peat margins (as at Holme Moss), with a slope of 25–30°, represents a horizontal retraction of the peat edge by 5–6 cm each year; as the peat edge is in places up to 10 m back from the break of slope which presumably limited lateral extension in the past, a recession of this magnitude indicates that erosion, at its current rate, must have been proceeding for about 200 years. The middle and upper reaches of many of the Type II gullies within the peat blanket are incised to a depth of 1.0–1.5 m; if these are deepened mainly by solution of the peat on the gully floor, which is currently proceeding at a rate of 6–8 mm per year, the gullies must represent 150–250 years of solution. A very similar estimate of the duration of existence of one of these gullies (200–250 years) was reached by Tallis (1973a) from a consideration of the total amount of solid material carried down in suspension each year in the gully.

Some major drainage gullies are much more deeply incised in their lower reaches (to 3.5–4.5 m) and could, on the same reasoning, be 500–750 years old at least.

In a few places in the southern Pennines, the approximate age of the gullies is known, and these examples can be used to derive an independent estimate of the rate of incision. Where the bog surface has been accidentally burned, to expose considerable expanses of bare peat, a superimposed gully system quickly develops, and the age of this system is the time elapsing since the fire. Two areas have been investigated cursorily, Coldharbour Moor (SK 0792), probably burnt around 1890, and Robinson's Moss (SK 0499), almost certainly burnt in 1947. The depth of incision of gullies on the bare peat slopes here was measured 20, 40 and 60 m downstream from the gully heads, and compared with similar measurements from gullies in unburnt areas (nearby on Robinson's Moss and on the Snake Pass). The mean depths of incision were:

Robinson's Moss, burnt 1947:	24 cm
unburnt:	167 cm
Coldharbour Moor, burnt *c.*1890:	82 cm
Snake Pass, unburnt:	129 cm

The two burnt gullies give mean rates of incision of *c.*7 mm and 9 mm per year. On this basis, the two unburnt gullies are about 160 and 200 years old.

There are a number of flaws in calculations of this sort. Erosion may now be proceeding at a more rapid rate than at any time previously, because of the increased intensity of biotic pressures on the bog vegetation: the change from a mossy dwarf-shrub-dominated vegetation to one with almost no mosses and a predominance of tussock-forming cyperaceous plants, under the joint influences of heavy sheep grazing and high air pollution, has almost certainly led in the last 200 years to a bog surface in the southern Pennines with a lower inherent resistance to erosion. Secondly, it is clear from recent work that the bulk of material lost from an upland catchment (80% or more – Crisp and Robson, 1979), whether it be peat or soil, is removed during a small number of discrete, high-intensity, short-duration episodes of erosion, perpetrated usually by thunderstorms and by snow-melt (Tallis, 1973a; Harvey, 1974; Crisp and Robson, 1979; Burt and Gardiner, 1981). There were seven such episodes at Featherbed Moss, Derbyshire, in 1970–1, and twelve at Howgill Fells, Cumbria, in 1971–2. It is probable that the number of these major erosion-inducing events varies quite widely from year to year, and that, accordingly, measurements of erosion rates over a period of 1–3 years only are strongly influenced by chance (an unusually mild/severe winter, a high/low incidence of thunderstorms, etc.). Thirdly, certain abnormal weather events, of exceptional severity but with a low probability of occurrence, result in exceptionally high erosion. For example, intense localised rainfall in the northern Pennines on 6 July 1963 led to two peat-slides on Meldon Hill, Cumbria, in which it is estimated that nearly 850 tonnes dry weight of peat was removed from an area of less than 0.01 km^2 (Crisp *et al.*, 1964). This value is similar to the total annual loss per 1 km^2 on Featherbed Moss, Derbyshire (Tallis, 1973a).

The onset of erosion

The earliest reliable descriptions of the south Pennine moorlands, from the beginning of the last century, refer to gullies, groughs and deep channels crossing the moors in all directions (Mayfield and Pearson, 1972), so it is clear that the more deeply hagged areas have been in existence for at least 170 years. Various workers (Bower, 1960a; Radley, 1962; Johnson and Dunham, 1963; Mayfield and Pearson, 1972) have mentioned the possible use of man-made features of the moorlands (drainage ditches, boundary trenches, peat cuttings) to date erosion gullies, but few unambiguous examples have been found. On Tintwistle High Moor in the southern Pennines (SK 0399), a disused pipeline runs across the moorland for about 1 km, from Robinson's Spring westwards towards a small, now empty, reservoir on Tintwistle Low Moor; the pipeline and reservoir were probably built round 1850, as part of the Longdendale Waterworks scheme. The pipeline crosses the upper reaches of all the gullies on the south flank of the moor, and is carried across on elaborately built drystone culverts up to 1.5 m high. The culverts are built on the mineral floors of the gullies, and water flowing down the gullies originally passed through each culvert along a conduit pipe (Fig. 17.2). If, as seems likely, the conduit pipe was positioned just above the level of the gully floor, then the gullies must already have been incised to a depth of at least 1 m by 1850.

More tenuous evidence of a much earlier onset of erosion comes from the use of terms of Norse origin (hagg, grough, grain) for certain erosion features. It could be argued that these features were already apparent when Norse colonists settled the southern Pennine uplands in the tenth century AD (Bower, 1960a). If the time-scale is as long as this, then clearly documentary evidence is of little use, and pollen–stratigraphic data afford the best means of deducing the time of onset of peat erosion.

The pollen–stratigraphic record

The nature of the evidence

The pollen–stratigraphic record may be defined as the sum of the characteristics of frequency, arrangement and state of preservation of organic remains in a peat profile: both micro-remains (particularly pollen grains) and macro-remains (leaves, stems, seeds, etc.) are thus included. At the actual site of erosion, the

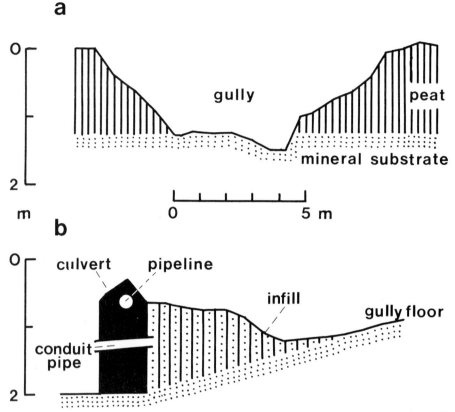

Fig. 17.2 The disused pipeline on Tintwistle High Moor: (a) cross-section of a gully
immeditely above the culvert; (b) longitudinal profile of the culvert and gully. The conduit
pipe is now blocked by infill material ponded up behind the culvert since its construction.

pollen–stratigraphic record has, by definition, been lost, so that almost invariably
evidence about peat erosion has to be inferred from *intact* peat profiles from
sites immediately adjacent to a gully or hagg. Such sites are nearly always better
drained as a result of erosion, so ideally these profiles should be compared with
undrained profiles from nearby uneroded areas. Improved drainage of the surface
peat layers in the southern Pennines today often leads to an increased abundance
of plants such as *Empetrum nigrum* and *Vaccinium myrtillus*, or even (in the past)
the moss *Racomitrium lanuginosum*, while other plants intolerant of lowered
water tables, notably certain species of *Sphagnum*, die out. Improved drainage
also leads to an increased rate of decay of organic material, and hence to more
humified peat that accumulates more slowly. Such effects, however, are not specific
to eroding sites, and similar changes can be produced in an uneroded peat mass
by a prolonged period of drier climate. Distinguishing between climate-induced
and erosion-induced changes is not easy; indeed, it is rarely possible from a single
profile alone. The best possibilities for discrimination are afforded by a comparison
of a series of peat profiles from different topographic situations and with varying

types of erosion, but within the same general area. A particular climatic effect
might be expected to show up in all, or nearly all, profiles, but an erosion-induced
effect to be perhaps more site-specific.

Table 17.3 Pollen–stratigraphic horizons in pollen diagrams from the southern Pennines.
  (a) Horizons recognised by Tallis (1964) and Tallis and Switsur (1973); the original
  radiocarbon dates have been calibrated using the tables in Klein *et al.*, (1982).
  (b) Equivalence of these horizons with those used by Conway (1954).

(a)	Horizon	Approx. date (yr BP)	Pollen–stratigraphic features
	E	950	Marked rise in herbaceous pollen (to > 60% TLP*) towards top of pollen diagram; *Plantago* values rise sharply to > 10% APC*, while *Alnus* and *Corylus* values fall.
	D	1550	Upper limit of a period of increased herbaceous (50–60% TLP) and *Plantago* (5–10% APC) pollen, and separated from later horizon E by a period of lowered herbaceous and *Plantago* pollen values.
	C	2240	Lower limit of period of increased herbaceous and *Plantago* pollen.
	B	2800	Junction of upper peat (uncompacted, *Sphagnum*-rich) and lower peat (humified, compacted). High values of *Empetrum* pollen (2–5% TLP) typically occur immediately below horizon B.
	Ulm.	5700	The Elm Decline: *Ulmus* pollen declines abruptly from values of 5–7% APC, and grains of *Plantago* often occur for the first time.

(b) The Ulm. horizon is also recognised by Conway (1954), while her horizon M is closely
  equivalent to horizon E as defined above. As Conway does not show *Plantago* values
  on her pollen diagrams, direct positioning of horizons D and C is difficult, but can
  be done approximately using the *Pteridium* curve instead. Conway recognises two
  horizons (recurrence surfaces) where uncompacted peat overlies humified peat (a feature
  of horizon B above), and suggests that the upper of these (C3) represents the
  Grenzhorizont at *c.*600 BC (which is of similar date to horizon B). A much more consistent pattern of peat accumulation rates is produced, however, if Conway's lower
  horizon (C2) is equated with horizon B, and this is the equivalence adopted in Fig. 17.7.

  It is not easy to relate the pollen–stratigraphic horizons above to those employed
  by Hicks (1971), and her radiocarbon dates do not help. Accordingly, Hicks's pollen
  diagrams from the East Moors were not considered for Fig. 17.7.

* TLP = total land pollen; APC = total tree pollen + *Corylus*

Although vegetation changes consequent upon increased drainage may sometimes be registered in the pollen record (by, for example, increased abundance of
Empetrum and *Vaccinium* pollen – Tallis, 1965), it is the character of the macro-
remains that is usually a better guide. The pollen record, summarised in pollen
diagrams, then forms the template and time-scale for comparing macro-remain
profiles. Certain distinctive sequences of change have been found in nearly all pollen

diagrams from the southern Pennine moorlands (Tallis, 1964; Tallis and Switsur, 1973; Bartley, 1975), and can be assigned dates based on radiocarbon and other evidence (Hicks, 1971; Tallis and Switsur, 1973; Bartley, 1975; Livett *et al.*, 1979). The more useful of the horizons recognised by Tallis (1964) are the B, C, D, E and Ulm. horizons; the pollen–stratigraphic characteristics of these are summarised in Table 17.3, where an attempt is made to relate these horizons to those of Conway (1954).

The macrofossil evidence

The E horizon (Conway's M horizon) delimits the onset of a major phase of forest clearance and settlement in the southern Pennine uplands that was almost certainly associated with the Norse colonists of the tenth century AD – the radiocarbon date for this horizon, 1023 ± 50 years bp[1] (Tallis and Switsur, 1973), gives a calibrated date of *c.*1000 AD (Stuiver, 1982). The depth of peat formed above this horizon (i.e. during the last 1000 years approximately) varies markedly from site to site: 90 cm at Ringinglow C (Conway, 1954), 40 cm at Featherbed Moss (Tallis and Switsur, 1973) and 10 cm at Bleaklow (Conway, 1954), for example. *Sphagnum* is a prominent component of this recently-formed peat where an appreciable depth has accrued, and may persist to within a few centimetres of the present-day surface (Tallis, 1965). At sites with only a slow build-up of peat in the last 1000 years, however, no *Sphagnum* remains are detectable in the peat formed in this time-interval. The occurrence or otherwise of 'recent' *Sphagnum* peat is clearly related to the type and extent of erosion at a site:

1. Sites with uneroded peat: *Sphagnum* present. Ringinglow C (Conway, 1954), Dean Head Hill I (Tallis, 1964).

2. Type II gullies not yet incised to the peat base: *Sphagnum* present. Goyt Moss (Tallis, 1964), Wessenden I (Tallis, 1964), Featherbed Moss sites 1–3 (Tallis, 1965), Rishworth Moor (Bartley, 1975).

3. Deeply incised main drainage gullies: *Sphagnum* absent. Featherbed Moss site 4 (Tallis, 1965).

4. Type I dissection systems on plateaux and cols: *Sphagnum* absent. Kinder I and VI (Conway, 1954), Bleaklow (Conway, 1954), Kinder (Tallis, 1964), Wessenden II (Tallis, 1964).

5. Sites close to the peat margin: *Sphagnum* absent. Woodhead (Conway, 1954).

More precise evidence of the timing and causes of peat erosion on the southern Pennine moorlands comes from detailed pollen–stratigraphic studies at a single site only – Featherbed Moss, Derbyshire (Tallis, 1965, 1973a, 1985; Tallis and Switsur, 1973). Here, within an area of *c.*1 km^2, all the erosion types listed above are represented. Figure 17.3 shows the distribution of sites with and without 'recent' *Sphagnum* peat (i.e. peat formed during the last millennium) on Featherbed Moss. Recently-formed *Sphagnum* peat is consistently present over much of the Moss,

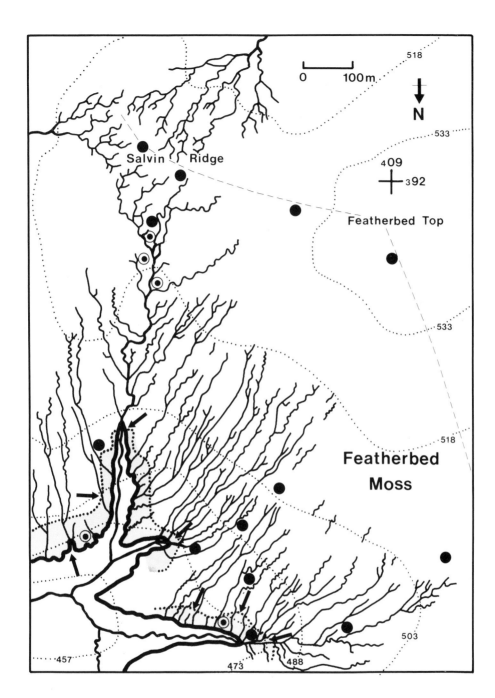

Fig. 17.3 Featherbed Moss, Derbyshire. Sites with (●) and without (◉) recently formed *Sphagnum* peat, and the probable extent (shaded) of peat affected by sliding. The heavy line marks the edge of the peat blanket. The gully system is mapped from a 1954 Air Ministry aerial photograph. ...., contours (m); ---, water-parting; ●●●●, upper limit of marginal peat showing evidence of mass movement; the arrows mark likely points of bog-bursting. A National Grid intersection is shown on Featherbed Top.

but is absent around the bog margins, along the sides of deeply incised drainage gullies, and over part of the closely gullied area occupying the saddle-shaped col on Salvin Ridge. At sites where recent *Sphagnum* peat is present, characteristic fluctuations in the abundance of *Sphagnum* can be recognised. Figure 17.4 shows the frequency of *Sphagnum* leaves at different depths in the peat at four sites – three sites along the sides of Type II gullies and one in uneroded peat *c.*50 m upstream of the gully heads. The position of horizon E, determined from localised pollen analyses, is indicated on the four *Sphagnum* profiles in Fig. 17.4. Three phases of reduced *Sphagnum* abundance occur in all profiles: a recent phase (c), probably attributable to air pollution; and two earlier phases (b and a), above and below horizon E, respectively. As these two earlier phases occur at both eroded and uneroded sites, it is more likely that they are climate-induced than erosion-induced – especially as Barber (1981) records two similar phases from his peat profiles at Bolton Fell, Cumbria (dated to 800–900 ad[1] and 1100–1300 ad), that he interprets as produced by periods of drier climate (the more recent of these two periods is the well-documented 'Little Climatic Optimum' (Lamb, 1977a)). At Featherbed Moss the onset of the phase of *Sphagnum* dominance separating these two drier phases, however, is progressively delayed down-gully (occurring latest in the lower reaches of the gully), while, conversely, the phase when very fresh-looking *Sphagnum* peat was forming between *c.*1300 and 1800 AD is rather better developed

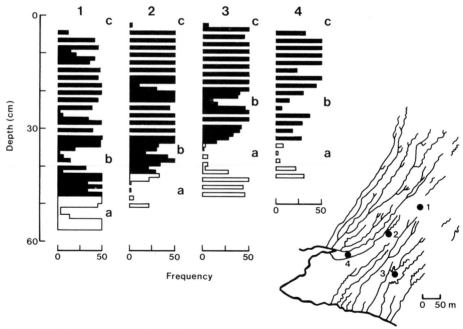

Fig. 17.4 Frequency (*ex* 50) of *Sphagnum* leaves in peat formed during the last millennium at four sites on Featherbed Moss. Histogram bars below horizon E (*c.*1000 AD) are unshaded; a, b and c designate periods of lowered *Sphagnum* abundance referred to in the text. The map (part only of the area shown in Fig. 17.3) shows the positions of the four sites.

down-gully (where it includes the aquatic species *S. cuspidatum*). These findings (as yet documented only for Featherbed Moss) might be interpreted as follows: the lines of at least the lower reaches of the present-day Type II gullies were established during an early period of active peat erosion, around 900 AD, and this erosion led to rather drier conditions developing around the bog margins (and particularly along the lower reaches of the gullies); subsequent *Sphagnum* growth after 900 AD, and particularly after 1300 AD, led to recolonisation of the gullies, but they persisted as 'water tracks' on the bog surface where wetter conditions prevailed (Ingram, 1967), and then entered a new phase of active erosion around 1750–1800 AD, when *Sphagnum* was killed off by air pollution.

These suggestions for the timing of gully erosion relate only to areas of the peat blanket where the gullies are peat-floored. Nearer to the bog margin these gullies typically converge towards a small number of steep-sided major gullies that are incised through the peat blanket into the underlying bedrock. Peat profiles from the sides of such major gullies typically lack any recently-formed *Sphagnum* peat, and the first prominent *Sphagnum* remains in the peat (at depths of 15–20 cm) date from the phase of active bog growth preceding the first dry phase (a). Streams linked to Type I gullies on cols and plateaux generally in the southern Pennines are of this form (see, for example, Conway's (1954) profiles for her Kinder VI and Bleaklow sites). Figure 17.5 shows peat profiles for six sites on the col of Salvin Ridge, Featherbed Moss, where a range of stages in the erosion of a former pool–hummock complex is represented. In addition to the network of deeply-incised gullies and intervening haggs characteristic of most Type I dissection, there are also extensive peat 'flats' (presumably the sites of former pools) and intervening drier hummocks, a microtopography that is best developed in the central part of the col most remote from the three streams draining the col (Figs. 17.5 and 17.6). Profiles 1 and 4 in in Fig. 17.5 are from sites bordering main gullies, profiles 2 and 3 from hummocks adjoining peat 'flats' and peat-floored gullies, and profiles 5 and 6 from high hummocks in the central part of the col. On these profiles are shown the positions (where appropriate) of the pollen–stratigraphic horizons E, D, C and Ulm. (Table 17.3), and also horizoin F (marked by a rise in *Plantago* and Cerealia pollen values to > 25% and > 10% APC, respectively, coupled with increases in *Fraxinus* and *Pinus* values), which can be dated on both radiocarbon and documentary evidence (Tallis, 1985) to 1550–1600 AD. Two major trends can be discerned in the profiles:

(1) A progressive reduction downstream (i.e. from sites 6 to 1) in the depth of peat built up between horizons D and E (400–1000 AD).
(2) A progressively earlier date downstream for the uppermost *Sphagnum* peat in the profiles – very recent at site 6, around 1700 AD at site 5, *c.*1000 AD at sites 2 and 4, and perhaps *c.*800 AD at sites 1 and 3.

At sites 1 and 3, the peat profiles record a phase of drier conditions just below horizon E (and presumably referable to phase (a) of Fig. 17.4) that is succeeded by highly humified peat containing leaves of the moss *Racomitrium lanuginosum*. At the other four sites, the return to wetter conditions on the bog surface at

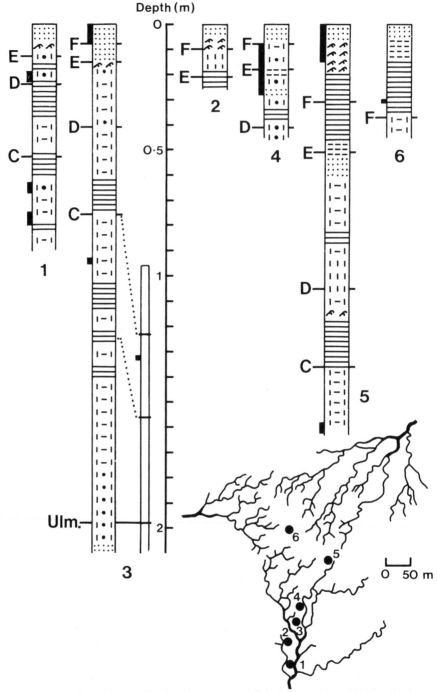

Fig. 17.5 Stratigraphic profiles for six sites on Salvin Ridge. C, D, E, F and Ulm. are pollen–stratigraphic horizons defined in Table 17.3 and in the text. ≡ , *Sphagnum* remains predominant in the peat; I-I, mixed bog with *Sphagnum* present; I·I, mixed bog with *Sphagnum* absent; II, *Eriophorum vaginatum* remains predominant; :::, highly humified or fibrous peat; ⌒ *Racomitrium* peat. The black bars on the left-hand side of each profile indicate regions where *Empetrum* pollen exceeds 3% total land pollen. The map showing the position of the six sites covers part only of the area mapped in Fig. 17.3. At site 3, corresponding levels (determined pollen-analytically) in profiles from a hummock (left) and an adjoining peat 'flat' are shown.

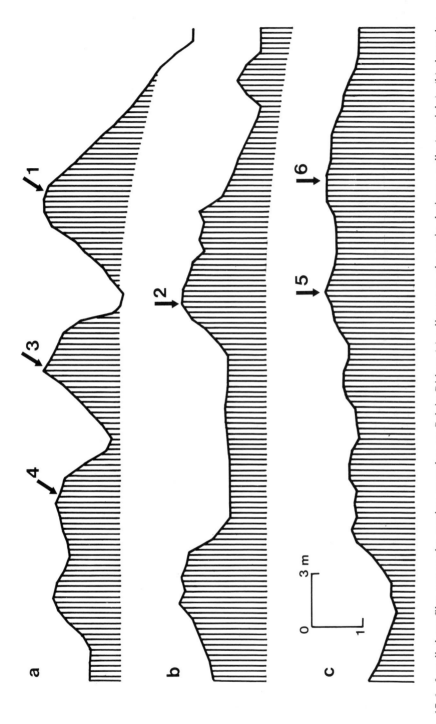

Fig. 17.6 Levelled profiles across the erosion complex on Salvin Ridge: (a) adjacent to the main drainage gully (on right); (b) through an area of hummocks and extensive peat flats; (c) in the central, low-relief, area of the hummock–hollow complex. The approximate positions of sites 1–6 are indicated. The depth of peat shown is approximate only.

*c.*900–1000 AD, documented in the *Sphagnum* profiles in Fig. 17.4, is recorded in the peat profiles from Salvin Ridge by a layer of *Sphagnum* peat. Dry phase (b) (during the early Mediaeval period of warm climate) is apparent as a super-posed layer of *Eriophorum* or *Eriophorum–Calluna* peat at sites 2 and 4, and this is overlain by highly humified peat (again with leaves of *Racomitrium lanuginosum* at site 2) above horizon F. At sites 5 and 6, active but localised hummock growth continued until 200–300 years ago, with lenses of very fresh-looking *Sphagnum fuscum* (site 5) or *S. imbricatum* (site 6) peat building up. The series of peat pro-files in Fig. 17.5 thus apparently records the progressive drawing off of water from the pool–hummock complex on Salvin Ridge over a period of about 1000 years, a process interrupted periodically at some sites by rejuvenated bog growth. The location of the sites with the earliest indications of erosion – along the upper reaches of the main stream draining northwards from the col – is in conformity with a process of erosion set in motion by headward stream erosion (the 'fluvial' hypothesis). That erosion was indeed acting on an existing pool–hummock complex is suggested by pollen analyses of two adjacent peat columns at site 3: one column from a dry hummock (total peat depth 218 cm), the other column from an adjoin-ing peat flat 3 m away (total peat depth 116 cm). Equivalent levels in the two peat columns, established from the pollen analyses (Tallis, 1985), are shown by dotted lines in Fig. 17.5, and suggest that a height-differential between pool and hummock of 0.5 m had developed by horizon C, and had probably been further accentuated by the time of onset of erosion.

Peat growth rates

Stratigraphic variation between peat profiles from different sites on Featherbed Moss, and probably the southern Pennine moorlands in general, is most marked above horizon E. This variation is expressed not only in the character of the macrofossils but also in the overall rate of peat build-up. Crude rates of peat build-up can be calculated from the pollen–stratigraphic data, using the various pollen–stratigraphic horizons to divide the peat column into time-segments of known duration. The depth of peat present in each time-segment at a particular site can be converted to average rates of peat build-up (as centimetres per century) in those time-intervals, and these rates then depicted in the form of a 'kite diagram'. Each superposed segment of the kite diagram is a time-interval (between two successive pollen–stratigraphic horizons), the vertical axis is a time-scale, and the width of each segment is proportional to the rate of peat build-up in that time-interval. 'Calibrated' radiocarbon dates (Klein *et al.*, 1982) should be used in the calculations, and, ideally, allowance should also be made for differences in bulk density within and between peat columns. In Fig. 17.7, where kite diagrams for thirteen sites are shown, calibrated dates have been used (see Table 17.3), but differences in bulk density have not been taken into account. The possible magnitude of this source of error is discussed in Tallis (1985).

Each kite diagram in Fig. 17.7 depicts the pattern of peat accumulation at that site over a time-period of *c.*5700 years. Rates of peat build-up during that time-

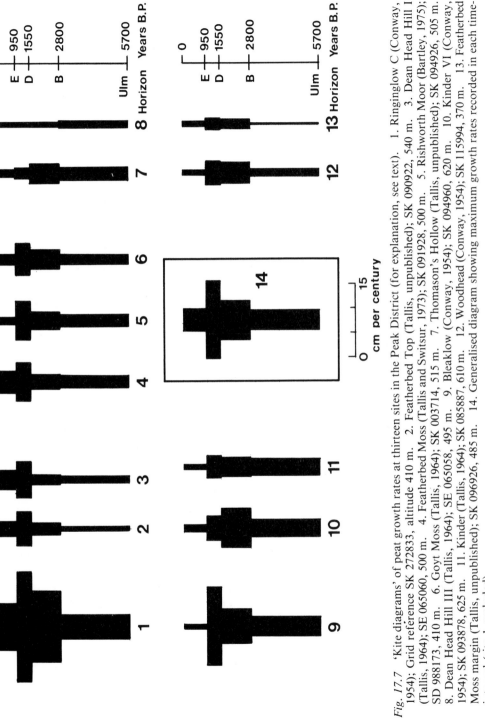

Fig. 17.7 'Kite diagrams' of peat growth rates at thirteen sites in the Peak District (for explanation, see text). 1. Ringinglow C (Conway, 1954); Grid reference SK 272833, altitude 410 m. 2. Featherbed Top (Tallis, unpublished); SK 090922, 540 m. 3. Dean Head Hill I (Tallis, 1964); SE 065060, 500 m. 4. Featherbed Moss (Tallis and Switsur, 1973); SK 091928, 500 m. 5. Rishworth Moor (Bartley, 1975); SD 988173, 410 m. 6. Goyt Moss (Tallis, 1964); SK 003714, 515 m. 7. Thomason's Hollow (Tallis, unpublished); SK 094926, 505 m. 8. Dean Head Hill III (Tallis, 1964); SE 065058, 495 m. 9. Bleaklow (Conway, 1954); SK 094960, 620 m. 10. Kinder VI (Conway, 1954); SK 093878, 625 m. 11. Kinder (Tallis, 1964); SK 085887, 610 m. 12. Woodhead (Conway, 1954); SK 115994, 370 m. 13. Featherbed Moss margin (Tallis, unpublished); SK 096926, 485 m. 14. Generalised diagram showing maximum growth rates recorded in each time-interval (site 1 excluded).

period commonly vary by four- to five-fold at any site (and as much as fourteen-fold at Bleaklow). Considerable differences in the pattern of accumulation are also apparent between sites, and some of these differences must be attributable to varying intensity and duration of erosion.

From these kite diagrams for specific sites, a generalised kite diagram can be constructed, showing the *maximum* rate of peat build-up observed for each time-interval at any site (Fig. 17.7, no. 14). Data for site 1 (Ringinglow Bog) have been ignored, as the rate of peat accumulation here is exceptionally fast. This generalised diagram shows clearly that peat build-up was most rapid in the time-segment DE (400–1000 AD), and depicts a pattern of change that is matched most closely in the diagrams from uneroded sites (nos. 1–3) and gully sides (nos. 4–6). Intensely eroded (9–11 and marginal sites (12 and 13) have rather different-looking kite diagrams.

For the deeper sites, but excluding Ringinglow Bog (i.e. sites 2–7 and 9–12), differences in rates of peat build-up are most marked in the time segment DE. Maximum and minimum rates (in centimetres per century) for the various time-segments are as follows:

Ulm. to B horizon,   1.0 and 4.0

B to D horizon,   2.0 and 8.0

D to E horizon,   1.8 and 15.8

E to present day,   1.1 and 4.5

As peat build-up at uneroded sites in this time-interval was high, it is likely that the lowered rates at other sites were caused by the onset of erosion. Subsequent rates of build-up (i.e. during the last millennium) at all but two of the sites where erosion is currently proceeding were very low (1.5 cm per century or less), and these data support the idea that erosion had set in at these sites by 1000 AD. The two sites with faster peat growth rates (3–5 cm per century) in the last millennium (sites 4 and 6) are both from the middle reaches of Type II gullies, where substantial recolonisation by *Sphagnum* may have occurred in the Middle Ages.

The kite diagrams for peat columns from deeply-incised gullies (sites 7 and 8) indicate that erosion here (and particularly at site 8) may have set in rather earlier.

Discussion

The evidence discussed in the preceding pages suggests that the blanket peat erosion visible on the southern Pennine moorlands today is the end-product of a complex series of events extending over at least 1000 years. The pollen–stratigraphic studies on Featherbed Moss point towards two periods of active peat erosion: a recent period covering the last 200–300 years and a much earlier period around 900 AD. There is good documentary evidence of marked contrasts in climate during the 800 years separating these two periods of erosion, such that the overall abundance of *Sphagnum* over large areas of the bog surface was considerably affected. Nevertheless, by 1700 AD the extent of erosion on the moorlands may have been substantially less than it was in 1000 AD, as a result of recolonisation by *Sphagnum* of the upper reaches of previously open gullies and some drier hummocks and peat 'flats'. It is probable that the recent phase of erosion was set in motion by the death of *Sphagnum* from air pollution. That being so, the situations with the greatest previous abundance of *Sphagnum* would have been most affected, and would then probably have become particularly vulnerable to erosion. Pools, 'flats' and water tracks are places on bog surfaces elsewhere in Britain that are characterised by abundant *Sphagnum* and also by excess surface water, and thus it is possible that some of the recolonised situations in the southern Pennines became lines of weakness following the death of *Sphagnum*.

These lines of weakness may well have been established during the first, much earlier, phase of erosion. The pollen–stratigraphic studies suggest that erosion affected the major stream channels and the pool–hummock complexes, but it is conceivable that some incision along the courses of the present Type II gullies also occurred. The pollen diagrams place the initial stratigraphic evidence of erosion *before* the major forest clearance of the tenth and eleventh centuries, so that biotic pressures in the uplands could have been low. However, even below horizon E there is a small presence of cereal and *Plantago* pollen, some evidence of burning (as charcoal and burnt twigs in the peat), and a steady decrease in *Betula* pollen values, and it is possible that the uplands were already being utilised for grazing before the hillslope forests were cleared. Hence biotic disturbance cannot be ruled out entirely as a cause of this first phase of erosion. Nevertheless, other causal mechanisms are at least as plausible.

It is easy to envisage, in simplistic terms, how progressive headward extension of major drainage streams might lead to the onset of erosion. It is also easy to see how natural headward extension might result from climatic change. The earliest peats to form in the southern Pennines were almost certainly in localised 'water-collecting' sites – sites in shallow basins, troughs and channels. One such suite of water-collecting sites would be the upper reaches of the existing stream courses, where the stream channels, provided the flow of water in them was only slow and intermittent, would be occupied by peat-forming 'flush' vegetation. Increased water flow along the stream courses, for example during periods of wetter climate, would cause the demise of these communities and the cessation of peat accumulation. The form of the peat-growth diagrams in Fig. 17.7 suggests an increasingly wetter environment in the uplands from the time of the Elm Decline (Ulm. horizon) onwards. Any peat in and adjacent to these 'pre-peat' stream courses (as at sites 7 and 8) would accordingly become increasingly liable to erosion.

A more catastrophic explanation can also be suggested. The first episode of erosion is placed within a period of cooler wetter climate, extending perhaps from 200 to 1000 AD (Barber, 1981). Rapid build-up of uncompacted *Sphagnum* peat occurred widely in the southern Pennines at this time, and appears to have been terminated by equally widespread erosion, affecting may parts of the peat blanket. 'Explanations' of the unusually high incidence of erosion in southern Pennine blanket peats have sought to isolate as causes any special features of the southern

Pennine environment that are not found elsewhere. Biotic pressures, and particularly air pollution, are such features, and can be used to 'explain' satisfactorily the more recent phase of erosion. Another special feature, only recently recognised, is the very early date of peat initiation on the southern Pennine uplands. This has been attributed to the exploitation of the uplands as grazing land for wild game by Mesolithic hunters prior to 3000 bc, leading to the gradual degeneration of the upland forest under the combined pressures of grazing and regular burning (Jacobi *et al.*, 1976), and a takeover of the flatter ground by peat-forming communities. In other upland areas of Britain without these pressures, a persistent upland forest cover may have effectively prevented the spread of blanket peat until much later (Moore, 1973; Bostock, 1980). In the southern Pennines, the present-day areal extent of blanket-peat had been almost achieved by 3000 bc (Fig. 17.8). In the Berwyn Mountains, Mid-Wales, the best investigated of other upland areas, only two of the seventeen sites studied palynologically had peat forming before 3000 bc; peat on the bog slopes probably did not develop until after 500 bc (Bostock, 1980). In the southern Pennines, the areal extent of the peat blanket at higher altitudes is largely limited by topography: oversteepened slopes present an obstacle which generally cannot be overcome, but which comes under increasing strain as the more central parts of the peat blanket continue to grow vertically. In simple terms, this strain is manifested by a general tendency for the bog margins to be pushed downslope by the weight of peat on the bog slope, and for 'exit points' of water transmitted laterally towards the headwaters of the main drainage streams to be enlarged. If resistance is overcome, bog-slides and bog-bursts result.

It seems reasonable to suggest that a process of this sort may have occurred widely in the southern Pennines during the period 400–1000 AD, with the actual timing (and perhaps also the intensity) of the event varying considerably from place to place. Once bog-bursting and sliding had occurred, the whole complex of erosion features that are recognisable today might have come into being rather suddenly, though in muted form: Type I dissection by the drawing off of water from a pool–hummock complex; Type II dissection by the extension upslope of gullies originating as 'nicks' in the peat margin (an extension involving, in part at least, underground 'pipes'); eroded margins to the peat blanket produced by sliding; and deeply-incised main streams gouged out by the bog-burst. Some of these features were subsequently recolonised and stabilised during the Middle Ages, only to suffer renewed erosion in recent centuries; other features were not, so that at some sites in the southern Pennines (e.g. Kinder, Bleaklow) bare peat has probably been continuously present for a millennium.

What corroborative evidence is there for this hypothesis? Only one small area of the southern Pennines has been examined in detail so far (Featherbed Moss, Derbyshire), but that one area appears to be typical, in terms both of topography and of erosion patterns, of large parts of the moors. The Moss (Fig. 17.3) consists of gently-rolling peat-covered slopes, rising to the south to the dome-shaped summit of Featherbed Top (544 m altitude), and abutting to the north on the deeply-incised headwaters of the River Ashop. Within the peat blanket are areas of Type I and Type II dissection and uneroded peat. The peat edge on the north side terminates

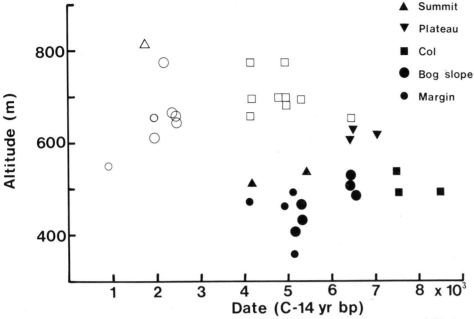

Fig. 17.8 Approximate dates of peat initiation at sites in the southern Pennines (filled symbols) and the Berwyn Mountains, Wales (unfilled symbols). Each symbol on the chart represents a site for which a pollen diagram of the basal deposits is available. In each area, pollen diagrams from the various sites have been compared with a radiocarbon-dated 'standard' diagram from the same area, and an approximate date assigned to the basal deposits. The Berwyn pollen data are from Bostock (1980); the southern Pennine pollen data are from Conway (1954), Bartley (1975), and Tallis (published and unpublished).

not in a single continuous erosion cliff but in a broad region (up to 50 m wide) where the bog surface is irregularly corrugated and terraced, and with frequent exposures of bare peat (Fig. 17.9); this microtopography is accentuated where major indentations ('nick' points) occur in the peat margin. Downslope from the main peat mass there are isolated blocks of peat on impossibly steep slopes (up to 30°) for peat accumulation. On aerial photographs of the Moss, or viewed from the moors to the north, a clear 'scarpline' of peat haggs is visible, some distance upslope from the edge of the peat blanket, and this scarpline defines the upper limit of the region of irregular topography.

These features at the peat edge on Featherbed Moss are best interpreted as the result of bog-bursting at a limited number cf points along the bog margin (the 'nick' points), accompanied by more extensive slumping and sliding of the marginal peats. The detailed pollen–stratigraphic evidence (Tallis, 1985) indicates that erosion on Featherbed Moss set in shortly before 1000 AD.

It would clearly be premature (and unwise) to attribute all peat erosion in the southern Pennines to this type of process. Extensive fieldwork is needed to establish how widespread is the evidence for bog bursting and sliding. The fieldwork needs to be backed up by further detailed pollen–stratigraphic studies in selected areas

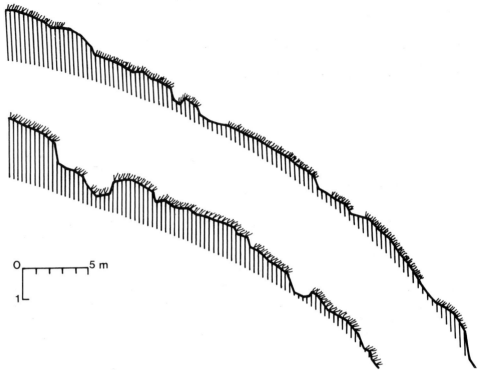

Fig. 17.9 Levelled profiles across the slumped peat at the margins of the blanket bog at two sites on Featherbed Moss. The depth of peat shown is approximate only.

(with varying overall intensities of erosion), to provide a firmer-based sequence of events and time-scale. A change in attitudes to peat erosion is also desirable. Certain attitudes have perhaps proved stultifying in the past: the belief that peat erosion is an unnatural feature of the upland landscape; the tendency to classify all erosion features within the rather rigid framework erected by Bower; and the belief that peat erosion is the result of specific *agencies*. The search for agencies is a natural corollary of the view that erosion in an upland landscape is a sign of poor management. If agencies can be identified, it may then be possible to formulate appropriate control measures. The evidence discussed in this chapter suggests that peat erosion may be the result of *processes* rather than of agencies – processes which are an inevitable consequence of the upland environment; and that, moreover, peat erosion may have been a feature of the southern Pennine landscape for many centuries. That it is not so prominent in other upland areas of Britain may merely be because the blanket peats there have not yet reached 'critical instability'. If this view is correct, then intensified erosion might well be expected in these areas in the future.

Note

1   ad, bp, and bc (rather than AD, BP, and BC) are used to denote uncalibrated radiocarbon dates.

Geomorphology and urban development in the Manchester area

Introduction

Approaching Manchester from the south across the Cheshire plain from Crewe or along the M6 motorway, the visitor gains little impression of the vigorous landscape of steep hills, fast flowing streams and glacial and periglacial deposits which surrounds the city to the north and east. These hills and streams which provided the original power source for Manchester's industrial expansion also produce a number of geomorphic problems. The streams descend rapidly to the plains which they traverse in broad meandering channels. Flooding, especially from summer thunderstorms, continues to cause local problems. Engineering works have to take account of the varied character of the glacial deposits. Since Stephenson built the Liverpool to Manchester railway across Chat Moss, a succession of ingenious solutions to the geomorphic problems of the Manchester area have been adopted.

River dynamics and urban growth

South of Manchester, the River Mersey is formed by the confluence of the River Tame and Goyt at Stockport. Downstream of Stockport the river has a meandering course through a wide floodplain whose margins are marked by a series of terraces. Throughout the floodplain area, historical records demonstrate a record of frequent inundation, early 1:10 560 maps indicating much of the floodplain as liable to flood. As early as 1841, the Bridgewater Canal which crosses the floodplain south of Stretford had to be protected by a flood overflow channel, with an off-take weir upstream of the canal and a 2.2 km-long overflow channel passing beneath the canal and the adjacent Chester Road (A56) at Eye Platt Bridge.

At Stockport, the Merseyway shopping precinct was constructed in 1965 on a deck over the river, confining the stream which has a long record of previous floods in the town. On 12 July 1828 the river rose over 6 m above its normal level, and on 8 June 1896 it caused widespread flooding in the town (Astle, 1971).

Flooding was widespread throughout the Greater Manchester section of the Mersey. In 1750 it was noted that in winter, and often in summer, flooding made the river impassable, causing loss of life and delays to the post on the Chester Road, part of the London to Manchester Post Road (Hainsworth, 1983). In the 1920s severe flooding through a breach in the embankment near Crossford Bridge persisted due to failure of farmers and the Ministry of Agriculture to agree about sharing the costs of embankment repairs. In July 1930 bank erosion during a flood took away 20 m of garden at Fernbank House, Ashton-upon-Mersey.

A 24-hour rainfall of 75 mm in July 1973 caused the river to rise 4.5 m above normal at Stockport Sewage Works, producing what the *Stockport Express* called the worst floods in living memory. There was some damage to parts of the Merseyway shopping precinct, but more severe damage occurred further downstream at Brinksway.

Downstream of Stockport the floodplain has been altered considerably by the construction of the M63 motorway which crosses the river four times between Stockport and Urmston. The river channel has been embanked throughout this section and is regulated by a series of flood-basins at Didsbury, Chorlton and Sale. North of Gatley part of a meander was cut off and the channel shortened. At Northenden the potential of the motorway embankment to obstruct the return of flood-water to the river is alleviated by a series of culverts which allow egress of water from the upstream to the downstream side of the motorway. Escape of the water to the river is regulated by sets of flood-gates.

At Chorlton and Sale the gravel pits used to supply aggregates for the motorway works have been converted into recreational lakes, or water parks, which have, within their embankments, storage for flood-waters which can be released from the river by lifting off-take weir gates upstream of the water parks. The long series of works modifying the Mersey channel, culminating in the construction of the motorway, have led to a reduction in sinuosity (Table 18.1). Several meanders have been cut off and reaches straightened (Fig. 18.1), thereby increasing the hydraulic efficiency of the channel, maintaining adequate depth and velocity to evacuate all but the severest flood flows without spillage over the embankments onto the floodplain.

Table 18.1   Change in sinuosity of River Mersey from Crossford Bridge, Sale, to Wheat Hey, Ashton-upon-Mersey, 1945–1971.

Sinuosity is chanel length divided by reach length.

Date	Sinuosity
1845	1.77
1904	1.68
1927	1.67
1954	1.50
1971	1.28

However, such river training works can lead to dramatic channel adjustments downstream of protected reaches. In the reach discussed in Table 18.1 and Fig. 18.1 the newest training works end 0.5 km upstream of Old Eea Brook at Urmston. Immediately downstream of the regulated reach, the river makes a bend with a radius of approximately 50 m, about twice the channel width. Severe bank erosion and channel migration has occurred in the period 1979–84, with the stream reducing its channel length and tending to straighten its course. Such bank erosion, with a series of rotational slumps in the stratified floodplain sediments, is similar to

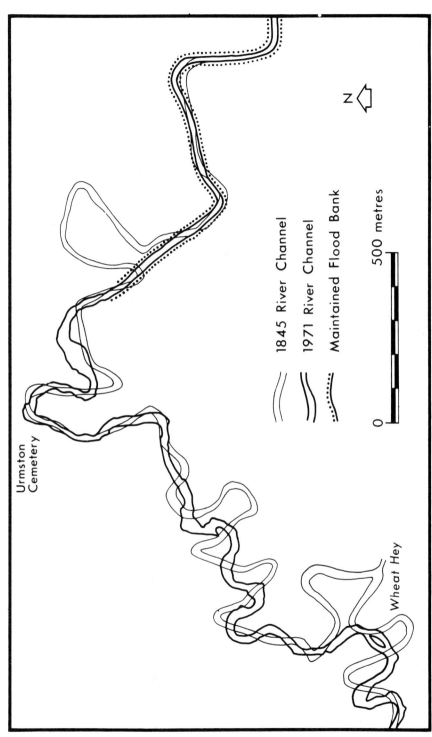

Fig. 18.1 The changing course of the River Mersey, south-west of Manchester between Urmston and Ashton-upon-Mersey, showing the limits of the maintained floodbank and the location of severe bank erosion near Urmston Cemetery.

Urmston
Cemetery

Wheat Hey

1845 River Channel
1971 River Channel
Maintained Flood Bank

N

0   500 metres

that which has occurred on many regulated rivers. Richards (1982) notes that a major requirement for stable channels in alluvium is that slope is not steepened by shortening the river course drastically and that channel bends are therefore maintained. On the Mississippi severe bank attack occurs if the bend radius is less than 2.5 times the channel width.

Such channel changes are widespread on the streams which debouch on to the Cheshire–Lancashire plain from the Pennines. Sometimes a change in shape may depend on major floods which cause channels to shift.

The River Irk underwent a dramatic change near Chadderton in December 1964 when the stream developed a new chute channel across the neck of a meander. The meander had developed after the flow of the river had been deflected by a small retaining wall built to prevent undercutting of the valley side below a new housing estate. The meander became entrenched in the south side of the valley, leading to oversteepening of the slope and a large supply of debris to the channel. During the flood, the sediment-laden meander channel restricted the rate of downvalley flow. Water piled up upstream of the meander, leading to a rise in level adequate for flow to begin across the meander neck, thus creating the chute (Johnson and Paynter, 1967). Such responses to minor channel modifications have occurred in several places in Greater Manchester. The channel of the River Bollin in Cheshire was stable between 1872 and 1935, after which year channel instability and meander cutoffs occurred, leading to a decrease in sinuosity from 2.34 to 1.37 (Mosley, 1975a). Although this change in shape was triggered by large floods in the 1930s, it may also have been a response to higher peak discharges as a result of agricultural land drainage and urban development.

Industrial and urban development had drastic effects on the rivers that flow through the centre of Manchester, especially the Irwell. By the early 1860s the bed of the Irwell in Manchester had been so raised by refuse and cinders dumped in the chanel that the main sewers and mill drains became blocked at high discharges and serious flooding occurred after rain (Aspin, 1969). At Warrington, 29 km downstream of the nearest culpable factory, the Mersey channel was obstructed by bars of cinders. Riverside factories often had chutes down which ashes were ejected directly into the river. In 1862, Thomas Coates estimated that more than 75 000 tonnes per year of cinders alone were discharged into the Irwell (Fig. 18.2).

The bed of the Irwell between Albert Bridge and Throstle Nest Weir, immediately downstream of the centre of Manchester, rose by an estimated 4 cm y^{-1} between 1862 and 1869 (Bancroft, 1881). Groynes were built in the river to improve navigation by trapping the gravels and cinders and reduce the flood discharge by narrowing the cross-sectional area. Formation of bars and shoals in the river at that time was sufficient to delay navigation for several days after flood events.

Such dramatic reductions of channel capacity made flooding all the more likely. In addition, the rapid expansion of the built-up areas of all the towns in the Irwell catchment accelerated the rate of runoff, with more drains and impermeable surfaces. To increase production to feed the expanding urban population, farmers on the surrounding hill lands undertook drainage works for pasture improvement. These drains added to the increased peak discharges after storms.

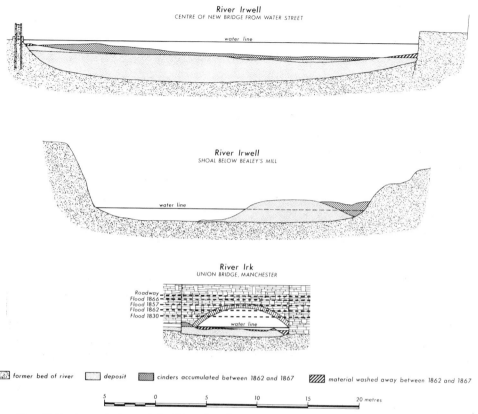

Fig. 18.2 Aggradation of the River Irwell and River Irk in the 1860s (based on data in the First Report of the Commissioners appointed in 1868 to inquire into the best means of preventing the pollution of rivers (Mersey and Ribble Basins), 1870).

The Irwell has a long history of major floods, with 29 occurring between 1816 and 1970, an average of one every 15 years. One of the worst floods was that of November 1866 when over 800 ha, covered by crowded tenements, dwelling houses and factories, were flooded to a depth of up to 2 m. In the four succeeding years another four floods caused similar, but somewhat smaller problems. The 1866 flood represented a discharge of 1.31 m³ km^{-2}s^{-1} of the Irwell catchment above Agecroft bridge (Bateman, 1884). The 1964 flood was equivalent to 1.06 m³ km^{-2}s^{-1}, yet still caused severe damage in the lower Kersal area. The channel capacity in 1946 was a discharge of only 0.63 m³ km^{-2}s^{-1}, some 0.43 m³ km^{-2}s^{-1} thus spilling over onto the floodplain (Glover, 1971; and see Fig. 18.3).

The tight meanders of the River Irwell and the encroachment of many structures into the river channel produce considerable variations in channel capacity and surface water slope. A model study suggested that the textural roughness of the sides and bed accounted for about half the total resistance to flow between Douglas Green Weir and Adelphi Weir, but only about one-third of that between Adelphi Weir and Hunt's Bank (Manchester Victoria Railway Station) (Allen and Shahwan,

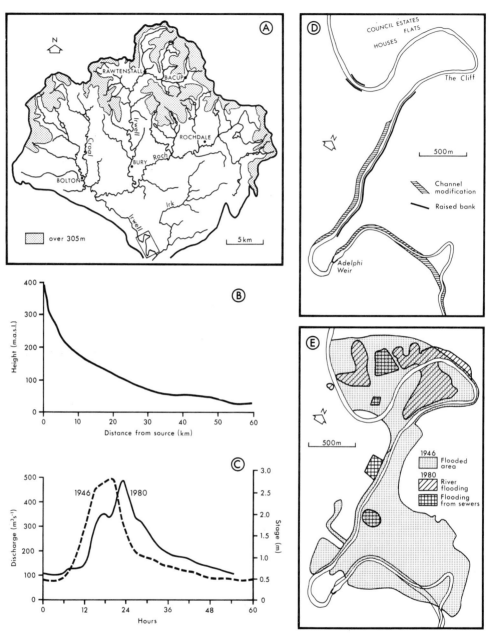

Fig. 18.3 The River Irwell and flood mitigation in Salford. (A) The Irwell catchment showing high ground and the area shown in more detail in D and E. (B) The long profile of the Irwell showing the steep upper course. (C) Hydrographs for the 1946 and 1980 floods at Adelphi Weir, Salford. (D) Flood mitigation works in the lower reaches of the Irwell showing the location of The Cliff and Lower Kersal. (E) Areas around the lower Irwell flooded in 1946 and 1980 (based on an unpublished dissertation by Gillian M. Carney).

1954). Conditions in the channel have been improved downstream of Adelphi Weir by the construction of the Sherburne Cut which eliminated a large meander. As the Irwell discharges into the Manchester Ship Canal 3.5 km further downstream, these river training works have not produced the type of channel modifications found on the Mersey at Urmston.

Nevertheless, the nature of the Irwell catchment, with a fan-like pattern of tributaries draining all the hills to the north of Manchester, is such that storms can produce rapid rises of river level. Despite the number of reservoirs in the catchment and various flood protection works, another major flood occurred at Lower Kersal in October 1980, where new private houses in St. Aidans Grove and the council houses which had been flooded before in 1946 were affected. Subsequently a new embankment was constructed in the vicinity.

Historical data on the Irwell and similar rivers are patchy, but analyses for the 1800–90 decade show that high suspended sediment concentrations occurred during flood flows in all the major tributaries, especially those draining industrial areas. By 1977 (Table 18.2) the suspended sediment concentrations had fallen but could still be sufficient to cause aggradation further downstream.

The Irwell, Mersey and Bollin all flow into the Manchester Ship Canal. Each river carries sufficient suspended sediment to cause problems for canal maintenance, dredging being required at the start of the canal in Pomona and Manchester Docks, at the confluence of the Mersey and where the Bollin enters the canal. While the rivers carry only fine silt at low flows, during floods they carry considerable quantities of sand which accumulates in the canal immediately below each confluence. Samples taken from the bottom of the canal below the Mersey confluence reveal layers of sand 10 cm or more thick alternating with layers of silt usually less than 2.5 cm thick (Milne, 1964). After severe floods the deposited material consists entirely of sand. Peak suspended sediment concentration recorded in the Mersey 1964–74 was 622 mgl^{-1}. If the flood is of short duration, the shoaling caused by the deposition of sand is confined to the immediate vicinity of the confluence, but prolonged flooding causes extensive accretion downstream for over a kilometre to the Partington Coaling Basin.

Dredged material removed from the canal is deposited over a large area between the canal and the old River Mersey channel just upstream of Warrington near Thelwall. The continued action of the rivers thus led to new people-made depositional landforms.

Cities and glacial deposits

Many cities are built on varied unconsolidated sandy, gravelly or clayey deposits which were laid down as the ice sheets of past glaciations retreated, or which resulted from the periglacial reworking of weathering mantles or glacial deposits. Foundation problems in such materials stem not only from the physical properties of the materials themselves, but also from their vertical and lateral variability. When piles were being sunk for the M56/M63 motorway interchange in the Mersey Valley

Table 18.2 Measures of suspended sediment in the Irwell catchment 1880 to 1977.

	Suspended sediment in mg 1^{-1}		
	1880–90	1964–74	1976–77
Irwell at Throstle's Nest Weir (1880–90), Salford University (1964–77)			
Mean	165	57	34
Minimum	26	11	10
Maximum	679	793	219
Medlock at Chester Road, immediately above Irwell confluence			
Mean	80	85	71
Minimum	–	15	30
Maximum	1190	2200	168
Irk at Red Bank, Manchester			
Mean	168	70	101
Minimum	–	9	46
Maximum	2380	576	570
Roch at Bury, above Irwell confluence			
Mean	2	38	67
Minimum	–	9	26
Maximum	1081	182	106
Croal immediately above Irwell confluence			
Mean	34	30	31
Minimum	–	5	6
Maximum	355	166	140

Sources: 1880–90 Davis and Davis (1890).

1964–74 Mersey and Weaver River Authority Annual Reports 1965 to 1974.

1977   North West Water Authority (1978).

south of Manchester, borehole investigations had revealed two tills, but the situation was complicated by buried river channels and variations in lithology. The glacial deposits of the area are sandwiched between the overlying river alluvium and bedrock of Keuper Waterstones or Upper Mottled Sandstones (Bunter Sandstone). Distinguishing the glacial sands and gravels from the underlying sandstones was not always easy. Careful trials showed that some of the piles would have to be driven deeper than originally thought, but careful investigations meant that only 18 per cent of the piles required the additional work. Potential problems were solved by detailed study of the glacial deposits (Tate, 1976).

Not every engineering problem posed by glacial deposits is so relatively easily overcome. Much of the Manchester area is mantled with lodgement till overlain by supraglacial tills and fluvio-glacial deposits. North of Manchester a well-defined ridge of fluvio-glacial outwash sands, in many places strongly current-bedded, extends from Swinton through Pendlebury, across the Irwell Valley to Prestwich and Cheetham Hill (Jones, 1924). Where the river meanders against this ridge in the great bend between Lower Kersal and Higher Broughton it has undercut a steep slope, known locally as The Cliff.

The undercutting of this sequence of sands and clays at The Cliff has caused a series of mass movements since 1882 (see Harrison and Petch, Chapter 19). Between 1931 and 1955 subsidence of 30 to 40 cm occurred in several places between the top of the cliff and Bury New Road to the east and as a consequence Salford City Council has refused planning permissions for extensions to premises on Bury New Road and has made The Cliff district into a conservation area to restore some of its former stately residential character on the attractive high bluff overlooking the river. The Cliff illustrates how attractive residential locations with extensive views are often not the best in terms of geomorphological site stability.

Further from the city centre in Greater Manchester and adjacent areas, the demand for housing land has been so great that new estates have been developed on steep valley-side slopes previously avoided by home builders. Many of these slopes are mantled with clayey tills and periglacial colluvium. In the foothills of the Pennines to the east and north of the conurbation these tills overlie the sandstones and clays of the Coal Measures. Subsurface water movement in the bedrock is essentially downward through the sandstones and then laterally along the surface of the clay to a spring or seepage in the valley side. The till has often blocked these outlets, diverting the water downslope at the base of the till which thus becomes potentially unstable, often leading to a series of small-scale slumps or creep phenomena on valley sides.

Development of such potentially unstable slopes has to be undertaken with great care. Adequate drainage of the slope is essential if foundations and footing are to be stable. One development, near Lower Fold, Marple Bridge, on the north-facing slope of a small valley draining to the River Goyt, has involved extensive cut-and-fill operations right down to the bed of a small stream over a slope which had been mapped as unsuitable for housing development (Fig. 18.4). The drainage works on the slope and the use of the steepest slopes as gardens seems to have avoided the potential for mass movement which alarmed the mapping team.

However, at Ewood Bridge in the Irwell Valley, 25 km upstream of the city centre, a housing site on the steep valley side was abandoned when construction was partially complete. Here platforms for the semi-detached houses and their gardens were cut into the till-mantled slope, with the debris being dumped downslope of the cuts to extend the platforms. At every cut below the summit, widespread seepage of subsurface water occurs with consequent small-scale slumping of material. This in turn caused subsidence of the material above, causing foundations to warp. With better drainage, the slope could be restored to housing development, but as adjacent undeveloped slopes all show signs of mass movement, it would be preferable to avoid using such land for residential purposes.

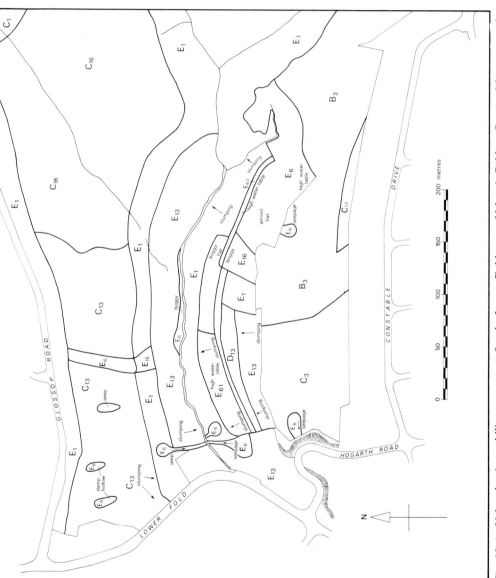

Fig. 18.4 Urban land capability assessment for the Lower Fold area of Marple Bridge, Greater Manchester (from a field survey by students of geography, University of Manchester). The classification codes relate to Land Capability (letters A–E); Landform Slope and terrain <del>(numbers). (see also Douglas, 1983, pp. 177–2).</del>

Access to Manchester across the Pennines and the development of the city's water supplies involves coping with the complex series of landslides, earthflows and rockfalls on valley sides on the alternating sandstones and shales of the Millstone Grit formations (Johnson, 1981). When the water reservoirs were being constructed in Longdendale in the mid nineteenth century, the old landslides were reactivated. Some of these appear to pre-date 7000 BP when vegetation limits were below the present, and others appear to relate to changes in slope stability caused by prehistoric forest clearance (Tallis and Johnson, 1980). One slip, at Rhodes Wood, had moved intermittently previously, the nearby turnpike road having slipped about a metre in 13 years some time before 1850. The slip was apparently stable again when a large channel 11.5 m wide and 3.3 m deep was cut across it. By April 1851 the slide was moving about 2 centimetres per day. Although 260 m long, the slide was only 6 m deep.

The next year, 1852, exceptionally heavy rain set in motion an old landslip of about 16 ha in extent, down the valley from Rhodes Wood. At this locality, where no signs of previous movement had been detected, a contractor's village, New Yarmouth, had been erected. On the night of 6 February, the village was moved about 20 cm downslope and the adjacent overflow channels from the Rhodes Wood reservoir were crushed and disturbed. Careful construction of the embankment and buttressing of the valley side stabilised the movement (Bateman, 1884).

Cities and subsidence

The term 'subsidence' embraces the features, phenomena and processes associated with human-induced lowering of the ground surface by removal of fluids and earth-materials, and compaction of sediments. Abstraction of groundwater, oil and gas, removal of salt, coal and other minerals, and pumping of water from underground structures and mines can all lead to subsidence. In cities, the hydrological changes caused by roofing and paving the land surface result in less replenishment of groundwater by infiltration and thus further lower water table levels.

Changes in ground level and consequent damage to buildings similar to that caused by subsidence occur in expansive soils, clays which alternately swell and shrink with the seasons. Unlike mining subsidence where *horizontal strains* cause trouble, the damage to buildings by clay shrinkage is due to differential *vertical settlement* (National Coal Board Mining Department, 1975). The urban problems of expansive soils are discussed in a later section.

One of the effects of removing coal or other mineral matter from a seam is to create a basin-like depression at the surface, unless measures to prevent subsidence are taken. The depression is caused by the strata overlying the seam settling into the space which was occupied by the material extracted. The settlements vary from zero at the limits of the basin to a maximum over the centre of the workings, but owing to the bulking of the strata over the seam, the maximum settlement is never as great as the thickness of the seam. The surface area affected, however, is greater than the area of the underground workings (Ministry of Works, 1951). The subsidence moves outwards from the centre of the depression affecting an increasingly large area. As buildings begin to come within the limits of the basin, they suffer tension stresses which tend to increase their length, but those closer to the centre of the depression will suffer compressional forces which will tend to buckle and shorten them. Drainage from mining subsidence is therefore expressed in terms of changes of length of a structure (Table 18.3).

Table 18.3   National Coal Board classification of subsidence damage.

Change of length of structure (m)	Class of damage	Description of typical damage
Up to 0.03	1. Very slight or negligible	Hair cracks in plaster. Perhaps isolated slight fracture in the building, not visible on outside.
0.03–0.06	2. Slight	Several slight fractures showing inside the building. Doors and windows may stick slightly. Repairs to decoration probably necessary.
0.06–0.12	3. Appreciable	Slight fracture showing on outside of building (or one main fracture). Doors and windows sticking: service pipes may fracture.
0.12–0.18	4. Severe	Service pipes disrupted. Open fractures requiring rebonding and allowing weather into the structure. Window and door frames distorted; floors sloping noticeably; walls leaning or bulging noticeably. Some loss of bearing in beams. If compressive damage, overlapping of roof joints and lifting of brickware with open horizontal fractures.
More than 0.18	5. Very severe	As above, but worse, and requiring partial or complete rebuilding. Roof and floor beams lose bearing and need shoring up. Windows broken with distortion. Severe slopes on floors. If compressive damage, severe buckling and bulging of the roof and walls.

In the Manchester area many instances of subsidence in coal mine areas have been recorded. In one severe incident in April 1945 an old shaft near Abram which had been plugged opened up and a train of mining wagons which was reversing over the area disappeared into the hole. Although the engine driver braked hard, the weight of the wagon pulled him and his engine down into the pit.

A large part of the Hulton area of Bolton was severely affected by subsidence in the 1950s. Several houses had to be rebuilt and many others underwent extensive repairs. Many structures in the Swinton area, including St. Luke's Catholic Church, had to be designed to withstand the subsidence that had affected older nearby buildings. The West Meade and Campbell Road areas of Swinton were particularly severely affected. In Leigh many areas, especially Plank Lane, suffered badly from subsidence, but perhaps the most significant geomorphic feature created by coal mining subsidence is the nearby Pennington Flash, a 57 ha lake which has developed since 1900.

Perhaps the most notorious subsidence in Manchester itself is that caused by the exploitation of the Roger Seam at Bradford Colliery to the northeast of the city centre. Over an area of 150 ha from Cheetham Hill Road to Ashton Old Road and from Queen's Road almost to Great Ancoats Street, subsidence of the shale above the coal seam caused irregular ground surface movements in the early 1960s. Many houses and factories were affected with cracking and lteral tension movements. A 283 000 m³ gas holder was so tilted that only half its storage capacity could be used. The bed of the abandoned Rochdale Canal was lowered, while similar disruptions of the River Medlock led to fears of local flooding. When Manchester Corporation built new housing estates at Miles Platting and Collyhurst, it had to spend £3 million on special precautions to prevent such damage. The problem was alleviated when economic conditions forced the closure of the mine in 1968.

Sewer collapse and ground conditions

By the end of the 1970s, Greater Manchester had begun to measure the magnitude of sewer collapses by the number of double-decker buses which could occupy the hole. However, some of these collapses are intimately related to geomorphic conditions. A classic example is provided by the subsidence at Fylde Street, Farnworth, on 12 September 1957, where at 0710 hours, after prolonged heavy rain, a small hole appeared in the street. The hole grew in size during the day, and by evening movement ceased with a crater 40 m long, 6 m wide and 4 m deep. Ground movement had occurred over a roughly semi-circular area within a radius of 75 m from the original hole (Hale and Dyer, 1963). Seventeen houses were damaged beyond repair and 121 houses had to be evacuated (Clark, 1958).

The hole was created by the collapse of a sewer following the line of the filled former Farnworth Hall Clough, a small first-order right-bank tributary of the River Croal. The stream was incised into a 3.5 m thick bed of glacial silt in the delta deposits of the Croal overflow channel (Tonks *et al.*, 1931a) between a layer of sand and gravel above and boulder clay below. The silt has many of the characteristics of an expansive soil. When dry it is stiff and can bear a considerable load, but when wet it becomes plastic and then changes to a silt of individual minute particles colloidally suspended in water.

The sewer runs parallel to the bed of the former Clough, crossing it in three places. The groundwater table was high, near the surface after rain. Some groundwater moved through the old streambed gravels and washed fine particles through the matrix, thus weakening the support of the sewer. With this failure of support, storm runoff in the sewer escaped through longitudinal joints and further weakened support. The jet of escaping water caused the adjacent silt to liquefy and so the hole enlarged. This event was the result of both the modification of the natural drainage system and the properties of the glacial deposits.

Subsidence from water and salt abstraction

In Greater Manchester, south Lancashire, Merseyside and north Cheshire, ground-water is abstracted from Permo-Triassic sandstone aquifers. The Permian Collyhurst sandstone, overlain by the Manchester Marl in the conurbation area, and the Sherwood sandstone form the main acquifers. Groundwater abstraction in the Trafford Park area of Manchester has led to a cone of depression in which the water table falls to 10 m below sea level. In the Liverpool area groundwater abstraction beneath the city and in the Widnes and Warrington areas has led to water table levels up to 40 m below sea level and penetration of salt water from the Mersey estuary. However, subsidence has not been reported as significant.

Much of the Cheshire plain, southwest of Manchester, is marked by small lakes, some being natural hollows in the glacial deposits, others being water-filled mediaeval marl pits, and yet others the result of subsidence due to the exploitation of the rock salt in the Triassic rocks of the Cheshire Basin. The roofs of many of the mine workings eventually collapsed and caused the development of large surface depressions which quickly filled with water, as happened during the 'Great Subsidence' at Northwich in 1881 (Wallwork, 1956). Such events produced the present water bodies of Neumann's and Ashton's Flashes north of Northwich (Fig. 18.5).

Brine pumping has had widespread effects on property. Northwich itself has experienced an overnight development of a subsidence crater near the Wheatsheaf Hotel. Such craters resemble collapse dolines. Howell and Jenkins (1976, 1984) describe these surface subsidences as a series of landforms related to salt karst. They recognise a suite of landforms which progress from dolines, including the crater subsidences, some 10 to 200 m in diameter to linear hollows up to 240 m wide and 8 km long. These craters are of particular concern as they develop rapidly, one 25 m in diameter being created in one week. Simulation of crater develop-ment in the UMIST 100G centrifuge led Howell and Jenkins (1984) to conclude that where the rock-salt beds are overlain by brecciated mudstones and uncon-solidated glacial deposits, loss of support by solution and pumping below can produce the propagation of a plume of disturbed material which moves towards the surface. When such a plume breaks the surface, the unconsolidated material is rebulked and a crater develops rapidly.

Such rapid movements can help to explain the sudden creation of surface depres-sions and the irregular rate of subsidence. Nevertheless, some problems are persistent. The main Crewe to Manchester railway line continues to subside at about 200 mm yr^{-1} near Sandbach, having fallen about 5 m between 1892 and 1956. Trains travel over the affected area slowly and constant ballasting is necessary to maintain the level of the track. The River Weaver suffered greatly in the main period of subsidence from 1870 to 1920 but is now virtually stable. However, legal cases claiming that brine pumping has damaged houses have arisen as close to Manchester as Knutsford.

Planning for Winsford New Town had to take account of subsidence. Detailed geological maps of the Cheshire basin showing a fivefold subdivision of the Keuper

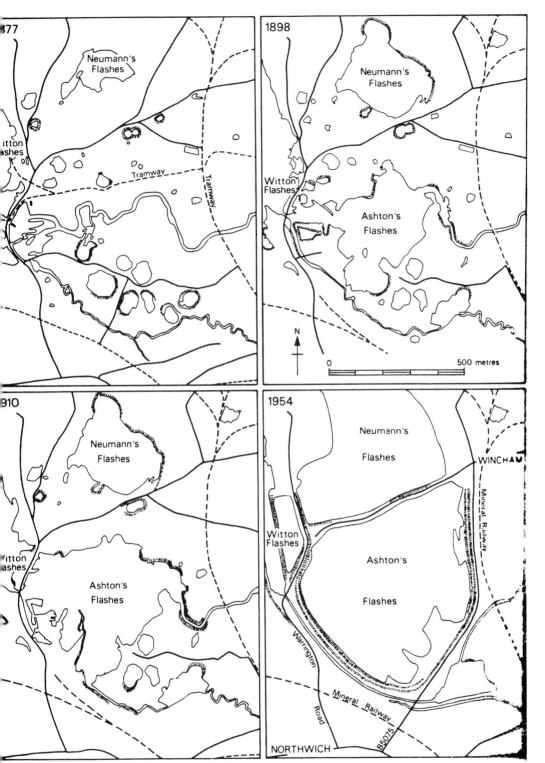

Fig. 18.5 Changes in landforms due to salt workings north of Northwich, Cheshire.

rocks enabled planners to locate land uses in the new town according to the subsidence risk and to allow for special foundation precautions to be taken where
such risk could not be avoided. An average extra 20 per cent is added to the cost
of building where full foundation precautions against subsidence have to be taken
and similar additional investment is needed to protect piped services and other
underground assets. Elsewhere the possibility of subsidence and the presence of
salt underground had previously been enough to persuade authorities that new
urban development should not occur (Woodland, 1968).

Conclusion

The growth of Greater Manchester and the surrounding industrial towns has
involved the solution of many geomorphological problems. In some cases, as in
the subsiding railway line near Sandbach, the solution is only partial, as continuous
adjustment, ballasting and regrading of the line is necessary. Elsewhere, protection
of urban areas has simply shifted the problem further downstream, as with the
channelisation of the Mersey downstream to Ashton-upon-Mersey (Fig. 18.1). In
other instances, continued urban development has introduced new problems,
particularly through the changing rainfall–runoff relationships in small catchments.
Economic pressures have forced the development of less suitable land for housing,
sometimes leading to slope stability problems as at Ewood Bridge. Other problems
arise because the engineering works of the nineteenth century are now reaching
the end of their life. Leaks from brick-lined sewers located in unsuitable materials,
as at Farnworth, can trigger off major subsidence events. Old culverts may now
have to take discharges far in excess of those for which they were designed. Much
of the nineteenth century infrastructure is now at risk, yet financial resources for
its systematic replacement are limited.

   Against this set of geomorphic problems must be set the dramatic series of landscape improvements undertaken by the Joint Land Reclamation Team of the
Greater Manchester and Lancashire County Councils and similar bodies in
removing some of the most unsightly industrial landforms, especially coal waste
tips such as the so-called 'Wigan Alps', and in reclaiming the slopes of former
industrial river valleys. Despite constraints of finance and adequate access to all
the land involved, reclamation has removed many of the major sediment sources,
has stabilised stream channels and established a new, but more natural, geomorphology. Application of the techniques of urban geomorphology and land
capability evaluation would add to these efforts to avoid the geomorphic hazards
which are still significant in the region.

Ground movements in parts of Salford and Bury, Greater Manchester – aspects of urban geomorphology

Introduction

This paper sets out a reappraisal, by the authors, of a geomorphological problem which has long been familiar to the inhabitants of Salford – the landslide phenomenon known locally as the Cliff landslip. An attempt is made to look behind the immediate causes of the landslip, to construct a larger view of its occurrence, and in so doing to place within a general framework not only this phenomenon but others, which are of a geomorphological origin and which are no less pressing.

The post-glacial valley of the River Irwell, which is cut into the glacial and fluvio-glacial deposits of the Manchester embayment, is a relatively immature landscape feature which provides the sites for many types of geomorphological adjustment in the form of gully erosion and mass movement. The largest such feature, where the River Irwell flows against the valley side slope, is the Cliff landslide in Broughton (Fig. 19.1).

The landslide is a complex mass movement feature in unconsolidated glacial deposits which overlie Permo-Triassic rocks to a depth of about 30 m at this point (Fig. 19.2). The general sequence is of silty laminated clays with lenses of silt and sand overlain by stiff brown clays with stone inclusions. These in turn are overlain by stiff clay with stone, silt and sand inclusions, with loose to very loose silty, medium to fine sands on the top. Mass movement in the lower clays takes the form of flow lobes which undercut the sand slopes above. These upper slopes show frequent small-scale failures and occasional, rapid, large-scale adjustments. There are no eye-witness reports of failures taking place, but it is presumed that in general block structures are not preserved, and failure develops into a type of flow. The top of the slope appears to be the site of active tension cracks which extend into Bury New Road and seem to be early manifestations of the slumping which has occurred at the present face.

The history of the landslip site

A clear record of events since the late nineteenth century is found in the files of the City Engineer's Department of the City of Salford. In February 1882 a substantial slip was reported, which did not extend as far as Great Clowes Street, but left it in a precarious position. The somewhat unwise response was to tip fill at the head of the slope, providing additional stress to the driving moment. By May of the same year cracks had appeared in the surface of Great Clowes Street, and

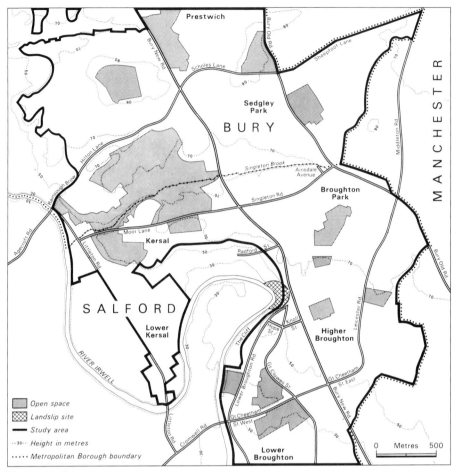

Fig. 19.1 Topography.

later in the year a rudimentary system was installed to drain the slope. Further slips were reported in 1886, April 1887 and June 1888, and in 1892 it became necessary to insert timber supports in the roadway. In April 1925 the tram service along Great Clowes Street was discontinued and the endangered section of the road was closed to all mechanically propelled vehicles in January 1926. A further major slip occurred in July 1927, and by July 1933 the problem had become so acute that the adjacent section of road was closed altogether. Between 1927 and 1967 there appears to have been slow subsidence of the top half of the slope, with slips recorded in November 1945, April 1947 and August 1948. Since 1948 no substantial movements have been recorded, but continual small changes persist.

The other principal effects of subsidence have been on housing. Numbers 402 to 416, Great Clowes Street, which were built in 1910, directly overlook the site of the slide (Fig. 19.3). These were the subject of a special clause in the Salford Corporation Act (1933) which permitted the City Engineer to evict without notice

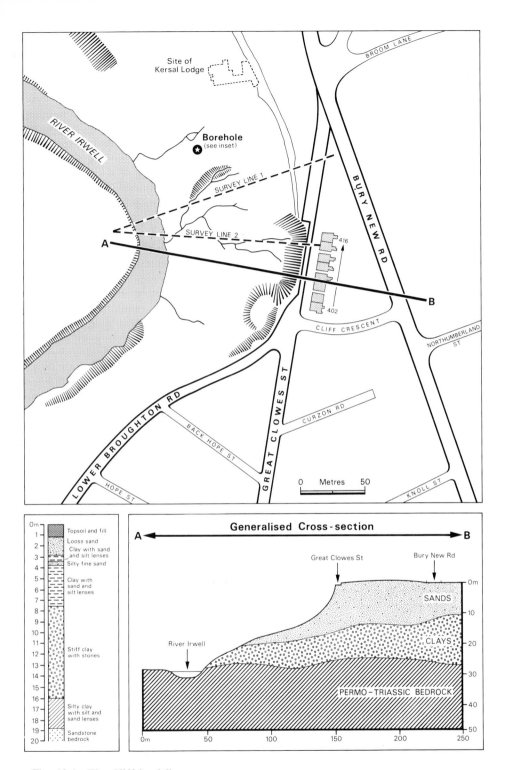

Fig. 19.2 The Cliff landslip.

if the houses became in imminent danger of collapse. The houses remain standing today but are less than 10 m from the slope crest.

Kersal Lodge, which was located on the valley slope, was allowed to fall into disrepair and was demolished after the Second World War. The Council refused to accept any responsibility for it or the access road which joined it to Great Clowes Street at the Cliff because of the natural hazard which was being allowed to take its course. The Council considered there was little they could do to prevent any damage. This indeed has been the pragmatic approach to the problem taken by local government in all questions arising from the mass movement.

Recently the Cliff Area Residents' Association (CARA) has sought action to stop the passage of large vehicles through the area because they were presumed to be causing damage to buildings. It has also expressed some concern about the mass movement and the fabric of houses in a report and questionnaire to the City Engineer, and has received the same pragmatic response. In short, the Council accepts no responsibility for a situation beyond its control.

Where the Council does have responsibility, in operating planning permissions, it has refused these for developments in the area of the Cliff along the valley slope at Radford Street (Fig. 19.1). In the Cliff area itself between the valley slope and Great Clowes Street there have been several building developments, presumably because these areas were thought not to be in any danger.

The causes of the landslip

At first glance there are few problems in explaining the presence of large scale and continuing mass movement at the Cliff. Many of the usual characteristics of mass movement sites are present: it is a slope composed mainly of sands overlying clays and having a river at the base. The slope materials have relatively low strengths and exhibit perched water tables.

Previous engineering analyses of the site and slope stability ascribe the mass movement to continuing erosion by the river. At this point in the bend of the river, the solid rock is exposed, part of a buried ridge which trends SW–NE. The river, it is postulated, is exposing the ridge and following its line to the northeast.

The history of events at the Cliff since the late nineteenth century indicates three or four periods when the upper slope was particularly active. What was happening to the lower slope can only be presumed. A possible determinant of this instability is erosion by the river of the toe of the slope. This mechanism was outlined by consultant engineers in 1948 and seems a rational explanation. However, map evidence from successive Ordnance Surveys since 1848 shows that very little movement of the channel has taken place, and there is no clear trend of erosion into the slope or along the Permo-Triassic exposure. It is possible that instability could arise as a result of temporary incursions by the river during flood, but this would hardly explain continuing instability when the position of the river channel remained virtually unchanged.

The Cliff lies close to the line of the Ardwick Fault, but there is no evidence to suggest that slope movement is triggered by tectonic movements. Records of

Fig. 19.3   The crest of the Cliff landslip.

earthquake activity in the Manchester area show no correspondence with periods of movement at the Cliff. Thus, although there could be some connection between instability and river or fault activity, there is little reason to think so, and prospects of establishing links seem thin. Moreover, such activities have little bearing on the wider aspects of the problem which have emerged.

Sub Soil Surveys (1967) estimated the factor of safety of the slope to be close to 1.0 along a number of hypothetical slip surfaces. Their analyses indicate that a slip could be either (a) in the upper sand slope, or (b) more deep-seated, passing through the lower glacial clay, or (c) a combination of (a) and (b). One proposed remedy, which would involve lowering the angle of slope, would necessitate either extending the crest to the east of Bury New Road, or placing material at the toe. In the latter case, Sub Soil Surveys (1967) still anticipate the possibility of instability in the overlying sand. The report continues:

Apart from the possibility of a general slip of the slope there could also be internal erosion of the sand deposit by the flow of water and sand from the deposit. The ground water is issuing from the base of the sand deposit at a number of points, and it is possible that due to internal 'piping' there is also sand being carried away with the ground water during very wet weather conditions. As this sand is eroded from the base it will tend to produce a subsidence of the sand slope.

A sub-soil drainage system was advised in order to prevent this removal of sand from the slope.

Standard penetration tests show a mean N value (blows/ft) of 4 (the penetration tests were often as low as 2 blows/ft), indicating a loose sand which has an angle of shearing resistance ϕ_d of 30°. Slip circles through the sand had a factor of safety of 0.93. Thus the sand is susceptible to slippage. However, this analysis has no bearing on its susceptibility to 'piping', except that we know that it is loose.

In 1965 the City Engineer decided to establish surveys of the slope in order to ascertain the nature of the movement. Since the Autumn of 1966, periodic, but not regular, surveys have been made of two profiles down the slope, and a third one was added in 1978. The surveys present some problems of interpretation in that marker pegs have been replaced, which eliminates continuity. Nevertheless, the surveys show that movement has taken place on both the upper and lower sections of the slope since 1966 (Fig. 19.4).

There is evidence of continued movement along cracks at the crest of the slope. Photographic evidence from 1978 and 1981 shows that between these two dates the pavement and fence at the crest had been reinstated and suffered subsequent deformation. Repeated observations have shown continual small-scale movements near the crest. Whether the upper slope movements are independent of those in the lower slope is not known, but close inspection of the slope surveys seem to show that there is some independence of movement. The lateral and downward movements in the sands may have led to deposition lower down, but it is possible that lower 'accumulations' could be due to bulging of wet clays.

Wider evidence of movement at the Cliff

The area known as the Cliff extends to the east and south of the landslide. It

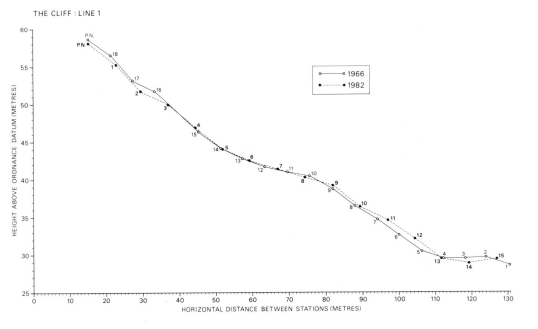

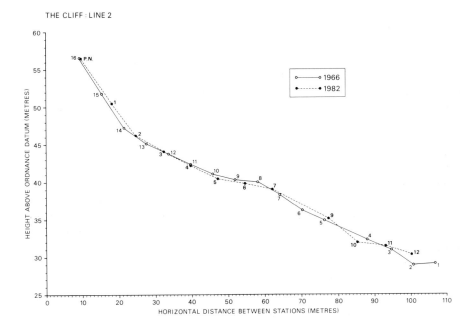

Fig. 19.4   Surveyed profiles across the Cliff landslip, 1966–82.

occupies the crest of the slope overlooking the River Irwell and extends southwards towards Lower Broughton where the valley slope disappears altogether.

It is well known that the area has experienced many problems of damage to buildings and roads. For instance, one of the major complaints of the CARA was about damage to houses caused by heavy vehicles. Road repairs to Lower Broughton Road, Great Clowes Street, and their tributary streets are an almost perpetual occurrence. Sewer repairs in Knoll Street and Hope Street lasted several months and required deep excavation.

In the light of such conditions in the area of the slip it was decided to examine the extent of housing and road damage in order to ascertain whether these effects could be attributed to the mass movement, or whether the mass movement and damage have a common origin.

Visual evidence of damage to houses and roads was collected in surveys by the authors and students from the Department of Geography of the University of Salford. The evidence was taken as being clearly visible external distortions to house walls and roofs, and severe irregularities in road surfaces. Examples of the severity of such damage are shown in Figs. 19.5 and 19.6. The survey was not undertaken by experienced building surveyors and therefore only the most obvious signs of damage were noted.

What emerged from this preliminary survey was:

1. The Cliff area generally was subject to damage to houses and roads, and the area beyond showed similar features.

2. There seemed to be no spatial correspondence between the damage and any possible mass movement feature.

The implications of this small survey are clear. Housing damage near the Cliff and perhaps even within the zone affected by landsliding is not due to the landslip but is caused by other movement and is possibly due to what heretofore has been considered the secondary process of internal erosion within the sand.

If this is the case then there is reason to expect that the area subject to such damage may have a much wider distribution. Moreover, since a causal mechanism can be put forward for such damage (the internal erosion of loose sand) then these areas may coincide with the areas of glacial deposits which have an arenaceous character. It was therefore decided to investigate more widely the problem which the Cliff area posed, in order to test this association.

The geological map of the area around the Cliff (Fig. 19.7) shows the principal divisions of the glacial and fluvial deposits. The former consist of sands and gravels, and clays. There is an apparently simple pattern of distribution, with the sands and gravels overlying the clays. In general, the sands and gravels occupy the higher ground in the north and the clays the lower ground in the south and along the lower slopes of the Irwell valley. The sand and gravel deposits are thought to have been laid down under pro-glacial conditions in what is, in fact, stratigraphically a very complex area. Boreholes drilled for the construction of the M62 motorway to the north show that the area underwent a period of deglaciation of consider-

Fig. 19.5 Ground movement damage, Hope Street.

able variations in deposition, such that the correlation of borehole sequences over even very short distances would appear to be a highly dubious procedure. It is clear that the 1910 drift map can only be used to indicate very generally where sand may be found at the surface. However, these sand areas were used to investigate whether they reveal the same sorts of problem encountered in the area of the Cliff.

Evidence of ground movement in areas of fluvio-glacial sands

The significance of this study derives from the fact that the area which was suspected of being affected by ground movement is built-up. But, this also means that much of the evidence that might be obtainable to establish the extent and timing of these movements is not available.

There are severe economic implications in designating an area, and more particularly a site, as being susceptible to ground movement. Therefore no survey could be contemplated which involved acquiring information from owners of property.

For any area there is an enormous amount of information on housing conditions, but this is almost entirely confidential and is in the hands of estate agents, surveyors, building societies and banks. It was decided therefore to collect information from Local Authority records which might have a bearing on ground movements. The survey area covered about 10 square kilometres around the Cliff site, including parts of Salford 7 and 8 and the Prestwich area of Bury. Metropolitan District boundaries were used because of record sources, and the northern and southern boundaries were dictated by logistical considerations. The study area was also limited by the extent of coal mining. It lies mainly outside the area within which the National Coal Board accepts liability for subsidence damage.

The Local Authority information takes the form of data on road repairs (including pavement repairs) and sewer repairs. It was thought that an exploratory analysis of the location of such repairs might reveal that the damage evident near the Cliff has expression in road and sewer damage, and that the wider incidence would also be indicated. The locations by street of repairs to roads and sewers were extracted from the records of the Engineers' Departments of Salford and Bury councils. The periods covered are 1977–81 for Salford and 1976–81 for Bury.

Road and sewer repairs can be undertaken for a large number of reasons, only one of which may be ground movement. Furthermore, ground movement can take place for a number of reasons, one of the more important being heavy traffic. However, if ground movement is taking place and is manifest in damage to roads and sewers and this movement is related to fluvio-glacial sands, then there should be some spatial correspondence between the two.

Maps were prepared of the frequency of road and sewer repairs per km of road for the study area. The area was divided into squares of $\frac{1}{3}$ km side length for analysis. The size of the square was the nearest simple ratio of map square size which would exhibit coherent patterns, which are discussed below. Other sizes were considered but rejected because of coarseness of scale or lack of resolution.

Fig. 19.6 Misalignment of terraced houses, Great Clowes Street.

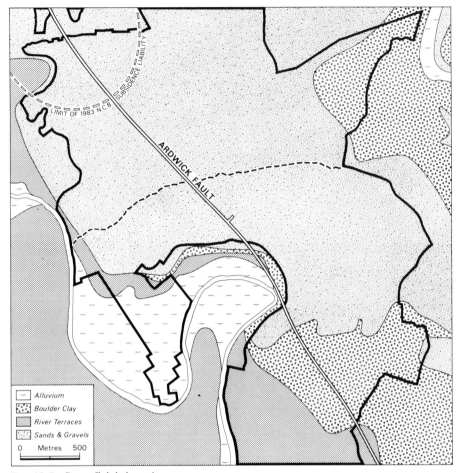

Fig. 19.7 Superficial deposits.

There is some degree of incompatibility between the data from Salford and Bury
which arises partly from the time periods involved and partly because repair
recording procedures and Council repair policies differ. For the part of Salford
studied, individual squares produce values ranging from 1 to 24 repairs per km
of road, with a mean value of 7. For the part of Bury studied, the equivalent values
are 4 and 41 with a mean of 19. Strictly, two separate maps of repair frequency
should be presented, but a comparison would show predominantly the higher
general repair frequency for Bury. For spatial continuity, the records have been
reduced to a standard form. For each of the two areas a frequency distribution
of repairs per km (by square) was plotted and divided into quartiles, which were
then mapped (Fig. 19.8). In this way the highest repair frequencies in both Districts
may clearly be seen. It cannot, however, be assumed that high repair frequency
squares in Salford have the same absolute damage levels as equivalent squares in
Bury, although the housing damage evidence presented below and the contiguity

of upper quartile squares in the area of the common local authority boundary may indicate that this indeed is the case.

A second survey was undertaken, with respect to housing damage. As with road and sewer repairs the object was to investigate the spatial association between such damage and sand deposits. Housing damage was measured by inspecting house fronts and noting clear evidence of major cracking, distortion, and walls out of plumb (Fig. 19.9). As with road and sewer repairs, the existence of housing damage can be due to a number of reasons, not least of which is the age and type of house. However, in this exploratory survey no distinction between house types was made as the areal extent of major damage was of paramount interest.

Surveys were made of a stratified, randomly selected sample of streets. These were collected by random points in each $\frac{1}{3}$ km square. For each, the percentage of houses exhibiting clear external evidence of damage was recorded. The survey was carried out in the second half of 1983, and therefore reveals damage existing

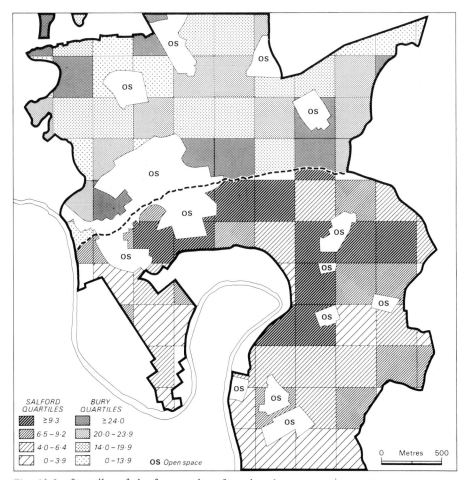

Fig. 19.8 Quartiles of the frequencies of road and sewer repairs per km.

Fig. 19.9 Ground movement damage, Ainsdale Avenue.

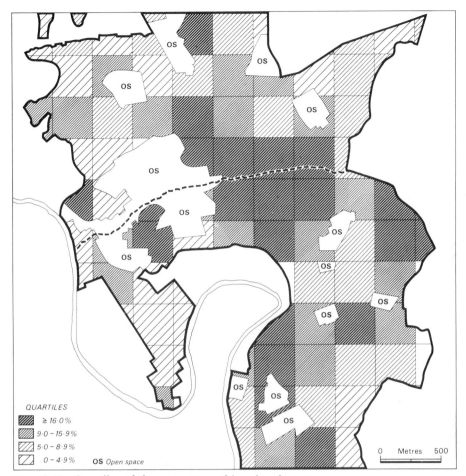

Fig. 19.10 Quartiles of the percentage of housing damage.

at that time, but not the period when each incidence occurred. The samples comprise approximately 4200 houses.

The samples in each square were aggregated and a frequency distribution of percentage damage by squares constructed. The quartiles of this distribution are presented in map form (Fig. 19.10) in order that they may be compared with the road and sewer repair maps. The housing damage data are also presented as an isoline map constructed from the individual street samples (Fig. 19.11). Since there are some small samples, this map must be viewed with some caution as far as the detail is concerned.

Discussion

The distributions of road and sewer repair frequency and housing damage are not random. The patterns of distribution are not simple but they are susceptible to interpretation. First, it is important to note the close correspondence between the patterns of high damage to houses and high road and sewer repair frequencies. This is irrespective of type of house or road. There are minor differences in the distribution but the major high value areas coincide.

High damage is recorded in the areas of the Cliff, the northern end of Leicester Road, at squares along the Irwell valley slope and in a band on either side of the Singleton Brook, along the boundary between Salford and Bury. High housing damage is also recorded in the Great Cheetham Street area of Lower Broughton, an area of high traffic density with mainly late nineteenth century housing located close to the road.

The Cliff area appears on each map as an area of high damage. Data for individual squares and streets indicate that this is an area of very intense damage. Moreover, the damage occurs in streets well away from the landslide. Heavy traffic follows Great Clowes Street, but this street does not stand out as severely damaged.

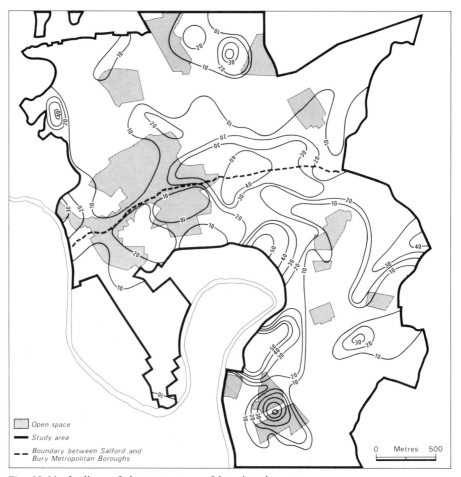

Open space
Study area
Boundary between Salford and
Bury Metropolitan Boroughs

0   Metres   500

Fig. 19.11 Isolines of the percentage of housing damage.

Some sections of other streets in the Cliff area show almost 100 per cent housing damage.

Some of the roads in this area have been subjected to very substantial repair. One aspect of repair not recorded is the magnitude of the engineering works. It is worth noting, for example, that recent (1976–8) repairs to Knoll Street and Hope Street sewers involved excavation of the entire roadway to depths of 4 or 5 metres, replacement with non-settling backfill and flexible joints in pipes.

Much of the area is built of terraced houses and other property of a relatively low quality and it might be the case that recorded damage is largely a function of house construction. However, if this were the case then road damage would not show spatial association and similar housing areas would show similar levels. Neither type of evidence shows a pattern related to house type and the area detailed above, irrespective of building condition, shows values in the upper quartile. Clear physical evidence of ground movement is provided by faults in walls, depressions in roads and misalignment of entire buildings.

Thus the earlier inference is borne out that the Cliff landslide is but a local manifestation of a general instability. Here, local geography has produced landsliding where elsewhere ground movement has occurred.

The correspondence of ground movement damage with areas of fluvio-glacial sands, as recorded by the Geological Survey, is borne out by the map evidence. However, there is no simple correspondence. Rather, some areas of sand are subject to damage while others are not. At present no independent evidence of the causes of ground movement is available and so the following interpretation is submitted.

There is considerable circumstantial evidence that ground movements occur and that they are related to some sort of movement in sands. Excessive settlement does not occur when structures are built, but changes occur at considerable intervals after construction and seem to continue for long periods of time. More specific evidence of collapse is found as holes in roads, reports of sewer engineers about large cavities around sewer pipes and brickwork, and from engineering reports of building and road subsidence. Indirect evidence of the possibility of sub-surface movement is provided in borehole records of running sands in addition to the surveys reported here.

Three mechanisms may be appropriate to explain ground movement. The first, which may be dismissed immediately, is internal erosion by translocation of depositional material. This requires a strongly bimodal particle size distribution (Hallsworth, 1963; Statham, 1977) in the sand-to-clay size range. However, the sands in this area contain very little clay and are distinctly unimodal, making translocation most unlikely.

The second mechanism is by collapse following piping. Terzaghi and Peck (1967) recognise two types of piping, due to either sub-surface erosion or heave. The former occurs by a process of headward erosion which starts at springs and proceeds upstream along the base of structures or along bedding planes. Piping due to heave occurs as a result of flow concentration, due to structures, such that the hydraulic gradient is sufficient for the seepage pressure to exceed the effective weight of the soil. In this condition there is the sudden rise of a large body of soil and the evacuation of a pipe along the flow lines as soil is removed.

The glacial sands of the area (Sub Soil Surveys, 1967; Lancashire County Council, 1964) are probably susceptible to both sorts of piping. In general terms the greatest susceptibility is presented by soils which are either '(i) poorly compacted though well graded cohesionless material with practically no binder or (ii) very uniform fine cohesionless sand, even though well compacted' (Krynine and Judd, 1957). Standard penetration tests show that the sands at the Cliff landslip have a mean N (number of blows/ft) of 4 and a range of 2–10. This places them in the range of very loose (0–4) or loose (4–10) soils (Terzaghi and Peck, 1967). Here, and in the area in general, particle size distributions show very well graded soils and measured bulk densities are in the range 96.8 to 120.0 lb/cu ft. Furthermore, the deposits are complex and as Terzaghi and Peck (1967) report, 'the greatest difficulties associated with preventing subsurface erosion are encountered in sedimentary deposits in which layers of inorganic silt are in direct contact with layers of clean coarse sand or gravel' (p. 619). Such situations are common in borehole records. Thus the soils are likely to exhibit sub-surface erosion or heave, but one problem remains in accepting these second mechanisms. There is at present no real evidence of the external features associated with piping. Heave has not been recorded in the area and there are no indictions of high sediment discharges in springs or along Singleton Brook. Neither are there any external depositional features which could be attributed to internal erosion except on a very small scale at the Cliff landslide.

One possibility is that conditions for sub-surface erosion are enhanced by buried structures and that the erosion is very localised. Existing cavities around footings, pipes and other structures may provide depositional sites, but this needs to be established independently.

The third mechanism is by collapse. This can occur where fine sands or silts of low density and with no binder are subjected to a rising water table either generally or even locally in response to impermeable barriers being inserted into soils. The strength of the unsaturated sand, which previously was in part dependent on capillary forces of attraction, is reduced on saturation and the open granular structure is destroyed. The very loose nature of the sands is indicated by many penetration tests. Loose and loose-to-medium sands are encountered along motorway sections both in the top metre where N values vary from 2 to 7 and at depths up to 7 m with values as low as 8 (Lancashire County Council, 1964). The motorway records contain very few sand density measurements but some bulk density measurements of very loose sands below 4 m show values less than 100 lb/cu ft. These values approach the range for deposits susceptible to collapse, although the very small number of samples leaves open to doubt the possibility of collapse as an explanation of widespread ground movement.

One fact remains clear. Whether sub-surface erosion, heave or collapse is responsible for ground movement, the distribution of damage indicates a strong dependence on particular topographic situations. It remains to be investigated whether the mechanism which is operating relates to ground water levels or flow rates. What is also clear is that there is no evidence of ground movement or sub-surface erosion not associated with structures, except at the Cliff landslide, and it must be assumed at this stage that damage to buildings, roads, pipes and sewers arises indirectly as a consequence of construction.

Conclusions

Parts of Salford and Bury are subject to housing and road damage which can only reasonably be explained by ground movements. Ground movement is probably caused by piping or collapse of sands of fluvio-glacial origin. Evidence of damage indicates that there are sites of greater movement in particular topographic situations. Circumstantial evidence indicates that ground movement is highly localised. It follows the construction of sewers, pipes and foundations rather than occurring naturally.

This initial study raises a number of problems. First, what processes cause ground movement? Secondly, how can the existence and extent of ground movement be predicted? Thirdly, how can it be prevented or controlled? Each of these questions requires a greater understanding than exists at present of the mechanisms of such movement and the factors which control its rate. Fourthly, how could such knowledge be used in short- and long-term planning?

P

Bibliography

Adams, A. E. (1980) 'Calcrete profiles in the Eyam Limestone (Carboniferous) of Derbyshire: petrology and regional significance', *Sedimentology*, **27**, 651–60.

Adams, A. E. (1984) 'Development of algal–foram–coral reefs in the Lower Carboniferous of Furness, northwest England', *Lethaia*, 17, 233–249.

Agassiz, L. J. (1840/1) 'Glaciers and the evidence of their having once existed in Scotland, Ireland and England', *Proc. Geol. Soc.*, **3**, 327–32.

Allen, J. and Shahwan, A. (1954) 'The resistance to flow of water along a tortuous stretch of the River Irwell (Lancashire) – an investigation with the aid of scale-model experiments, *Proc. Inst. of Civ. Engrs.*, 17, 144–65.

Allen, J. R. L. (1960) 'The Mam Tor Sandstones: a "turbidite" facies of the Namurian deltas of Derbyshire, England', *J. Sedim. Petrol.*, **30**, 193–208.

—— (1965) 'Late Quaternary Niger delta and adjacent areas: sedimentary environments and lithofacies', *Bull. Am. Ass. Petrol. Geol.*, **49**, 547–600.

Allen, P. (1975) 'Wealden of the Weald: a new model', *Proc. Geol. Assoc.*, **88**, 389–437.

Allen, P., Keith, M. L., Tan, F. C. and Deines, P. (1973) 'Isotopic ratios and Wealden environments', *Palaeontology*, **16**, 607–21.

Al Sabti, G. (1977) *'Morphometry of the limestones in a part of the Peak District of Derbyshire'*, unpublished M.Sc. thesis, University of Birmingham.

Al Saigh, N. H. (1977) *'Geophysical investigations of glacial sediments in the region of Madeley, Staffordshire'*, unpublished M.Sc. thesis, University of Keele.

Anderson, P. and Shimwell, D. (1981), *Wild Flowers and other Plants of the Peak District – an Ecological study* (Moorland Press, Ashbourne).

Anderson, P. and Tallis, J. (1981) 'The nature and extent of soil and peat erosion in the Peak District – field survey', in Phillips, J., Yalden, D. and Tallis, J. (eds.), *Moorland Erosion Study, Phase 1, Report*, (Peak Park Joint Planning Board, Bakewell), pp. 52–64.

Anderson, S. S. (1972) 'The ecology of Morecambe Bay. II. Inter-tidal invertebrates and factors affecting their distribution', *J. appl. Ecol.*, **9**, 161–78.

Anderton, R., Bridges, P. H., Leeder, M. R. and Sellwood, B. W. (1979), *A Dynamic Stratigraphy of the British Isles*, (Allen and Unwin, London).

Andrew, R. and West, R. G. (1977) 'Appendix. Pollen analyses from Four Ashes, Worcs.', *Phil. Trans. R. Soc.*, Lond. **280B**, 242–6.

Andrews, J. T., King, C. A. M. and Stuiver, M. (1973) 'Holocene sea-level changes, Cumberland coast, north-west England: eustatic and glacio-isostatic movements', *Geol. Mijnb.*, **52**(1) 1–12.

Arthurton, R. S. and Wadge, A. J. (1981) 'Geology of the country around Penrith', *Inst. Geol. Sci.*, (HMSO, London).

Ashmead, P. (1970) 'Notes on the sediments of Kirkhead Cave', *Archaeol. News Bull. for Northumb., Cumb., Westm.*, **8**, 5–6.

Ashmead, P. (1974) 'The caves and karst of the Morecambe Bay area', in A. C. Waltham (ed.), *The Limestones and Caves of North West England*, (David and Charles, Newton Abbot), pp. 201–26.

Ashworth, A. C. (1972) 'A late-glacial insect fauna from Red Moss, Lancs., England',

Entomol. Scand., **3**, 211–24.

Aspin, C. (1969) *Lancashire, the First Industrial Society* (Helmshore Local History Society, Helmshore).

Astle, W. (ed.) (1971) *History of Stockport* (with a new introduction by J. Christie-Miller), (S.R. Publishers, Wakefield).

Atkinson, T. C., Harmon, R. S., Smart, P. L., and Waltham, A. C. (1978), 'Palaeoclimatic and geomorphic implications of ^{230}Th/^{234}U dates on speleothems from Britain', *Nature*, **272**, 24–8.

Audley-Charles, M. G. (1970) 'Triassic palaeogeography of the British Isles', *J. Geol. Soc. Lond.*, **126**, 49–89.

Bakker, J. P. and Levelt, Th. W. M. (1964) 'An inquiry into the probability of a polyclimatic development of peneplains and pediments (etchplains)', in Europe in the Senonian and Tertiary period. *Publication van de Universsitat van Amsterdam*. No. 4.

Ball, D. F. (1982) The sand and gravel resources of the country south of Wrexham, Clwyd. Description of 1:25 000 sheet SJ 34 and parts of SJ 24. *Mineral Assessment Rep. Inst. Geol. Sci.*, 106.

Bancroft, H. (1881) *Report to the Irwell Floods Committee representing the inhabitants of Strangeways and Lower Broughton.*

Barber, K. E. (1981) *Peat Stratigraphy and Climatic Change.* (Balkema, Rotterdam).

Barke, F. (1920) 'The evolution of river valleys', *Trans. N. Staffs. Field Club*, **54**, 17–27.

—— (1929) 'The old course of the River Churnet', *Trans. N. Staffs. Field Club*, **63**, 90–7.

Barnes, B. (1975) '*Palaeoecological studies of the late Quaternary period in Northwest Lancashire*', unpublished Ph.D. thesis, University of Lancaster.

Barnes, F. and Hobbs, J. L. (1951) 'Newly discovered flint-chipping sites in the Walney Island locality', *Trans. Cumb. and Westm. Antiq. Archaeol. Soc.*, N.S. **50**, 20–29.

Barnes, F. A. (1963) 'Peat erosion in the southern Pennines: problems of interpretation', *E. Midld. Geogr.*, **3**(4), 216–22.

Barrett, E. C. (1964) 'Local variations in rainfall trends in the Manchester region', *Trans. Inst. Br. Geogr.*, **35**, 55–72.

Bartley, D. D. (1975) 'Pollen analytical evidence for prehistoric forest clearance in the upland area west of Rishworth, West Yorkshire', *New Phytol.*, **74**, 375–81.

Bateman, J. F. La T. (1884) *History and Description of the Manchester Waterworks*, (Day, Manchester and Spon, London).

Battiau-Queney, Y. (1980) 'Contribution à l'étude Géomorphologique du Massif Gallois', Thèse, l'Université de Bretagne Occidentale.

Beaujeau-Garnier, J. (1975) 'Variety of the natural elements', in J. M. Houston (ed.), *France* (Longman, London), pp. 4–19.

Beaumont, P., Turner, J. and Ward, P. F. (1969) 'An Ipswichian peat raft in glacial till at Hutton Henry, Co. Durham', *New Phytol.*, **68**, 797–805.

Beaver, S. H. and Turton, B. J. (1979) *The Potteries: a description of the O.S. 1:50 000 sheet 118.* (British landscape through maps. No. 19), (Geographical Assoc., Sheffield).

Beck, J. S. (1980) '*Aspects of speleogenesis in the Carboniferous Limestone of North Derbyshire*', unpublished Ph.D. thesis, University of Leicester.

Beckett, S. C. (1981) 'Pollen diagrams from Holderness, North Humberside', *J. Biogeogr.*, **8**, 177–98.

Belderton, R. H. (1964) 'Holocene sedimentation in the western half of the Irish Sea', *Marine Geol.*, **2**, 147–63.

Bell, F. G. (1969) 'The occurrence of Southern steppe and halophyte elements in Weichselian floras from southern Britain', *New Phytol.*, **68**, 913–22.

Binney, E. W. (1848) 'Sketch of the drift deposits of Manchester and its neighbourhood', *Mem. Manchr. Lit. Phil. Soc.*, Ser. 2(8), 195–234.

Birks, H. J. B. (1964) 'Chat Moss, Lancs.', *Mem. Proc. Manchr. Lit. Phil. Soc.*, **106**, 1–45.

—— (1965a) 'Late-glacial deposits of Bagmere, Cheshire and Chat Moss, Lancs.', *New Phytol.*, **64**, 270–85.

—— (1965b) 'Pollen analytical investigations at Holcroft Moss, Lancs. and Lindow Moss, Cheshire', *J. Ecol.*, **53**, 299–314.

—— (1982) 'Mid-Flandrian forest history of Roudsea Wood National Nature Reserve, Cumbria', *New Phytol.*, **90**, 339–54.

Bishop, W. W. and Coope, G. R. (1977) 'Stratigraphical and faunal evidence for Late-Glacial and Early Flandrian environments in south-west Scotland', in J. M. Gray and J. J. Lowe (eds.), *The Scottish Late-Glacial Environment*, (Pergamon, Oxford), pp. 61–88.

Blatt, H., Middleton, G. and Murray R. (1972), *Origin of Sedimentary Rocks*, (Prentice Hall, New Jersey).

Boardman, J. (1984) 'Red Soils and Glacial erosion in Britain', *Quat. Newsl.*, **42**, 21–4.

Bolton, J. (1862) 'On a deposit with insects, leaves etc. near Ulverston', *Proc. Geol. Soc. Lond.*, **18**, 274–7.

Bonsall, J. C. (1976) 'Monk Moors, Eskmeals', *Archaeological Excavations, 1975*, (HMSO, London), p. 35.

—— (1981) 'The coastal factor in the Mesolithic settlement of north-west England', in B. Gramsch (ed.), *Mesolithikum in Europa*, (VEB Deutscher Verlag der Wissenschaften, Berlin), pp. 451–72.

—— (1982) 'Eskmeals, Cumbria. SD 0892', *Proc. Prehist. Soc.*, **48**, 504–7.

—— (1983) 'Eskmeals, Cumbria. SD 0892', *Proc. Prehist. Soc.*, **49**, 382.

Bonsall, J. C. and Mellars, P. A. (1977) 'Williamson's Moss, Monk Moors and Langley Park, Archaeology II', in M. J. Tooley (ed.), *The Isle of Man, Lancashire Coast and Lake District*, 10th INQUA Congress Excursion Guide (Geo Abstracts, Norwich), pp. 42–4.

Bostock, J. L. (1980) *'The History of the Vegetation of the Berwyn Mountains, North Wales, with Emphasis on the Development of the Blanket Mire'*, unpublished Ph.D. thesis, Univ. of Manchester.

Bott, M. H. P. (1964) 'Gravity measurements in the north-eastern part of the Irish Sea', *J. Geol. Soc. Lond.*, **120**, 369–96.

—— (1967) 'Geophysical investigations of Northern Pennine basement rocks. *Proc. Yorks. Geol. Soc.*, **36**, 139–68.

—— (1974) 'The geological interpretation of a gravity survey of the English Lake District and the Vale of Eden', *J. Geol. Soc. Lond.*, **130**, 309–31.

—— (1978) 'Deep structure', in F. Moseley, (ed.), *The geology of the Lake District, Yorkshire Geological Society*, Occ. Publ. 3, 25–40.

Bott, M. H. P. and Masson-Smith, D. (1957), 'The Geological interpretation of a gravity survey of the Alston block and the Durham coalfield', *J. Geol. Soc. Lond.*, **113**, 93–117.

Bott, M. H. P. and Young, D. G. G. (1971) 'Gravity measurements in the north Irish Sea', *J. Geol. Soc., Lond.*, **126**, 413–34.

Boulton, G. S. (1970) 'On the origin and transport of englacial debris in Svalbard glaciers', *J. Glaciol*, **9**, 213–28.

—— (1971) 'Till genesis and fabric in Svalbard, Spitzbergen', in R. P. Goldthwait (ed.) *Till: a Symposium*, (Ohio U.P., Columbus), pp. 41–72.

—— (1972) 'Modern Arctic glaciers as depositional models for former ice sheets', *J. Geol. Soc. Lond.*, **128**, 361–93.

—— (1977) 'A multiple till sequence formed by a Late Devensian Welsh ice-cap, Glanllynau, Gwynedd', *Cambria*, **1**(4), 10–31.

Boulton, G. S., Jones, A. S., Clayton, K. M. and M. J. Kenning (1977) 'A British Ice Sheet Model and Patterns of Glacial Erosion and Deposition in Britain', in F. Shotton (ed.), *British Quaternary Studies: Recent Advances*, (Oxford University Press), 231–246.

Boulton, G. S. and Jones, A. S. (1979) 'Stability of temperate ice caps and ice sheets resting on beds of deformable sediment', *J. Glaciol.*, **24**, 29–43.

Boulton, G. S. and Paul, M. A. (1976) 'The influence of genetic processes on some geotechnical properties of glacial tills', *Q. J. Engng. Geol.*, **9**, 159–94.

Boulton, G. S. and Worsley, P. (1965) 'Late Weichselian glaciation of the

Cheshire–Shropshire basin', *Nature*, **207**, 704–6.

Bowden, K. F. (1955) 'Physical oceanography of the Irish Sea', Min. Agric. Fish. and Food, *Fisheries Investigations.* Series II. **18**(8), 1–68.

Bowen, D. Q. (1973) 'The Pleistocene succession of the Irish Sea', *Proc. Geol. Assoc.*, **84**, 249–73.

—— (1978) *Quaternary Geology.* (Pergamon, Oxford).

Bowen, D. Q. and Sykes, G. A. (1984) 'Amino acid dating the British Pleistocene', Paper abstract, Section C (Geology), Brit. Ass. Adv. Sci., Norwich.

Bower, M. M. (1960a) 'Peat erosion in the Pennines', *Adv. Sci.*, **16**, 323–31.

—— (1960b) 'The erosion of blanket peat in the southern Pennines'. *E. Midld. Geogr.*, **13**, 22–33.

—— (1961) 'The distribution of erosion in the blanket peat bogs of the Pennines', *Trans. Inst. Br. Geogr.*, **29**, 17–30.

—— (1962) 'The cause of erosion in blanket peat bogs', *Scot. Geog. Mag.*, **78**, 33–43.

Bradshaw, M. (1982) 'Process, time and the physical landscape; geomorphology today', *Geography*, **67**, 15.28.

Bramwell, D. (1964) 'The excavations at Elder Bush Cave, Wetton, Staffordshire', *N. Staffs. J. Field Stud.*, **4**, 46–60.

Bremer, H. (1980) 'Landform development in the Humid Tropics, German Geomorphological research', *Z. Geomorph., Suppl.*, **36**, 162–75.

—— (1983) 'Albrecht Penck (1858–1945) and Walter Penck (1888–1923), two German Geomorphologists', *Z. Geomorph.*, N.F. **27**, 129–38.

Brenchley, P. (1968) 'An investigation into the glacial deposits at Thurstaston, Wirral', *Amateur Geologist*, **3**, 27–40.

Broadhurst, F. M. and Simpson, I. M. (1967) 'Sedimentary infillings of fossils and cavities in limestone at Treak Cliff, Derbyshire', *Geol. Mag.*, **104**, 443–8.

—— (1973) 'Bathymetry on a Carboniferous reef', *Lethaia*, **6**, 367–81.

—— (1983) 'Syntectonic sedimentation, rigs and fault reactivation in the Coal Measures of Britain', *J. Geol.*, **91**, 330–37.

Broadhurst, F. M., Simpson, I. M. and Hardy, P. G. (1980) 'Seasonal sedimentation in the Upper Carboniferous of England', *J. Geol.*, **88**, 639–51.

Brodrick, H. (1902) 'Martin Mere', *Ann. Rep. Southport Sci. Soc.*, **8**, 5–18.

—— (1903) 'Martin Mere', *Rep. Brit. Assoc. Adv. Sci.*, Trans. Section C, 656.

Brook, D. (1974) 'Cave development in Kingsdale', in A. C. Waltham (ed.), *Limestone and Caves of North-West England*, (David and Charles, Newton Abbott), pp. 310–34.

Brook, D., Coe, R. G., Davies, G. M. and Long, M. H. (1972–6) *Northern Caves*, (Dalesman, Clapham, Yorks.).

Brooks, M. (1973) 'Some aspects of the Palaeogene evolution of Western Britain', *J. Geol.*, **81**, 81–8.

Brown, E. H. (1979) 'The Shape of Britain', *Trans. Inst. Br. Geogr.*, New Ser. **4**, 449–62.

Brown, J. E. (1973) 'Depositional histories of sand grains from surface textures', *Nature*, **242**, 396–8.

Brunsden, D. (1980) 'Applicable models of longterm landform evolution', *Z. Geomorph., Suppl.*, **36**, 16–26.

Bryson, R. A. and Wendland, W. M. (1967) 'Tentative climatic patterns for some late glacial and postglacial episodes in central North America', in W. Mayer-Oakes (ed.), *Life, Land and Water*, (Manitoba), pp. 271–98.

Buckland, W. (1824) 'On the excavation of valleys by diluvial action', *Trans. Geol. Soc.* (second series), **1**, 95–102.

Büdel, J. (1957) 'Die Doppelten einebnungs-glachen in den feuchten Tropen', *Z. Geomorph.*, N.F. **1**, 201–28.

—— (1979) 'Reliefgenerationem und klimageschichte in Mitteleuropa', *Z. Geomorph., Suppl.*, **33**, 1–15.

—— (1982) *Climatic Geomorphology*, (Princeton University Press, New Haven).

Bull, A. J. (1940) 'Cold conditions and landforms in the South Downs', *Proc. Geol. Assoc.*, **51**, 63–71.

Bunting, B. T. (1964) 'Hill slope development and soil formation on some British sandstones', *Geogr. J.*, 130, 73–9.

Burek, C. V. (1977) 'The Pleistocene Ice Age and After', in T. D. Ford (ed.), *Limestone and Caves of the Peak District*, (Geo Abstracts, Norwich), pp. 87–128.

—— (1978) *'Quaternary deposits on the Carboniferous Limestone of Derbyshire'*, unpublished Ph.D. thesis, Univ. of Leicester.

—— (1982) 'An unusual occurrence of sands and gravels in Derbyshire', *Mercian Geol.*, **17**, 123–30.

Burek, C. V. and Cubitt, J. M. (1979) 'Trace element distribution in the superficial deposits of Northern Derbyshire, England', *Minerals and the Environment*, 1(3), 90–100.

Burke, K. and Dewey, J. F. (1973) 'Plume generated triple junctions. Key indicators in applying plate tectonics to old rocks', *J. Geol.*, **81**, 406–33.

Burt, T. P. and Gardiner, A. T. (1981) *Some aspects of the runoff response from a small peat-covered catchment*, (Huddersfield Polytechnic, Department of Geography).

—— (1982) 'The permanence of stream networks in Britain: some further comments', *Earth surf. processes and landf.*, **7**, 327–32.

Buurman, P. (1980) 'Palaeosols in the Reading beds (Palaeocene) of Alum Bay, Isle of Wight, U.K.', *Sedimentology.*, **27**, 593–606.

Carter, R. W. G. (1982) 'Sea-level changes in Northern Ireland', *Proc. Geol. Assoc.*, **93**, 7–23.

Caston, G. F. (1976) 'The floor of the North Channel, Irish Sea: a side scan sonar survey', *Rep. Brit. Geol. Surv. No. 76/7.*

Catt, J. A. and Hodgson, J. M. (1976) 'Soils and geomorphology of the chalk of south east England', *Earth surf. processes and landf.*, **1**, 181–93.

Challinor, J. (1978) 'The "Red Rock Fault", Cheshire: a critical review', *Geol. J.*, **13**(1), 1–10.

Chapman, S. B. (1965) 'The ecology of Coom Rigg Moss, Northumberland III, Some water relations of the bog system', *J. Ecol.*, **53**, 371–84.

Charlesworth, J. K. (1926) 'The readvance marginal kame moraines of the south of Scotland and some later stages of retreat', *Trans. R. Soc. Edinb.*, **67**, 25–50.

—— (1939) 'Some observations on the glaciation of north-east Ireland', *Proc. R. Ir. Acad.*, B, **45**, 253–95.

Cheetham, F. H. (1923) 'Blowick: the name and the place', *Trans. Hist. Soc. Lancs. and Cheshire*, N.S. **39**, 186–202.

Cherry, J. (1963) 'Eskmeals sand-dunes occupation sites – Phase I, flint workings', *Trans. Cumb. and Westm. Antiq. Archaeol. Soc.*, N.S. **63**, 31–52.

—— (1965) 'Flint-chipping sites at Drigg', *Trans. Cumb. and Westm. Antiq. Archaeol. Soc.*, N.S. **65**, 66–85.

—— (1967) 'Prehistoric habitation sites at Seascale', *Trans. Cumb. and Westm. Antiq. Archaeol. Soc.*, N.S. **67**, 1–16.

—— (1969) 'Early Neolithic sites at Eskmeals', *Trans. Cumb. and Westm. Antiq. Archaeol. Soc.*, N.S. **69**, 40–53.

—— (1977a) 'Williamson's Moss, Monk Moors and Langley Park Archaeology I', in *The Isle of Man, Lancashire Coast and Lake District*, M. J. Tooley (ed.), 10th INQUA Congress Excursion Guide (Geo Abstracts, Norwich), pp. 41–2.

—— (1977b) 'Drigg Archaeology', in *The Isle of Man, Lancashire Coast and Lake District*, M. J. Tooley (ed.), 10th INQUA Congress Excursion Guide (Geo Abstracts, Norwich), p. 47.

Cherry, J. and Cherry, P. J. (1973) 'Mesolithic habitation sites at St. Bees, Cumberland', *Trans. Cumb. and Westm. Antiq. Archaeol. Soc.*, N.S. **73**, 47–66.

Cheshire, S. G. and Bell, J. D. (1977) 'The Speedwell Vent, Castleton: A Carboniferous littoral cone', *Proc. Yorks. Geol. Soc.*, **41**, 173–84.

Chorley, R. J., Dunn, A. J. and Beckinsale, R. P. (1964) *The History of the Study of Land-forms. Volume one: Geomorphology before Davis*, (Methuen, London).

Christopher, N. S. J. (1980) 'A preliminary flood pilse study of Russett Well, Derbyshire', *Trans. Brit. Cave Res. Assoc.*, **7**(1), 1–12.

—— (1981) *'Hydrogeochemistry of the Carboniferous Limestone of North Derbyshire'*, unpublished Ph.D. thesis, University of Leicester.

Christopher, N. S. J. and Beck, J. S. (1977) 'A survey of Carlswark Cavern, Stoney Middleton, Derbyshire with geological and hydrological notes', *Trans. Brit. Cave Res. Assoc.*, **4**(3), 361–5.

Christopher, N. S. J., Trudgill, S. T., Crabtree, R. W., Pickles, A. M., and Culshaw, S. M. (1981) 'A hydrological study of the Castleton area, Derbyshire', *Trans. Brit. Cave Res. Assoc.*, **8**(4), 189–206.

Christopher, N. S. J. and Wilcock, J. D. (1981) 'Geochemical controls on the composition of limestone ground waters with special reference to Derbyshire', *Trans. Brit. Cave Res. Assoc.*, **8**(3), 135–58.

Clark, A. D. (1958) 'The Farnworth Disaster', *Inst. Municipal Eng. J.*, **85**, 88–96.

Clark, J. G. D. (1972) *'Star Carr: a Case Study in Bioarchaeology'* (Addison-Wesley, Reading, Massachusetts).

Clayton, K. M. (1953) 'The denudation chronology of part of the Middle Trent basin', *Trans. Inst. Brit. Geogr.*, **19**, 25–36.

—— (1966) 'The origin of the landforms of the Malham area', *Field Studies*, **2**, 359–84.

—— (1980) 'Geomorphology', in E. H. Brown (ed.), *Geography Yesterday and Tomorrow* (University Press, Oxford), pp. 167–80.

—— (1981) 'Explanatory description of the Landforms of the Malham Area', *Field Studies*, **5**, 389–423.

Clough, R. Mck. (1977) 'Some aspects of corrie initiation and evolution in the English Lake District', *Proc. Cumb. Geol. Soc.*, **3**, 209–32.

Coles, J. M. (1971) 'The early settlement of Scotland: excavations at Morton, Fife', *Proc. Prehist. Soc.*, **37**, 284–366.

Colhoun, E. A., Common, R. and Cruickshank, M. M. (1965) 'Recent bog flows and debris slides in the north of Ireland', *Sci. Proc. R. Dublin Soc.*, **2A**, 163–74.

Colhoun, E. A. and McCabe, A. M. (1973) 'Pleistocene glacial, glaciomarine and associated deposits of Mell and Tullyallen Townlands, near Drogheda, Eastern Ireland', *Proc. R. Ir. Acad.*, **73B**, 165–206.

Colter, V. S. and Barr, K. W. (1975) 'Recent developments in the geology of the Irish Sea and Cheshire Basins', in A. W. Woodland (ed.), *Petroleum and the Continental Shelf of Northwest Europe. I. Geology*, (Applied Science Publishers, London), pp. 61–73.

Colter, V. S. and Ebbern, J. (1978) 'The petrography and reservoir properties of some Triassic sandstones of the northern Irish Sea Basin, *J. Geol. Soc.*, Lond., **135**, 57–62.

Conway, V. M. (1954) 'Stratigraphy and pollen analysis of southern Pennine blanket peats', *J. Ecol.*, **42**, 117–47.

Coope, G. R. (1959) 'A late Pleistocene insect fauna from Chelford, Cheshire', *Proc. R. Soc. Lond.*, B, **151**, 70–86.

—— (1975) 'Mid Weichselian climatic changes in Western Europe, reinterpreted from coleopteran assemblages' in R. P. Suggate and M. M. Cresswell (eds.), *Quaternary Studies*, R. Soc. New Zealand, 101–8.

—— (1977a) 'Quaternary Coleoptera as aids in the interpretation of environmental history', in F. W. Shotton (ed.), *British Quaternary Studies: Recent Advances*, (University Press, Oxford), 55–68.

—— (1977b) 'Fossil coleopteran assemblages as sensitive indicators of climatic changes during the Devensian (Last) cold stage', *Phil. Trans. R. Soc. Lond.*, B, **280**, 313–40.

—— (1980) 'The climate of England during the Devensian glacial maximum: evidence from coleoptera', *Quat. Newsl.*, **30**, 11–13.

Coope, G. R. and Brophy, J. A. (1972) 'Late Glacial environmental changes indicated by a coleopteran succession from north Wales', *Boreas*, **1**, 97–142.

Coope, G. R. and Joachim, M. J. (1980), 'Late glacial environmental changes interpreted from fossil Coleoptera from St. Bees, Cumbria', in J. J. Lowe, J. M. Gray and J. E. Robinson (eds.), *Studies in the late-glacial of North West Europe*, (Pergamon, Oxford), pp. 55–68.

Coope, G. R. and Pennington, W. (1977) 'The Windermere Interstadial of the Late Devensian', *Phil. Trans. R. Soc.*, B, **280**, 337–9.

Cope, F. W. (1979) 'The age of the volcanic rocks in the Woo Dale borehole Derbyshire', *Geol. Mag.*, **116**, 319–20.

Corbel, J. (1957) *Les Karsts du Nord Ouest de l'Europe*, Institute des Etudes Rhodaniènnes de l'Université de Lyon, Memoirs et Documents 12.

Craddock, J. M. (1976) 'Annual rainfall in England since 1725', *Quart. J. R. Met. Soc.*, **102**, 823–40.

Crickmay, C. H. (1975) 'The hypothesis of unequal activity', in W. N. Melhorn and R. C. Flemal (eds.), *Theories of Landscape Development*, (Allen and Unwin, London), pp. 103–10.

Crisp, D. T. (1966) 'Input and output of minerals for an area of Pennine moorland: the importance of precipitation, drainage, peat erosion and animals', *J. of appl. Ecol.*, **3**, 327–48.

Crisp, D. T., Rawes, M. and Welch, D. (1964) 'A Pennine peat slide', *Geogr. J.*, **130**, 519–24.

Crisp, D. T. and Robson, S. (1979) 'Some effects of discharge upon the transport of animals and peat in a North Pennine headstream', *J. Appl. Ecol.*, **16**, 721–36.

Crompton, E. (1956) 'The environmental and pedological relationships of peaty gleyed podzols', *Sixième Congrès de la Science du sol.*, Paris, 155–61.

Cronin, T. M. (1982) 'Rapid sea level and climatic change: evidence from continental and island margins', *Quaternary Sci. Rev.*, **1**, 177–214.

Cubbon, A. M. (1957) 'The Ice Age in the Isle of Man', *Proc. Isle of Man Nat. Hist. Antiq. Soc.*, **5**, 499–512.

Cundill, P. R. (1976) 'Late Flandrian vegetation and soils in the Carlingill Valley, Howgill Fells', *Trans. Inst. Br. Geogr., New Ser.*, **1**, 301–9.

Cunningham, F. F. (1965) 'Tor theories in the light of South Pennine evidence', *E. Midld. Geogr.*, **3**, 424–33.

Dale, E. (1900) *Geology and Scenery of the Peak*, (Sampson Low, London).

Daley, B. (1972) 'Some problems concerning the Early Tertiary of Southern Britain', *Palaeogeogr. Palaeoclimatol. Palaeoecol.*, **11**, 177–90.

Dalton, A. C. (1958) 'The distribution of dolerite boulders in the glaciation of N.E. Derbyshire', *Proc. Geol. Assoc.*, **68**, 278–85.

Dansgaard, W., Johnson, S. J., Clausen, H. B. and Langway, C. C. (1971), 'Climate record revealed by the Camp Century ice core', in K. K. Turekian (ed.), *The Late Cenozoic Glacial Ages*, (Yale U.P., New Haven), pp. 37–56.

Davies, J. L. (1964) 'A morphogenic approach to world shorelines', *Z. Geomorph.*, N.F. **8**, 127*–42*.

Davis, G. E. and Davis, A. R. (1890) *The River Irwell and its Tributaries: A Monograph on River Pollution*, (Heywood, Manchester and London).

Davis, W. M. (1895) 'The development of certain English rivers', *Geogr. J.*, **5**, 127–46.

Dean, V. (1950) 'The Age and Origin of the Sale Wheel Gorge and of the deviation of the Starling Brook in the same locality', *J. Manchr. Geol. Assoc.*, **2**, 1–6.

—— (1953) 'Some unrecorded overflow channels in north east Lancashire', *L'pool and Manchr. Geol. J.*, **1**, 153–60.

Dearden, J. (1963) 'Derbyshire limestone: its removal by man and by nature', *E. Midld. Geogr.*, **3**(4), 199–205.

Denton, G. H. and Hughes, T. J. (1981) *The Last Great Ice Sheets*, (Wiley, New York), pp. 437–67.

De Rance, C. E. (1869) 'On the surface geology of the Lake District', *Geol. Mag.*, Decade one, **6**, 489–94.

—— (1875) 'The geology of the country around Blackpool, Poulton and Fleetwood', *Mem. Geol. Surv. G.B.*

—— (1877) 'The superficial geology of the country adjoining the coasts of southwest Lancashire', *Mem. Geol. Surv. G.B.*

Derbyshire, E. (1961) 'Sub-glacial col channels and the deglaciation of the north east Cheviots', *Trans. Inst. Br. Geogr.*, **29**, 31–46.

Derbyshire, E., Gregory, K. J. and Hails, J. R. (1979) *Geomorphological Processes*, (Dawson, Folkestone), pp. 187–286.

Dewey, J. F. (1982) 'Plate tectonics and the evolution of the British Isles', *J. Geol. Soc.*, Lond. **139**, 371–412.

Dickinson, W. (1973) 'The development of the raised bog complex near Rusland in the Furness District of North Lancashire', *J. Ecol.*, **61**(3), 871–86.

—— (1975) 'Recurrence surfaces in Rusland Moss, Cumbria (formerly North Lancashire)', *J. Ecol.*, 913–35.

Dickson, C. A., Dickson, J. H. and Mitchell, G. F. (1970) 'The Late Weichselian flora of the Isle of Man', *Phil. Trans. R. Soc. Lond.*, **B258**, 31–79.

Dobson, M. R. (1977a) 'The geological structure of the Irish Sea', in C. Kidson and M. J. Tooley (eds.), *The Quaternary History of the Irish Sea Geol. J. Special Issue*, **7**, (Wiley, Chichester), pp. 13–26.

—— (1977b) 'The history of the Irish Sea basins', in C. Kidson and M. J. Tooley (eds.), *The Quaternary History of the Irish Sea, Geol. J., Special Issue*, **7**, (Wiley, Chichester), pp. 93–8.

Douglas, I. (1964) 'Intensity and periodicity in denudation processes with special reference to the removal of material in solution by rivers', *Z. Geomorph.*, N.F. **8**, 453–73.

—— (1983) *The Urban Environment* (Arnold, London).

—— (1984) 'How tectonic movements have changed the landscape', Dynamic Earth 8: the Forces that Created Britain', *The Geog. Mag.* **56**(6), 304–11.

Duckworth, J. A. and Seed, G. (1969) 'Bowland forest and Pendle floods', *1969 Yearbook, Assoc. of River Authorities*, 81–90.

Dunham, K. C. (1972) 'The evidence for subsidence. The regional setting', *Phil. Trans. R. Soc. Lond.*, A.**272**, 81–6.

—— (1973) 'A recent deep borehole near Eyam, Derbyshire', *Nature*, **241**, 84–5.

Dunkley, P. N. (1981) 'The sand and gravel resources of the country north of Wrexham. Description of 1:25 000 resource sheet SJ 35 and part of SJ 25. *Mineral Assessment Rep. Inst. Geol. Sciences*, 61.

Dury, G. H. (1983) 'Geography and geomorphology: the last fifty years', *Trans. Inst. Brit. Geogr.*, New Series, **8**, 90–9.

Early, K. R. and Skempton, A. W. (1972) 'Investigations of the landslide at Walton's Wood, Staffordshire', *Quart. J. Engng. Geol.*, **5**, 19–41.

Earp, J. R., Magraw, D., Poole, E. G., Land, D. H. and Whiteman, A. J. (1961) *Geology of the Country around Clitheroe and Nelson*, Mem. Geol. Surv. G.B. (HMSO, London).

Eastwood, T., Hollingworth, S. E., Rose, W. C. C. and Trotter, F. M. (1968), *Geology of the Country around Cockermouth and Caldbeck*, Mem. Inst. Geol. Sci., (HMSO, London).

Edmunds, W. M. (1971) *Hydrogeochemistry of groundwaters in the Derbyshire Dome with special reference to trace constituents*, Inst. Geol. Sci. Rept. number 71/7.

Edwards, O., Martin, G. and Scharf, A. (1972) *The Age of Revolutions*, (Open Univ. Press, Milton Keynes).

Embleton, C. and Thornes, J. (1979) *Process in Geomorphology*, (Arnold, London).

Emiliani, C. (1961) 'Cenozoic climatic changes, as indicated in the stratigraphy and chronology of deep sea cores', *New York Acad. Sci.*, **95**, 521–36.

Erdtmann, G. (1925) 'Pollen statistics from the Curragh and Ballaugh, Isle of Man', *Proc. L'pool. Geol. Soc.*, **14**, 158–63.

Evans, G. H. (1970) 'Pollen and diatom analyses of Late Quaternary deposits in the Blelham Basin, north Lancashire', *New Phytol.*, **69**, 821–74.

Evans, I. S. and Clough, R. McK. (1977) 'Cirques of Mungrisdale: Bowscale and Bannerdale', in M. J. Tooley (ed.), *The Isle of Man, Lancashire Coast and Lake District*, 10th INQUA Congress Excursion Guide (Geo Abstracts, Norwich), pp. 52–5.

Evans, J. G. (1975) *The Environment of Early Man in the British Isles*, (Elek, London).

Evans, R. (1977) 'Overgrazing and soil erosion on hill pastures with particular reference to the Peak District', *J. Brit. Grassland Soc.*, **32**, 65–76.

Evans, W. B. and Arthurton, R. S. (1973) 'Northwest England', in G. F. Mitchell, L. F. Penny, F. W. Shotton and R. G. West (eds.), *A correlation of the Quaternary deposits of the British Isles. Geol. Soc. Lond. Special Rep.* 4.

Evans, W. B., Wilson, A. A., Taylor, B. J. and Price, D. (1968), *Geology of the country around Macclesfield, Congleton, Crewe, and Middlewich*, Mem. Geol. Surv. G.B. 2nd ed. (HMSO, London).

Exley, C. S. (1970) 'Observations on the geology of the Keele district', *North Staffs. J. Field Studies*, **10**, 49–63.

Eyles, N. (ed.) (1983) *Glacial Geology*, (Pergamon, Oxford).

Eyles, N. (1983) 'Glacial geology: a landsystems approach' in N. Eyles (ed.), *Glacial Geology*, (Pergamon, Oxford), pp. 1–18.

Eyles, N. and Sladen, J. A. (1981) 'Stratigraphy and geotechnical properties of weathered lodgement tills in Northumberland, England', *Quart. J. Engng. Geol.*, **14**, 129–41.

Fairbridge, R. W. and Finkel, C. W. (1980) 'Cratonic erosional unconformities and peneplains', *J. Geol.*, **88**, 69–86.

Fearnsides, W. G. (1932) 'The valley of the Derbyshire Derwent', *Proc. Geol. Assoc.*, **43**, 153–78.

Fell, C. I. (1971) 'Committee for Prehistoric Studies – unpublished records for Westmorland and Lancashire north of the Sands, 1940–May 1970', *Trans. Cumb. Westm. Antiq. Archaeol. Soc.*, N.S. **71**, 1–11.

Ferguson, R. I. (1981) 'Channel form and channel change', in J. Lewin, (ed.), *British Rivers*, (Allen and Unwin, London), pp. 90–125.

Firman, R. J. (1978) 'Intrusions', in F. Moseley (ed.), *The Geology of the Lake District*, Yorkshire Geological Society (Occ. Publ. No. 3), pp. 146–63.

Fishwick, H. (1907) 'The history of the parish of Lytham in the County of Lancaster', *Cheetham Soc.*, N.S. **60**, 1–118.

Fitch, F. J., Miller, J. A. and Thompson, D. B. (1966) 'The palaeogeographic significance of isotopic age determinations on detrital micas from the Triassic of the Stockport–Macclesfield district, Cheshire, England', *Palaeogeogr. Palaeoclimatol. Palaeoecol.*, **2**, 281–312.

Fletcher, T. W. (1982) 'The agrarian revolution in arable Lancashire', *Trans. Lancs. Ches. Antiq. Soc.*, **72**, 93–122.

Flint, R. F. (1929) 'Stagnation and dissipation of the last ice sheet', *Geog. J.*, **19**, 256–89.

—— (1970) *Glacial and Quaternary Geology.* (Wiley, New York).

Fookes, P. G., Gordon, D. L. and Higginbottom, I. E. (1978) 'Glacial Landforms, their deposits and engineering characteristics', in *The Engineering Behaviour of Glacial Materials.* (Geo Abstracts, Norwich), pp. 18–51.

Ford, D. C. (1980) 'Threshold and limit effects in karst geomorphology', in D. R. Coates and J. D. Vitek (eds.), *Thresholds in Geomorphology*, (Allen and Unwin, London), pp. 345–62.

Ford, D. C. and Schwarez, H. P. (1981) 'Uranium series disequilibrium dating methods', in A. S. Goudie (ed.), *Geomorphological Techniques*, (Allen and Unwin, London), pp. 284–7.

Ford, D. C. and Stanton, W. I. (1968) 'The geomorphology of the South Central Mendip Hills', *Proc. Geol. Assoc.*, **79**, 401–27.

Ford, T. D. (1963) 'The dolomite tors of Derbyshire', *E. Midld. Geogr.*, **3**(19), 148–53.

—— (1964) 'Fossil karst in Derbyshire', *Proc. Brit. Speleol. Assoc.*, **2**, 50–62.

—— (1966) 'The underground drainage systems of the Castleton area, Derbyshire and their evolution', *Cave Sci.*, **5**(39), 369–96.

—— (1972) 'Evidence of early stages in the evolution of the Derbyshire karst', *Trans. Cave Res. Grp. G.B.*, **14**(2), 73–7.

—— (1977a) *'Limestone and Caves of the Peak District*, (Geo. Abstracts, Norwich).

—— (1977b) 'Sands and solution in the Tertiary era', in T. D. Ford (ed.), *Limestone and Caves of the Peak District*, (Geo Abstracts, Norwich), pp. 79–86.

Ford, T. D. and Burek, C. V. (1976) 'Anomalous limestone gorges in Derbyshire', *Mercian Geol.*, **6**(1), 59–66.

Ford, T. D., Burek, C. V. and Beck, J. S. (1975) 'The evolution of Bradwell Dale and its caves, Derbyshire', *Trans. Brit. Cave. Res. Assoc.*, **2**(3), 133–40.

Ford, T. D., Gascoyne, M. and Beck, J. S. (1983) 'Speleothem dates and Pleistocene chronology in the Peak District of Derbyshire', *Trans. Brit. Cave Res. Assoc.*, **10**(2), 103–15.

Ford, T. D. and Gill, D. W. (1979) *Caves of Derbyshire*, (Dalesman, Clapham, Yorks).

Ford, T. D. and Rieuwerts, J. H. (1975) *Lead Mining in the Peak District*, (Peak Park Planning Board, Bakewell).

Ford, T. D. and Worley, N. E. (1977) 'Mineral veins and cave development', *Proc. 7th Int. Spe. Congr.*, Sheffield, 192–3.

Francis, E. A. (1978) *Field Handbook: Annual field meeting 1978*, (*Keele*), (Quaternary Research Assoc., Keele).

—— (1980) 'The Limit of the Last Glaciation in England: A consideration of its definition with special reference to the West Midlands, the Southwest Pennines and the Vale of York', *Quat. Newsl.*, 30, 1–4.

Franks, J. W. and Johnson, R. H. (1964) 'Pollen analytical dating of a Derbyshire landslide', *New Phytol.*, **63**, 209–16.

Franks, J. W. and Pennington, W. (1961) 'The Late-glacial and Post-glacial deposits of the Esthwaite Basin, north Lancashire', *New Phytol.*, **60**, 27–42.

Freeman, T. W., Rodgers, H. B. and Kinvig, R. H. (1966) *Lancashire, Cheshire and the Isle of Man*, (Nelson, London).

Frenzel, B. (1973) *Climatic Fluctuations of the Ice Age*, (Case Western University Press, Cleveland, U.S.A.).

Furness, R. R. (1978) *Soils of Cheshire*, Soil Surv. G.B. Bull. No. 6.

Furness, R. R. and King, S. J. (1972) *Soils in Westmorland. I., Soil Survey Record No. 10, Sheet SD 58 (Sedgwick)*, (Soil Survey England and Wales, Harpenden).

Gale, S. J. (1981a) 'The geomorphology of the Morecambe Bay karst and its implications for landscape chronology', *Z. Geomorph.*, N.F. **25**, 457–69.

—— (1981b) *'Karst Palaeoenvironments'*, unpublished Ph.D. thesis, University of Keele.

Gale, S. J. and Hunt, C. O. (1985) 'The stratigraphy of Kirkhead Cave, a Later Upper Palaeolithic site in northern England', Proc. Prehist. Soc., 51 (in press).

Gale, S. J., Hunt, C. O. and Southgate, G. A. (1984) 'Kirkhead Cave: biostratigraphy and magnetostratigraphy', *Archaeometry*, **26**(2), 192–98.

Garnett, A. (1956) 'Climate', in D. L. Linton (ed.) *Sheffield and its Region: a scientific and historical survey'*, (Brit. Ass. Adv. Sci. Sheffield), pp. 44–69.

Garrard, R. A. (1977) in C. Kidson and M. J. Tooley (eds.) *The Quaternary History of the Irish Sea, Geol. J. Special Issue*, **7**, (Wiley, Chichester), pp. 25–38.

Gascoyne, M. (1981) 'Rates of cave passage entrenchment and valley lowering determined from speleothem age measurements', *Proc. 8th Int. Speleol. Congr.*, Bowling Green, 99–100.

Gascoyne, M., Ford, D. C. and Schwarcz, H. P. (1983a) 'Rates of cave and landform development in the Yorkshire Dales from speleothem ages', *Earth surf. processes and landf.*, **8**(6), 557–68.

Gascoyne, M., Schwarcz, H. P. and Ford, D. C. (1983b) 'Uranium-series ages of speleothem from northwest England: Correlation with Quaternary Climate', *Phil. Trans. R. Soc., Lond.*, B, **301**, 143–64.

Gates, W. L. (1976) 'Modelling the ice-age climate', *Science*, **191**, 1138–44.

Geike, A. (1865) *Scenery of Scotland*, (London).

—— (1868) 'On denudation now in progress', *Geol. Mag. Decade one,* **5**, 249–54.

Gemmell, A. D. and George, P. K. (1972) 'The Glaciation of the West Midlands: A review of recent research', *N. Staffs. J. Field Studies*, **12**, 1–20.

George, T. N. (1974) 'Prologue to a geomorphology of Britain', in E. H. Brown and R. S. Waters (eds.), *Progress on Geomorphology*, Inst. Br. Geogr. Spec. Pub. 7, 113–25.

Gibson, W. and Wedd, C. B. (1905) *Geology of the North Staffordshire Coalfields*, Mem. Geol. Surv. G.B. (HMSO, London).

Gibson, W., Wedd, C. B. and Scott, A. (1925) *Geology of the country around Stoke upon Trent*, Mem. Geol. Sur. G.B. (HMSO, London).

Glover, H. A. (1971) 'River Irwell Improvement Scheme', *Association of River Authorities Yearbook and Directory*, 17–24.

Glover, R. R. (1974) 'Cave development in the Gaping Gill System', in A. C. Waltham (ed.), *Limestone and Caves of Northwest England*, (David and Charles, Newton Abbot), pp. 343–84.

Godwin, H. (1943) 'Coastal peat beds of the British Isles and North Sea', *J. Ecol.*, **31**(2), 199–247.

—— (1945) 'Coastal peat-beds of the North Sea region, as indices of land- and sea-level changes', *New Phytol.*, **44**, 29–69.

—— (1956) *The History of the British Flora*. (Cambridge University Press).

—— (1959) 'Studies of the post glacial history of British vegetation. XIV. Late-glacial deposits at Moss Lake, Liverpool', *Phil. Trans. R. Soc. Lond.* B. **242**, 127–49.

—— (1975) *The History of the British Flora* (2nd edn), (Cambridge University Press).

—— (1977) 'Quaternary history of the British flora', in F. W. Shotton (ed.), *British Quaternary Studies, Recent Advances*, (University Press, Oxford), pp. 107–18.

—— (1978) *Fenland: its ancient past and uncertain future*, (University Press, Cambridge).

Godwin, H. and Clifford, M. H. (1938) 'Studies of the post-glacial history of British vegetation II. Origin and Stratigraphy of deposits in southern Fenland', *Phil. Trans. R. Soc.* B. **229**, 363–406.

Goldie, H. S. (1973) 'Limestone pavements of Craven', *Trans. Cave Res. Grp. G.B.*, **15**(3), 175–89.

—— (in press) 'Human influence on limestone pavements in N.W. England', in M. M. Sweeting and K. Paterson (eds.), *New directions in Karst*, (Geo Abstracts, Norwich).

Good, T. R. and Bryant, I. B. (1985) 'Fluvio-aeolian sedimentation; an example from Banks Island, N.W.T. Canada', *Geogr. Annal.* **67A**, in press.

Goodchild, J. G. (1889/90) 'The history of the Eden and some rivers adjacent', *Trans. Cumb. and Westm. Assoc.*, **14**, 73–90.

—— (1890) 'Notes on some observed rates of weathering of limestones, *Geol. Mag.*, **17**, 463–6.

Goudie, A. S. (1977) *Environmental Change*, (University Press, Oxford).

Goulden, C. E. (1964) 'The history of the Cladoceran fauna of Esthwaite Water (England) and its limnological significance', *Archiv für Hydrobiologie*, **60**, 1–52.

Grayson, R. (1972) *'The buried bedrock topography between Manchester and the Mersey Estuary'*, unpublished M.Sc. thesis, University of Manchester.

Greenwood, G. (1857) *Rain and Rivers*, (London).

Gregory, K. J. (1971) 'Drainage density changes in south-west England', in K. J. Gregory and W. D. L. Ravenhill, *Exeter Essays in Geography* (Reading University Press), pp. 33–53.

Gresswell, R. K. (1951) 'The Glacial Geomorphology of the south-eastern part of the Lake District', *L'pool and Manchr. Geol. J.*, **1**, 57–70.

—— (1953) *Sandy Shores in South Lancashire, The geomorphology of south-west*

Lancashire (Liverpool University Press).

—— (1957) 'Hillhouse coastal deposits of south Lancashire', *L'pool and Manchr. Geol. J.*, **2**, 60–78.

—— (1958) 'The Post-glacial raised beaches in Furness and Lyth', *Trans. Inst. Br. Geogr.*, **25**, 79–103.

—— (1962) 'The glaciology of the Coniston Basin', *Geol. J.*, **3**, 83–96.

—— (1964) 'The origin of the Mersey and Dee Estuaries', *Geol. J.*, **4**, 77–86.

—— (1967a) *Physical Geography*, (Longman, London).

—— (1967b) 'The geomorphology of the Fylde', in R. W. Steel and R. Lawton (eds.), *Liverpool Essays in Geography: a Jubilee Collection*, (Longman, London), pp. 25–42.

Gresswell, R. K. and Lawton, D. (1964) *Merseyside*, (Geog. Assoc. Sheffield).

Grieve, W. and Hammersley, A. D. (1971) 'A re-examination of the Quaternary deposits of the Barrow area', *Proc. Barrow Nat. Field Club*, **10**, 5–25.

Gunn, J. (1974) 'A model of the karst percolation system of Waterfall Swallet, Derbyshire', *Trans. Brit. Cave Res. Assoc.*, **1**(3), 159–64.

—— (in press) 'Solute processes and karst landforms', in S. T. Trudgill, (ed.), *Solute Processes*, (Wiley, London).

Hageman, B. P. (1969) 'Development of the western part of the Netherlands during the Holocene', *Geologie Mijnb.*, **48**(4), 373–88.

Hainsworth, V. (1983) *Looking Back at Sale*, (Willow Press, Timperley).

Hale, H. T. and Dyer, E. A. (1963) 'The subsidence of Fylde Street, Farnworth', *Proc. Inst. Civ. Engng.*, **24**, 207–22, **26**, 178–84.

Hall, A. M. (1983) *'Preglacial landscape evolution in Scotland'*, unpublished Ph.D. thesis, Univ. of St. Andrews.

Hall, A. R. (1980) 'Late Pleistocene deposits at Wing, Rutland', *Phil. Trans. R. Soc. Lond.*, **B**, **289**, 135–64.

Hall, B. R. (1954) 'Borehole records from the mosses of south-west Lancashire', *Soil Survey of England and Wales*, (Harpenden).

Hall, B. R. and Folland, C. J. (1967) *Soils of Lancashire*, Mem. Soil. Surv. G.B. (Harpenden).

Hallam, J. S., Edwards, B. J. N., Barnes, B. and Stuart, A. J. (1973) 'The remains of a Late Glacial elk associated with barbed points from High Furlong, near Blackpool, Lancashire', *Proc. Prehist. Soc.*, **39**, 100–28.

Halliwell, R. A. (1979) 'Influence of contrasted rock types and geological structure on solutional processes in northwest Yorkshire', in A. F. Pitty (ed.), *Geographical Approaches to Fluvial Processes*, (Geobooks, Norwich), pp. 51–71.

—— (1980) 'Karst waters of the Ingleborough area, North Yorkshire', *Proc. Univ. Bristol Spel. Soc.*, **15**(3), 183–205.

Hallsworth, E. G. (1963) 'An examination of some of the factors affecting the movement of clay in an artificial soil', *J. Soil Sci.*, **14**, 360–71.

Hamblin, R. J. O. (1973) 'The Haldon gravels of south Devon', *Proc. Geol. Assoc.*, **84**, 459–76.

Harland, W. B., Cox, A. V., Llewellyn, P. G., Pickton, C. A. G., Smith, A. G. and Walters, R. (1982) *A Geologic Time Scale*, (Cambridge University Press).

Harmsworth, R. V. (1968) 'The developmental history of Bleham Tarn (England) as shown by animal microfossils, with special reference to the Cladocera', *Ecological Monographs*, **38**, 223–41.

Hartley, J. J. (1932) 'The volcanic and other igneous rocks of Great and Little Langdale, Westmorland', *Proc. Geol. Assoc.*, **43**, 32–69.

Harvey, A. M. (1974) 'Gully erosion and sediment yield in the Howgill Fells, Westmorland', in K. J. Gregory and D. E. Walling (eds.), *Fluvial Processes in Instrumented Watersheds*, Inst. Brit. Geogr. Spl. Publ. 6, pp. 45–58.

—— (1977) 'Event frequency in sediment production and channel change', in K. J. Gregory (ed.), *River Channel Changes*, (Wiley, Chichester), pp. 301–15.

—— (1984) 'Seasonality of processes on eroding gullies: a twelve year record of erosion rates', (Abstracts Geog. Union Congress, Paris 1984).

—— (in press) 'Geomorphic affects of a 100 year storm in the Howgill Fells, N.W. England', *Z. Geomorph.*

Harvey, A. M., Alexander, R. W. and James, P. A. (1984) 'Lichens, soil development and the age of Holocene valley floor landforms, Howgill Fells, Cumbria', *Geogr. Annlr.*, **66A**, 353–66.

Harvey, A. M., Hitchcock, D. H. and Hughes, D. J. (1979) 'Event frequency and the morphological adjustment of fluvial systems in upland Britain', in D. D. Rhodes and C. P. Williams (eds.), *Adjustments to the Fluvial System*, (Dubuque, Iowa), pp. 139–67.

Harvey, A. M., Oldfield, F., Baron, A. F. and Pearson, G. W. (1981) 'Dating of post-glacial landforms in the central Howgills', *Earth surf. processes and landf.*, **6**, 401–12.

Haszeldine, R. S. and Anderton, R. (1980) 'A braidplain facies model for the Westphalian B. Coal Measures of northeast England', *Nature*, **284**, 51–3.

Haworth, E. Y. (1969) 'The diatoms of a sediment core from Blea Tarn, Langdale', *J. Ecol.*, **57**, 429–39.

Hays, J. D., Imbrie, J., and Shackleton, N. J. (1976) 'Variations in the earth's orbit: Pacemaker of the Ice Ages', *Science*, **194**, 1121–32.

Hays, J. D. and Pitman, W. C. (1973) 'Lithospheric plate motion, sea level change and climatic and ecological consequences', *Nature*, **246**, 18–22.

Heyworth, A. (1978) 'Submerged Forests', in J. Fletcher (ed.), *Dendrochronology in Northern Europe*, Br. Archaeol. Rep. Internat. Ser. (Suppl) S-51 (London), 279–88.

Hibbert, F. A., Switsur, V. R. and West, R. G. (1971) 'Radiocarbon dating of Flandrian pollen zones at Red Moss, Lancashire', *Proc. R. Soc. Lond.*, B, **177**, 161–76.

Hicks, S. P. (1971) 'Pollen-analytical evidence for the effect of prehistoric agriculture on the vegetation of north Derbyshire', *New Phytol.*, **70**, 647–67.

Hildreth, S. P. (1836) 'Observations on the Bituminous coal deposits of the valley of the Ohio', *American J. Sci.*, **29**, 291–335.

Hind, W. (1906) 'Speculations on the evolution of the river Trent', *Trans. N. Staffs. Field Club*, **41**, 93–100.

Hitchcock, D. H. (1977a) '*Channel pattern changes in divided streams*', unpublished Ph.D. thesis, Univ. of Liverpool.

—— (1977b) 'Channel pattern changes in divided reaches: an example in the coarse bed material of the Forest of Bowland', in K. J. Gregory (ed.), *River Channel Changes*, (Wiley, Chichester), pp. 207–20.

Hodgson, E. (1862) 'On a deposit containing Diatomaceae, leaves, etc. in the iron-ore mines near Ulverston', *J. Geol. Soc. Lond.*, **19**, 19–31.

—— (1867) 'The moulded limestones of Furness', *Geol. Mag.*, **4**, 401–6.

Hodgson, J. M., Catt, J. A. and Weir, A. H. (1967) 'The origin and development of clay with flints and associated soil horizons on the south downs', *J. of Soil Sci.*, **18**, 85–102.

Hollingworth, S. E. (1929) 'The evolution of the Eden drainage in the south and west', *Proc. Geol. Assoc.*, Lond., **40**, 115–38.

—— (1931) 'Glaciation of western Edenside and adjoining areas and the drumlins of the Edenside and Solway basin', *J. Geol. Soc., Lond.*, **87**, 281–359.

—— (1935) 'High level erosional platforms in Cumberland', *Proc. Yorks. Geol. Soc.*, **23**, 159–77.

Holmes, P. W. (1968) 'Sedimentary studies of Late Quaternary material in Windermere Lake (Great Britain)', *Sediment Geol.*, **2**, 201–24.

Holyoak, D. T. (1983) 'The identity and origins of *Picea abies* (L.) Karston from the Chelford Interstadial (Late Pleistocene) of England', *New Phytol.*, **95**, 153–7.

Hooke, J. M. and Harvey, A. M. (1983) 'Meander changes in relation to bend morphology and secondary flows', in J. D. Collinson and J. Lewin (eds.), *Modern and Ancient Fluvial Systems*, Int. Assoc. Sediment. Sp. Publ. 6, 121–32.

Hooke, J. M. and Kain, R. J. (1982) *Historical Change in the Physical Environment: A guide to sources and techniques*, (Dawson, Folkestone).

Howell, F. T. (1973) 'The sub-drift surface of the Mersey and Weaver catchments and adjacent areas', *Geol. J.*, **8**(2), 285-96.

Howell, F. T. and Jenkins, P. L. (1976) 'Some aspects of the subsidences in the rock salt districts of Cheshire, England', Int. Assoc. Hydrol. Sci. Publn. **121**, 507-20.

—— (1984) 'Centrifuge modelling of salt subsidences', in W. H. Craig (ed.), *The application of centrifuge modelling to geotechnical design*, (Engineering Department, University of Manchester), pp. 196-205.

Huddart, D. (1971a) 'Textural distinction of Main Glaciation and Scottish Readvance tills in the Cumberland lowland', *Geol. Mag.*, **108**, 317-24.

—— (1971b) 'A relative glacial chronology from the tills of the Cumberland lowland', *Proc. Cumb. Geol. Soc.*, **3**, 21-32.

—— (1981) 'Pleistocene foraminifera from south-east Ireland – some problems of interpretation', *Quat. Newsl.*, **33**, 28-41.

Huddart, D. and M. J. Tooley (eds.) (1972) *The Cumberland Lowland Handbook* (Quat. Res. Assoc., Durham).

Huddart, D., Tooley, M. J. and Carter, P. A. (1977) 'The coasts of north west England', in C. Kidson and M. J. Tooley (eds.), *The Quaternary History of the Irish Sea*, Geol. J. Spec. Issue 7, (Wiley, Chichester), pp. 119-54.

Hudson, R. G. S. (1933) 'Scenery and Geology of North West Yorkshire', *Proc. Geol. Assoc.*, **44**, 228-55.

Hughes, M. (1901) 'Ingleborough', *Proc. Yorks. Geol. Soc.*, **14**, 125.

Hughes, M. J. (1978) 'Foraminifera from vibrocore G5 samples', in H. M. Pantin (ed.), Quaternary sediments from the north-east Irish Sea: Isle of Man to Cumbria, *Bull. Geol. Sur. Gt. Britain*, **64**, 1-43.

Hughes, T. Mck. (1886) 'On some perched blocks and associated phenomena', *J. Geol. Soc.*, Lond., **42**, 527-39.

Hull, E. (1864) *Geology of the country around Oldham, including Manchester and its suburbs*, Mem. Geol. Surv. G.B. London.

—— (1866) 'River denudation of valleys', *Geol. Mag.*, Decade one, **3**, 474-7.

—— (1868) 'Observations on the relative age of the leading physical features and lines of elevation of the Carboniferous district of Lancashire and Yorkshire', *Geol. Mag.*, **24**, 287-8.

Hull, E., Dakyns, J. R., Tiddeman, R. H., Ward, J. C., Gunn W. and De Rance, C. E. (1875), *The Geology of the Burnley Coalfield*, Mem. Geol. Surv. G.B., London.

Hunt, C. O. (1984) 'Erratic palynomorphs from some British tills', *J. of Micropalaeontology*, **3**, 71-4.

Hunt, C. O., Hall, A. R. and Gilbertson, D. D. (1984) 'The palaeobotany of the Late-Devensian sequence at Skipsea Whitow Mere', in D. D. Gilbertson, D. J. Briggs, J. R. Flenley, A. R. Hall, C. O. Hunt and D. Woodall (eds.), *Late Quaternary Environments and Man in Holderness*, (Br. Archaeol. Rept., 134, pp. 81-108.

Huntley, B. and Birks, H. J. B. (1983) *An Atlas of Past and Present Pollen Maps for Europe: 0-13000 Years Ago*, (University Press, Cambridge).

Hutton, J. (1788) 'Theory of the Earth', *Trans. R. Soc. Edinb.*, **1**, 209-34.

Imeson, A. C. (1971) 'Heather burning and soil erosion on the North Yorkshire Moors', *J. appl. Ecol.*, **8**, 357-542.

Ingham, J. K., McNamara, K. J. and Rickards, R. B. (1978) 'The Upper Ordovician and Silurian rocks', in F. Moseley (ed.), *The Geology of the Lake District*, Yorkshire Geological Society, (Occ. Publ. No. 3), pp. 121-45.

Ingham, J. K. and Rickards, R. B. (1974) 'Lower Palaeozoic rocks' in D. H. Rayner and J. E. Hemingay (eds.), *The Geology and Mineral Resources of Yorkshire,* (Yorkshire Geological Society, Leeds), pp. 29-44.

Ingram, H. A. P. (1967) 'Problems of hydrology and plant distribution in mires', *J. Ecol.*, **55**, 711-24.

Innes, J. B. and Tomlinson, P. R. (1983) 'Cultural implications of Holocene landscape evolution in the Merseyside region', *Amateur Geologist*, **10**, 3–17.

Iversen, J. (1960) 'Problems of the early post-glacial forest development in Denmark', *Danm. Geol. Unders.*, Ser. IV, **4**(3), 1–32.

Jackson, D. E. (1978) 'The Skiddaw Group', in F. Moseley (ed.), *The Geology of the Lake District*, Yorkshire Geological Society Occ. Publ. No. 3, pp. 79–98.

Jackson, J. W. (1953) 'Archaeology and Palaeontology' in C. H. D. Cullingford (ed.), *British Caving; An introduction to speleology*, (Routledge, London).

Jacobi, R. M. (1973) 'Aspects of the "Mesolithic Age" in Great Britain', in S. K. Kozlowski (ed.), *The Mesolithic in Europe*, (Warsaw), pp. 237–65.

Jacobi, R. M., Tallis, J. H. and Mellars, P. A. (1976) 'The Southern Pennine Mesolithic and the ecological record', *J. Archaeol. Sci.*, **3**, 307–20.

Jardine, W. G. (1975) 'Chronology of Holocene marine transgression and regression in south-western Scotland', *Boreas*, **4**, 173–96.

Jarvis, R. A., Bendelow, V. C., Bradley, R. I., Carrol, D. M., Furness, W. R., Kilgour, I. N. L. and King, S. J. (1984) *Soils and their Use in Northern England*, Bull. Soil Surv. G.B., (Harpenden).

Jelgersma, S., de Jong, J., Zagwijn, W. H. and van Regteren Altena, J. F. (1970) 'The coastal dunes of the western Netherlands; geology, vegetational history and archaeology', *Meded. Rijks geol. Dienst.*, N.S. **21**, 93–167.

Jelgersma, S., Oele, E. and Wiggers, A. J. (1979) 'Depositional history and coastal development in the Netherlands and the adjacent North Sea since the Eemian', in E. Oele, R. T. E. Schüttenhelm and A. J. Wiggers (eds.), *The Quaternary History of the North Sea*, Acta Univ. Ups, Symp, Univ. Ups. Annum Quingentesimum Celebrantis, (Uppsala), **2**, 115–42.

Johnson, G. A. L. and Dunham, K. C. (1963) *The Geology of Moor House*, (HMSO, London).

Johnson, P. A. (1971) 'Soils in Derbyshire I Sheet SK 17 (Tideswell)', *Soil Survey Record*, No. 4, (Harpenden).

Johnson, R. H. (1957a) 'Observations on the stream patterns of some peat moorlands in the southern Pennines', *Mem. Proc. Manchr. Lit. and Phil. Soc.*, **99**, 110–27.

—— (1957b) 'An examination of the drainage pattern of the eastern part of the Peak District of North Derbyshire', *Geog. Studies*, **4**, 46–55.

—— (1963) 'The Roosdyche, Whaley bridge: a new appraisal', *E. Midld. Geogr.*, **3**(3), (3)155–63.

—— (1965a) 'Glacial Geomorphology of the West Pennine slopes between Cliviger and Congleton', in J. B. Whittow and P. D. Wood (eds.), *Essays in Geography for A. A. Miller*, (Reading University Press), pp. 58–94.

—— (1965b) 'The origin of the Churnet and Rudyard Valleys', *N. Staffs. J. Field Studies*, **5**, 95–105.

—— (1967) 'The former course of the River Goyt', *Amateur Geologist*, **2**(2), 38–50.

—— (1968) 'Four temporary exposures of Solifluction Deposits Pennine hillslopes in North east Cheshire', *Mercian Geol.*, **2**(4), 379–87.

—— (1969a) 'The Derwent–Wye confluence re-examined', *E. Midld. Geogr.*, **4**(7), 421–6.

—— (1969b) 'The Glacial Geomorphology of the area around Hyde, Cheshire', *Proc. Yorks. Geol. Soc.*, **37**, 189–230.

—— (1969c) 'A reconnaisance survey of some river terraces in part of the Mersey and Weaver catchment', *Mem. Proc. Manchr. Lit. Phil. Soc.*, **112**, 1–35.

—— (1971) 'The last glaciation in North-West England: a general survey', *Amateur Geologist*, **5**(2), 18–37.

—— (1975) 'Some Late Pleistocene involutions at Dalton-in-Furness, Northern England', *Geol. J.*, **10**, 23–34.

—— (1980) 'Hillslope stability and landslide hazard – a case study from Longdendale, north Derbyshire, England', *Proc. Geol. Assoc.*, **91**(4), 315–25.

—— (1981) 'Four maps for Longdendale – a geomorphological contribution to Environmental Land Management in an Upland Pennine Valley', *Manchr. Geographer*, **2**(2), 6–29.

Johnson, R. H., Franks, J. W. and Pollard, J. E. (1970) 'Some Holocene faunal and floral remains in the Whitemoor meltwater channel at Bosley, East Cheshire', *N. Staffs. J. Field Studies*, **10**, 65–74.

Johnson, R. H. and Musk, L. F. (1974) 'A comment on Dr F. T. Howell's paper on the sub-drift surface of the Mersey and Weaver catchment and adjacent areas', *Geol. J.*, **9**(2), 209–10.

Johnson, R. H. and Paynter, J. (1967) 'The development of a cutoff on the River Irk at Chadderton, Lancashire', *Geography*, **52**, 41–9.

Johnson, R. H. and Rice, R. J. (1961) 'Denudation Chronology of the Southwest Pennines', *Proc. Geol. Assoc.*, **72**, 21–32.

Johnson, R. H., Tallis, J. H. and Pearson, M. (1972) 'A temporary section through Late-Devensian sediments at Green Lane, Dalton-in-Furness, Lancashire', *New Phytol.*, **71**, 533–44.

Johnson, R. H. and Walthall, S. (1979) 'The Longdendale Landslides', *Geol. J.*, **14**(2), 135–58.

Jones, D. K. C. (1974) 'The influence of the Calabrian transgression on the drainage evolution of south east England', in E. H. Brown and R. S. Waters (eds.), *Progress in Geomorphology*, Inst. Brit. Geogr. Sp. Publ. No. 7, 139–58.

—— (1981) *South East and Southern England: The Geomorphology of the British Isles series*, E. H. Brown and K. Clayton (eds.), (Methuen, London).

Jones, O. T. (1924) 'The origin of the Manchester Plain', *J. Manchester Geog. Soc.*, **39-40**, 89–124.

Jones, R. C. B., Tonks, L. H. and Wright, W. B. (1938), *The Geology of the Country around Wigan*, Mem. Geol. Surv. G.B. (HMSO, London).

Jowett, A. (1914) 'The glacial Geology of East Lancashire', *J. Geol. Soc. Lond.*, **70**, 199–231.

Jowett, A. and Charlesworth, J. K. (1929) 'The glacial geology of the Derbyshire Dome and the western slopes of the southern Pennines', *J. Geol. Soc. Lond.*, **85**, 307–34.

Jukes, J. B. (1862) 'On the mode of formation of some of the river valleys in the South of Ireland', *J. Geol. Soc. Lond.*, **18**, 378–403.

Kear, B. S. (1968) '*An investigation into soils developed on the Shirdley Hill Sand in south-west Lancashire*', unpublished M.Sc. thesis, University of Manchester.

Kendall, J. D. (1881) 'Interglacial deposits of west Cumberland and north Lancashire', *J. Geol. Soc. Lond.*, **37**, 29–39.

Kendall, P. F. (1894) 'On the glacial geology of the Isle of Man', *Yn Lioar Manninagh*, **1**, 397–437.

—— (1902) 'A system of glacier lakes in the Cleveland Hills', *J. Geol. Soc. Lond.*, **58**, 471–571.

Kenney, J. P. J. (1978) '*The pollen stratigraphy of the Moss Farm Bog near Eccleshall, Staffordshire – a preliminary study*', unpublished M.Sc. dissertation, North London Polytechnic and the Polytechnic of Central London.

Kent, P. E. (1975) 'The tectonic development of Great Britain and the surrounding seas', in A. W. Woodland (ed.), *Petroleum and the Continental Shelf of North West Europe*, (Applied Science, London), pp. 3–28.

Kerney, M. P., Brown, E. H. and Chandler, T. J. (1964) 'The Late-Glacial and Post-Glacial history of the Chalk escarpment near Brook, Kent', *Proc. R. Soc. Lond.*, **B248**, 135–92.

Kidson, C. (1966) 'Beaches in Britain', *Inaugural lecture, University College of Wales, Aberystwyth* (University of Wales Press).

—— (1968) 'Coastal Geomorphology', in R. W. Fairbridge (ed.), *The Encyclopedia of Geomorphology*, (Reinhold, New York), pp. 134–9.

—— (1977) 'Some problems of the Quaternary of the Irish Sea', in C. Kidson and M. J. Tooley (eds.), *The Quaternary History of the Irish Sea, Geol. J. Special Issue* 7, (Wiley, Chichester), pp. 1–12.

Kidson, C. and Heyworth, A. (1979) 'Sea "level" ', in K. Suguio, T. R. Fairchild, L. Martin and J. M. Flexor (eds.), *Proceeedings of the 1978 Symposium on Coastal Evolution in the Quaternary*, Universidade de São Paulo, Sao Páulo.

Kidson, C. and Tooley, M. J. (eds.) (1977) *The Quaternary History of the Irish Sea*, Geological Journal Special Issue No. 7, (Wiley, Chichester).

King, C. A. M. (1959) *Beaches and Coasts*, (Arnold, London).

—— (1960) 'The Churnet Valley', *E. Midld. Geogr.*, **14**, 33–40.

—— (1962) *Oceanography for Geographers*, (Arnold, London).

—— (1963) 'Some problems concerning marine planation and the formation of erosion surfaces', *Trans. Inst. Br. Geogr.*, **33**, 29–43.

—— (1966) *Techniques in Geomorphology*, (Arnold, London).

—— (1969) 'Trend surface analysis of central Pennines erosion surfaces', *Trans. Inst. Br. Geogr.*, **47**, 47–59.

—— (1976) *Northern England: The Geomorphology of the British Isles series*, E. H. Brown and K. M. Clayton (eds.), (Methuen, London).

—— (1977) 'The early Quaternary landscape with considerations of Neotectonics matters', in F. W. Shotton (ed.), *British Quaternary Studies*, (Oxford University Press), pp. 137–52.

King, L. C. (1953) 'Canons of landscape evolution', *Geol. Soc. Am. Bull.*, **64**, 721–52.

—— (1958) Correspondence on the problem of tors. *Geogr. J.*, **124**, 289–91.

Klatkowa, H. (1965) 'Vallons en berceau et vallées sèches aux environs de Lodz', *Acta Geographica Lodziensia*, **19**, 124–42.

Klein, J., Lerman, J. C., Damon, P. E. and Ralph, E. K. (1982) 'Calibration of radiocarbon dates: tables based on the consensus data of the Workshop on Calibrating the Radicarbon Time Scale', *Radiocarbon*, **24**, 103–50.

Knight, C. (1979) 'Urbanisation and natural stream channel morphology: the case of two English new towns', in G. E. Hollis (ed.), *Man's Impact on the Hydrological Cycle in the United Kingdom*, (Geobooks, Norwich), pp. 181–98.

Knight, D. J. (1977) 'Morecambe Bay feasibility study – sub-surface investigations', *Quart. J. Engng. Geol.*, **10**, 303–19.

Knighton, A. D. (1972) 'Changes in a braided reach', *Geol. Soc. Am. Bull.*, **83**, 3813–22.

—— (1973) 'River bank erosion in relation to streamflow conditions, River Bollin-Dean, Cheshire', *E. Midld. Geogr.*, **5**, 416–26.

—— (1975a) 'Variations in at-a-station hydraulic geometry', *Am. J. Sci.*, **275**, 186–218.

—— (1975b) 'Channel gradient in relation to discharge and bed material characteristics', *Catena*, **2**, 263–74.

Kroonenberg, S. B. and Melitz, P. J. (1983) 'Summit levels, bedrock control and the etchplain concept in the basement of Suriname', *Geologie en Mijnbouw*, **62**, 389–400.

Krynine, D. and Judd, W. (1957) *Principles of Engineering Geology and Geotechnics*, (McGraw-Hill, New York).

Lamb, H. H. (1965) 'The early medieval warm epoch and its sequel', *Palaeogeogr., Palaeoclimatol., Palaeoecol.*, **1**, 13–37.

—— (1970) 'Climatic fluctuations', in H. Flohn (ed.) *World Survey of Climatology Volume 2, General Climatology 2*, (Elsevier, Amsterdam), pp. 173–249.

—— (1977a) *Climate Present, Past and Future, Volume 2, Climatic History and the Future*, (Methuen, London).

—— (1977b) 'The late Quaternary history of the climate of the British Isles', in F. W. Shotton (ed.), *British Quaternary Studies: Recent Advances*, (Oxford University Press), pp. 283–98.

Lamb, H. H. and Johnson, A. I. (1959) 'Climatic variation and observed changes in the general wind circulation', Parts I and II, *Geogr. Annlr.*, **41**, 94–134.

Lamb, H. H. and Woodroffe, A. (1970) 'Atmospheric circulation during the last ice age', *Quat. Res.*, **1**, 29–58.

Lamplugh, G. W. (1903) *The Geology of the Isle of Man*, Mem. Geol. Surv. U.K.

—— (1911) 'On the shelly moraine of the Sefstrom Glacier and other Spitsbergen phenomena illustrative of British glacial conditions', *Proc. Yorks. Geol. Soc.*, **17**, 216-41.

Lancashire County Council (1964) Borehole data sheets, Lancashire-Yorkshire Motorway, Department of the Environment, Manchester.

Lawrence, G. R. P. (1967) 'Erosion surfaces in the Sow River Basin', *N. Staffs. J. Field Studies*, **7**, 34-43.

Lee, J. (1981) 'Atmospheric pollution and the Peak District blanket bogs', in J. Phillips, D. Yalden and J. Tallis (eds.), *Moorland Erosion Study. Phase I Report*, (Peak Park Joint Planning Board, Bakewell), pp. 104-8.

Lee, M. P. (1979) 'Loess from the Pleistocene of the Wirral Peninsula, Merseyside', *Proc. Geol. Assoc.*, **90**, 21-6.

Leeder, M. R. (1982) 'Upper Palaeozoic basins of the British Isles – Caledonian inheritance versus Hercynian plate margin processes', *J. Geol. Soc. Lond.*, **139**, 479-91.

Lewin, J. (1969) 'The formation of Chalk dry valleys: the Stonehill Valley, Dorset', *Biul. peryglac.*, **19**, 345-50.

—— (1981) 'Contemporary erosion and sedimentation', in J. Lewin (ed.), *British Rivers*, (Allen and Unwin, London), pp. 34-58.

Lewis, F. J. (1908) 'The British Vegetation Committee's excursion to the west of Ireland', *New Phytol.*, **7**, 253-60.

Lewis, H. C. (1894) *Papers and Notes on the Glacial Geology of Great Britain and Ireland*, (Longman, London).

Linton, D. L. (1950) 'Report on the joint meeting of sections C and E of the British Association for the Advancement of Science', *Nature*, **166**, 757.

—— (1955) 'The problem of Tors', *Geog. J.*, **121**, 470-87.

—— (1956) 'Geomorphology', in D. L. Linton (ed.), *Sheffield and its Region: A scientific and historical survey*, (Brit. Ass., Sheffield), pp. 24-43.

—— (1957) 'Radiating valleys in glaciated lands', *Tidjdschrift van het koninklijke Nederland. Aardrijkskundig*, **74**, 297-312.

—— (1963) 'The forms of glacial erosion', *Trans. Inst. Br. Geogr.*, **33**, 1-28.

—— (1964) 'The origin of Pennine tors: an essay in analysis', *Z. Geomorph.*, N.F. **8**, 5-24.

Livett, E. A., Lee, J. A. and Tallis, J. H. (1979) 'Lead, zinc and copper analyses of British blanket peats', *J. Ecol.*, **67**, 865-91.

Lockwood, J. G. (1979) 'Water balance of Britain, 50,000 yr. BP to the present day', *Quat. Res.*, **12**, 297-310.

—— (1982) 'Snow and ice balance in Britain at the present time and during the last glacial maximum and late glacial periods', *Journal of Climatology*, **2**, 209-31.

Longworth, D. (1984) '*An Investigation into the Glacial Deposits of the Lancashire Plain*', unpublished M.Sc. thesis, University of Manchester.

Lyell, C. (1830) *Principles of Geology*, (John Murray, London).

McArthur, J. L. (1981) 'Periglacial slope planations in the southern Pennines, England', *Biul. peryglac.*, **28**, 85-96.

McCabe, A. M. (1973) 'The glacial stratigraphy of eastern counties Meath and Louth', *Proc. R. Irish. Acad.*, B, **73**, 355-82.

McConnell, R. B. (1938) 'Residual erosion surfaces in mountain ranges', *Proc. Yorks. Geol. Soc.*, **24**, 31-59.

—— (1939) 'The relic surfaces of the Howgill Fells', *Proc. Yorks. Geol. Soc.*, **24**, 152-64.

Mackereth, F. J. H. (1966) 'Some chemical observations on post-glacial lake sediments', *Phil. Trans. R. Soc. Lond.*, B, **250**, 165-213.

Mackintosh, D. (1865a) 'A tourist's notes on the surface geology of the Lake District', *Geol. Mag.*, Decade one, **2**, 299-306.

—— (1865b) 'On the relative extent of atmospheric and oceanic denudation with a particular reference to certain rocks and valleys in Yorkshire and Derbyshire', *Geol. Mag.*, Decade one, **2**, 519-20.

Manabe, S. and Hahn, D. G. (1979) 'Simulation of the tropical climate of an ice age', *J. Geophys. Res.*, **82**, 3889-911.

Manley, G. (1959) 'The Late-Glacial climate of north-west England', *L'pool and Manchr. Geol. J.*, **2**, 188–215.

—— (1964) 'The evolution of the climatic environment', in J. Wreford Watson and J. B. Sissons (eds.), *The British Isles, A Systematic Geography*, (Nelson, London), pp. 152–76.

—— (1971) 'Manchester rainfall since 1765, *Mem. Proc., Manchr. Lit. and Phil. Soc.*, **114**, 70–89.

—— (1974) 'Central England temperatures: monthly means 1659 to 1973, *Quart. J. R. Met. Soc.*, **100**, 389–405.

—— (1975) 'Fluctuations of snowfall and persistence of snow cover in marginal-oceanic climates', in *Proc. WHO/IAMAP Symposium on Long-term Climatic Fluctuations*, (WHO, Geneva), pp. 183–88.

Marcussen, I. (1976) 'Supposed area-wasting of the Weichselian ice sheet in Denmark', *Boreas*, **6**, 167–73.

Marr, J. E. (1889) 'On the superimposed drainage of the English Lake District', *Geol. Mag.*, Decade one, V, 40–42.

—— (1906) 'The influence of the geological structure of the English Lakeland upon its present features', Anniversary address of the President, *J. Geol. Soc. Lond.*, **62**, LXVI.

—— (1916) *The Geology of the Lake District*, (Cambridge University Press).

Marr, J. E. and Fearnsides, W. G. (1909) 'The Howgill Fells and their topography', *J. Geol. Soc. Lond.*, **65**, 587–610.

Mayfield, B. and Pearson, M. C. (1972) 'Human interferences with the north Derbyshire blanket peat', *E. Midld. Geogr.*, **5**, 245–51.

Mellars, P. A. (1970) 'Flints from the Kirkhead Cavern', *Archaeol. News Bull. Northumberland, Cumberland and Westmorland*, **8**, 6.

Mercer, J. (1970) 'The microlithic succession in N. Jura, Argyll, W. Scotland', *Quaternaria*, **13**, 177–85.

Merritt, J. W. (1982) 'A possible interglacial or interstadial deposit near Oundle, Northamptonshire', *Quat. Newsl.*, **37**, 10–11.

Miall, L. C. (1880) 'The (Raygill) Cave and its contents', *Proc. Yorks. Geol. Soc.*, **7**, 207–8. (See also Earp *et al.*, op. cit.)

Miller, A. A. (1938) 'Pre-glacial erosion surfaces round the Irish Sea', *Proc. Yorks. Geol. Soc.*, **24**, 31–59.

Milliman, J. D. and Emery, K. O. (1963) 'Sea-level changes during the last 35,000 years', *Science*, **162**, 1121–3.

Millward, D., Moseley, F. and Soper, N. J. (1978) 'The Eycott and Borrowdale volcanic rocks', in F. Moseley (ed.), *The Geology of the Lake District*, Yorkshire Geological Society (Occ. Publ. No. 3), pp. 99–120.

Milne, D. C. (1964) 'Dredging at the Port of Manchester', *Proc. Inst. Civ. Eng.*, **27**, 17–46.

Ministry of Works (1951) 'Mining subsidence: effects on small houses', *National Building Studies Report*, 12, (HMSO, London).

Mitchell, G. F. (1960) 'The Pleistocene history of the Irish Sea', *Adv. Sci.*, **17**, 313–25.

—— (1963) 'Morainic ridges on the floor of the Irish Sea', *Irish Geogr.*, **4**, 335–44.

—— (1965) 'The Quaternary deposits of the Ballaugh and Kirkmichael districts, Isle of Man', *J. Geol. Soc. Lond.*, **121**, 359–81.

—— (1972) 'The Pleistocene history of the Irish Sea: a second approximation', *Sci. Proc. R. Dubl. Soc.*, **A4**, 181–99.

Mitchell, G. F., Penny, L. F., Shotton, F. W. and West, R. G. (1973) 'A correlation of the Quaternary deposits in the British Isles', *Geol. Soc. Lond., Spec. Rep.*, No. 4.

Moffatt, J. D. (1977) '*Risley Moss Draft Management Plan*', Warrington New Town Planning Department.

Montford, H. (1970) 'The terrestrial environment during Upper Cretaceous and Tertiary times', *Proc. Geol. Assoc.*, **81**, 181–204.

Moore, P. D. (1973) 'The influence of prehistoric cultures upon the initiation and spread of blanket bog in upland Wales', *Nature*, **241**, 350–3.

—— (1975) 'Origin of blanket mires', *Nature*, London, **256**, 267–9.

Moore, P. D. and Bellamy, D. J. (1974) *Peatlands*, (Elek Science, London).

Morgan, A. (1973) 'Late Pleistocene environmental changes indicated by fossil insect faunas of the English Midlands', *Boreas*, **2**, 173–212.

Morgan, A. V. (1973) 'The Pleistocene geology of the area north and west of Wolverhampton, Staffordshire, England', *Phil. Trans. R. Soc. Lond.*, B, **265**, 233–97.

Morgan, A. V., Duthie, H. C., Morgan, A., Fritz, P. and Rearden, E. J. (1975) 'The Stafford project; a multi-disciplinary analysis of a Late-Glacial sequence in the West Midlands', *I.N.Q.U.A.*, 309.

Morgan, A. V. and Morgan, A. (1977) 'The English Midlands. Keele to Birmingham excursion', *I.N.Q.U.A.*, Guide 2, 34–9.

Morrison, M. F. S. and Stephens, N. (1965) 'Stratigraphy and pollen analysis of the raised beach deposits at Ballyhalbert, Co. Down', *New Phytol.*, **59**, 153–62.

Moseley, F. (1961) 'Erosion surfaces in the Forest of Bowland', *Proc. Yorks. Geol. Soc.*, **33**, 173–96.

—— (1972) 'A tectonic history of North West England', *J. Geol. Soc. Lond.*, **128**, 561–98.

—— (1978) 'An introductory review', in F. Moseley (ed.), *The Geology of the Lake District*, Yorkshire Geological Society, (Occ. Publ. No. 3), 1–16.

Moseley, F. and Walker, D. (1952) 'Some aspects of the Quaternary Period in North Lancashire', *The Naturalist*, Hull, April–June, 41–54.

Moseley, M. P. (1972) 'Gully systems in the blanket peat, Bleaklow, North Derbyshire', *East Midld. Geogr.*, **5**, 235–44.

—— (1975a) 'Channel changes on the River Bollin, Cheshire, 1872–1973', *E. Midld. Geogr.*, **6**, 185–99.

—— (1975b) 'Meander cutoffs on the River Bollin, Cheshire in July 1973', *Rev. de Geomorph. Dyn*, **24**, 21–31.

Moss, C. E. (1904) 'Peat mosses of the Pennines: their age, origin and utilization', *Geog. J.*, **23**, 660–71.

—— (1913) *Vegetation of the Peak District*, (Cambridge University Press).

Moulson, J. R. (1967) '*Some Aspects of the Geomorphology of the Lune Basin*', unpublished M.A. thesis, University of Manchester.

Muller, P. (1979) 'Investigating the age of a Pennine landslip', *Mercian Geol.*, **7**(3), 211–18.

National Coal Board Mining Department (1975) *Subsidence Engineers' Handbook*, (National Coal Board, London).

Newson, M. D. (1981) 'Mountain streams', in J. Lewin (ed.), *British Rivers*, (Allen and Unwin, London), pp. 59–89.

Nickson, D. and MacDonald, J. H. (1956) 'A preliminary report on a Microlithic site at Drigg, Cumberland', *Trans. Cumb. Westm. Antiq. Archaeol. Soc.*, **55**, 17–29.

North West Water Authority (1978) *Water Quality Review* 1977 (Directorate of Scientific Services, Warrington).

Oakman, C. D. (1979) 'Derbyshire sough hydrogeology and the artificial drainage of the Stanton syncline near Matlock, Derbyshire', *Trans. Brit. Cave. Res. Assoc.*, **6**(4), 169–94.

Oldfield, F. (1960a) 'Late Quaternary changes in climate, vegetation and sea level in Lowland Lonsdale', *Trans. Inst. Brit. Geogr.*, **28**, 99–117.

—— (1960b) 'Studies in the Post-glacial history of British vegetation: Lowland Lonsdale', *New Phytol.*, **59**, 192–217.

—— (1963) 'Pollen-analysis and man's role in the ecological history of the south-east Lake District', *Geogr. Annlr.*, **45**(1), 23–40.

—— (1965a) 'Problems of mid-Post glacial pollen zonation in part of north-west England', *J. Ecol.*, **53**, 247–60.

—— (1966) 'The palaeoecology of an early Neolithic waterlogged site in north-western England', *Rev. Palaeo-bot, Palynol.*, **62**, 53–61.

—— (1969) 'Pollen analysis and the history of land use', *Adv. Sci.*, **25**, 298–311.

Oldfield, F. and Statham, D. C. (1963) 'Pollen analytical data from Urswick Tarn and

Ellerside Moss, North Lancashire', *New Phytol.*, **62**, 53–66.

—— (1965) 'Stratigraphy and pollen analysis on Cockerham and Pilling Moses, North Lancashire', *Mem. Proc. Manchr. Lit. Phil. Soc.*, **107**, 1–16.

Ollier, C. (1979) 'Evolutionary Geomorphology of Australia and Papua New Guinea', *Trans. Inst. Br., Geogr.*, NS4, 516–39.

—— (1981) *Tectonics and Landforms*, Geomorphology texts series, K. M. Clayton (ed.), (Longman, London).

Osborne, P. J. (1972) 'Insect faunas of late Devensian and Flandrian age from Church Stretton, Shropshire', *Phil. Trans. R. Soc. Lond.*, B, **263**, 327–67.

—— (1976) 'Evidence from the insects of climatic variation during the Flandrian period: a preliminary note', *World Archaeol.*, **8**, 150–8.

O'Sullivan, P. E. (1975) 'On the origin and age of Whixall Moss, Shropshire', *North Staffs. J. Field Studies*, **15**, 19–25.

Osvald, H. (1949) 'Notes on the vegetation of British and Irish mosses', *Acta Phytogeogr. Suecica*, **26**, 7–62.

Ovenden, J. C. and Gregory, K. J. (1980) 'The permanence of stream networks in Britain', *Earth Surf. processes and Landf.*, **5**, 47–60.

Owen, D. E. (1947) 'The Pleistocene history of the Wirral peninsular', *Proc. L'pool Geol. Soc.*, **19**, 210–39.

Palmer, J. and Neilson, R. (1962) 'The origin of granite tors of Dartmoor, Devonshire', *Proc. Yorks. Geol. Soc.*, **33**, 315–40.

Palmer, J. and Radley, J. (1961) 'Gritstone tors of the English Pennines', *Z. Geomorph.*, N.F. **5**, 37–52.

Pantin, H. M. (1975) 'Quaternary sediments of the north-eastern Irish Sea', *Quat. Newsl.*, **17**, 7–9.

—— (1977) in C. Kidson and M. J. Tooley (eds.), *The Quaternary History of the Irish Sea, Geological Journal*, Special Issue 7, (Wiley, Chichester), pp. 27–54.

—— (1978) 'The Quaternary sediments from the north-east Irish Sea: Isle of Man to Cumbria', *Bull. Geol. Surv. U.K.* **64**, 1–50.

Parker, W. R. (1975) 'Sediment mobility and erosion on a multibarred foreshore (southwest Lancashire, U.K.)', in J. Hails and A. Carr (eds.), *Nearshore Sediments Dynamics and Sedimentation*, (Wiley, London), pp. 151–80.

Parry, J. T. (1960) 'The erosion surfaces of the south western Lake District', *Trans. Inst. Br. Geogr.*, **28**, 39–54.

Parry, M. L. (1975) 'Secular climatic change and marginal agriculture', *Trans. Inst. Br. Geogr.*, **64**, 1–13.

Paterson, K. and Chambers, B. (1982) 'Techniques for the study of limestone pavements', *Teaching Geography*, **8**(1), 3–9.

Paul, M. A. (1983) 'The Supraglacial Landsystem', in N. Eyles (ed.), *Glacial Geology*, (Pergamon, Oxford), pp. 71–90.

Peake, D. S. (1961) 'Glacial changes in the Alyn river system and their significance in the glaciology of the north Welsh border', *J. Geol. Soc. Lond.*, **117**, 335–66.

—— (1978) 'The Llay till and the Padeswood kettle-drift Clwyd', *Quat. Newsl.*, **26**, 12–14.

Pearsall, W. H. (1918) 'The aquatic and marsh vegetation of Esthwaite Water V. The marsh and fen vegetation of Esthwaite Water', *J. Ecol.*, **6**, 53–74.

—— (1941) 'The "mosses" of the Stainmore district', *J. Ecol.*, **29**, 161–75.

—— (1950) *Mountains and Moorlands*, (Collins, London).

Pécsi, M. (1964) *Ten years of Physicogeographic Research in Hungary*, Studies in Geography No. 1. Hungarian Acad. Sci. (Akademiai Kiado, Budapest), 30–56.

Pemberton, M. (1980) 'Shakeholes: a morphometric field project for sixth form geographers', *Geography*, **65**(3), 180–93.

Penck, W. (1924) *Die Morphologische Analyse*, (Engelhorns, Stuttgart).

Pennington, W. (1943) 'Lake sediments: the bottom deposits of the north basin of Windermere, with special reference to the diatom succession', *New Phytol.*, **42**, 1–27.

—— (1947) 'Studies of the Post-Glacial History of British vegetation VII. Lake sediments: pollen diagrams from the bottom deposits of the north basin of Windermere', *Phil. Trans. R. Soc. Lond.*, B, **233**, 137–75.

—— (1964) 'Pollen analyses from the deposits of six upland tarns in the Lake District', *Phil. Trans. R. Soc. Lond.*, B, **248**, 205–44.

—— (1965a) 'The interpretation of some post-glacial vegetation diversities at different Lake District sites', *Proc. R. Soc. Lond.*, B, **161**, 310–23.

—— (1965b) 'Pollen analysis at a microlithic site at Drigg', *Trans. Cumb. West. Antiq. Archaeol. Soc.*, N.S. **65**, 82–5.

—— (1970) 'Vegetation history in the north-west of England: a regional synthesis', in D. Walker and R. G. West (eds.), *Studies in the Vegetational History of the British Isles*, (Cambridge University Press), 41.79.

—— (1973) 'Absolute pollen frequencies in the sediments of lakes of different morphometry', in H. J. B. Birks and R. G. West (eds.), *Quaternary Plant Ecology*, (Blackwell, Oxford), pp. 79–104.

—— (1974) *The History of British Vegetation*, 2nd edn. (English University Press, London).

—— (1975a) 'A chronostratigraphic comparison of Late-Weichselian and Late-Devensian subdivisions, illustrated by two radiocarbon-dated profiles from Western Britain', *Boreas*, **4**, 157–71.

—— (1975b) 'The effect of Neolithic man on the environment in north-west England: the use of absolute pollen diagrams', in J. G. Evans, S. Limbrey and H. Cleere (eds.), *The Effect of Man on the Landscape: the Highland Zone*, Council for British Archaeol. Res. Rept. 11, (London), pp. 74–86.

—— (1977) 'The late-glacial flora and vegetation of Britain', *Phil. Trans. R. Soc. Lond.*, B, **280**, 247–71.

—— (1978) 'Quaternary Geology', in F. Moseley (ed.), *The Geology of the Lake District*, Yorkshire Geological Society, Occ. Publ. 3, pp. 207–25.

—— (1979) 'The origin of pollen in lake sediments: an enclosed lake compared with one receiving inflow streams', *New Phytol.*, **83**, 189–213.

Pennington, W. and Bonny, A. P. (1970) 'Absolute pollen diagram from the British lake-glacial', *Nature*, **226**, 871–3.

—— (1977) 'Blelham Bog', in M. J. Tooley (ed.), *The Isle of Man, Lancashire Coast and Lake District*, 10th INQUA Congress Excursion Guide A 4, (Geo Abstracts, Norwich), pp. 50–2.

Pennington, W. and Lishman, J. P. (1971) 'Iodine in lake sediments in northern England and Scotland', *Biol. Rev.*, **46**, 279–313.

Penny, L. F. (1964) 'A review of the last glaciation in Great Britain', *Proc. Yorks. Geol. Soc.*, **34**, 387–411.

Penny, L. F., Coope, G. R. and Catt, J. A. (1969) 'Age and insect fauna of the Dimlington Silts, E. Yorks.', *Nature*, **224**, 65–7.

Pentecost, A. (1981) 'The tufa deposits of the Malham District, North Yorkshire', *Field Studies*, **5**, 365–87.

Pethybridge, G. H. and Praeger, R. L. (1904–5) 'The vegetation of the district lying south of Dublin', *Proc. R. Irish Acad.*, **25**, 124–80.

Petts, G. E. (1980a) 'Implications of the fluvial process–channel morphology interaction below British reservoirs for stream habitats', *Sci. of the total envt.*, **16**, 149–63.

—— (1980b) 'Morphological changes of river channels consequent upon headwater impoundment', *J. Inst. Water Engr. and Sci.*, **34**, 374–82.

Petts, G. E. and Lewin, J. (1975) 'Physical effects of reservoirs on river systems', in G. E. Hollis (ed.), *Man's Impact on the Hydrological Cycle in the United Kingdom*, (Geobooks, Norwich), pp. 79–91.

Phillips, B. A. M. (1967) 'The post-glacial raised shoreline around the northern plain of the Isle of Man', *Nth. Univ. Geog. J.*, **8**, 56–63.

—— (1970a) 'The significance of *inheritance* in the interpretation of marine and lacustrine coastal histories', *Lakehead University Review*, 3(1), 36–45.

—— (1970b) 'Effective levels of marine planation on raised and present rock platforms', *Rev. Géogr. Montr.*, 24(3), 227–40.

Phillips, J., Yalden, D. and Tallis, J. H. (1981) *'Peak District Moorland Erosion Study, Phase 1 Report*, (Peak Park Joint Planning Board, Bakewell).

Pigott, C. D. (1962) 'Soil Formation and development on the Carboniferous Limestone of Derbyshire (I) Parent Materials', *J. Ecol.*, **50**, 145–56.

—— (1965) 'The structure of limestone surfaces in Derbyshire', *Geogr. J.*, **131**, 41–4.

Pigott, C. D. and Huntley, J. P. (1980) 'Factors controlling the distribution of *Tilia cordata* at the northern limits of its geographical range II. History in north-west England', *New Phytol.*, **84**, 145–64.

Pill, A. L. (1948) 'The Perryfoot–Castleton system – an experiment in quantitative hydrology', *Cave Sci.*, **3**, 92–7.

Pinchemel, P. (1969) *France: A Geographical Survey*, (Bell, London).

Pitman, W. C. and Talwani, M. (1972) 'Sea floor spreading in the North Atlantic', *Bull. Geol. Soc. Amer.*, **83**, 619–46.

Pitty, A. F. (1966) *An approach to the study of karst water*, Univ. Hull Occ. Papers in Geogr., 5.

—— (1968a) 'The scale and significance of solutional loss from the limestone tract of the Southern Pennines', *Proc. Geol. Assoc.*, **79**, 153–77.

—— (1968b) 'Some features of calcium hardness fluctuations in two karst streams and their possible value in geohydrological studies', *J. Hydrol.*, **6**, 202–8.

—— (1968c) 'Calcium carbonate content in karst water in relation to flow-through time', *Nature*, **217**, 939–40.

—— (1971a) 'Rate of uptake of calcium carbonate in underground karst water', *Geol. Mag.*, **108**(6), 537–43.

—— (1971b) 'Biological activity and the uptake and re-deposition of calcium carbonate in lake waters', *Environ. Letters*, **1**, 103–9.

—— (1971c) 'Observations of tufa deposition', *Area*, **3**, 185–9.

—— (1972) 'The contrast between Derbyshire and Yorkshire in the average value of calcium carbonate in their cave and karst waters', *Trans. Cave Res. Grp. G.B.*, **14**(2), 151–2.

—— (1974) 'Karst water studies in and around Ingleborough Cave', in A. C. Waltham (ed.), *Limestones and Caves of North West England*, (David and Charles, Newton Abbot), pp. 127–39.

Pitty, A. F., Bracewell, J. L. and Halliwell, R. A. (1977) 'Calcium hardness fluctuations in the show-cave section of White Scar Cave, Ingleton, Yorkshire', *Proc. 7th Int. Speleol. Congr.*, Sheffield, 359–62.

Plant, J. (1866) 'On the existence of a sea beach on the limestone moors near Buxton', *Trans. Manchr. Geol. Soc.*, **5**, 272.

Pocock, T. I. (1906) *The Geology of the Country around Macclesfield, Congleton, Crewe and Middlewich*, Mem. Geol. Surv. G.B., 1st edn. (HMSO, London).

Poole, E. G. (1968) 'The age of the Upper Boulder Clay Glaciation in the Midlands', *Nature*, **217**, 1137–8.

Poole, E. G. and Whiteman, A. J. (1961) 'The glacial drifts of the southern part of the Shropshire–Cheshire Basin', *J. Geol. Soc. Lond.*, **117**, 91–130.

—— (1966) *Geology of the Country around Nantwich and Whitchurch*, Mem. Geol. Sur. (HMSO, London).

Prentice, I. C. (1983) 'Postglacial climatic change: vegetation dynamics and the pollen record', *Progress in Physical Geography*, **7**, 273–86.

Prentice, J. E. and Morris, P. G. (1959) 'Cemented screes in the Manifold valley, North Staffordshire', *E. Midld. Geogr.*, **2**(11), 16–19.

Price, D., Wright, W. D., James, R. C. B., Tonks, L. H. and Whitehead, T. H. (1963)

Geology of the Country around Preston, Inst. Geol. Sci., (HMSO, London).

Price, R. J. (1983) *Scotland's Environment during the last 30,000 years*, (Scotland Academic Press, Edinburgh).

Prus-Chacinski, T. M. and Harris, W. B. (1963) 'Standards for lowland drainage and flood alleviation and drainage of peat lands, with special reference to the Crossens Scheme', *Proc. Instn. civ. Engrs.*, **24**, 177–205.

Radley, J. (1962) 'Peat erosion on the high moors of Derbyshire and West Yorkshire', *E. Midld. Geogr.*, **3**, 40–50.

Radley, J. and Mellars, P. A. (1964) 'A Mesolithic structure at Deepcar, Yorkshire, England, and the affinities of its associated flint industries', *Proc. Prehist. Soc.*, **30**, 1–24.

Ragg, J. M., Beard, G. R., George, H., Heaven, F. W., Hollis, J. M., Jones, R. J. A., Palmer, R. C., Reeve, M. J., Robson, J. D., and Whitefield, W. A. D. (1984) *Soils and their use in Midland and Western England, Bull. Soil Surv. G.B.*, (Harpenden).

Raistrick, A. (1930) 'Some Glacial Features of the Settle District', *Proc. Univ. Durham Phil. Soc.*, **8**, 239–51.

—— (1933) 'Glacial and Post-Glacial periods in W. Yorkshire', *Proc. Geol. Assoc.*, **44**, 263–9.

Ramsay, A. C. (1846) 'The denudation of South Wales', *Mem. Geol. Surv. G.B.*, Vol. 1, London.

—— (1863) *The Physical Geology and Geography of Great Britain*, (Stanforth, London).

—— (1872) 'On the river courses of England and Wales', *J. Geol. Soc. Lond.*, **28**, 148–60.

Ramsbottom, W. H. C. (1973) 'Transgressions and regressions in the Dinantian: a new synthesis of British Dinantian Stratigraphy', *Proc. Yorks. Geol. Soc.*, **39**,, 567–607.

—— (1979) 'Rates of transgression and regression in the Carboniferous of northwest Europe', *J. Geol. Soc. Lond.*, **136**, 147–53.

Ratcliffe, D. A. (1977) *A Nature Conservation Review*, Volume 2 (University Press, Cambridge).

Ratcliffe, R. A. S. (1981) 'Variability of weather over approximately the last century', in A. Berger (ed.), *Climatic Variations and Variability: Facts and Theories*, (Reidel, Dordrecht), pp. 303–16.

Rayner, D. H. and Hemingway, J. E. (eds.) (1974) *The Geology and Mineral Resources of Yorkshire*, (Yorkshire Geological Society, Leeds).

Reade, T. M. (1883) 'The drift beds of North West England and North Wales', *J. Geol. Soc. Lond.*, **39**, 82–132.

—— (1894) 'An ancient glacial shore', *Geol. Mag.*, N.S. 1(2), 76–7.

—— (1895) 'Foraminiferal boulder clay at Great Crosby and at Blackpool', *Proc. L'pool Geol. Soc.*, **7**, 387–90.

—— (1904) 'Glacial and postglacial features of the lower valley of the River Lune and its estuary', *Proc. L'pool Geol. Soc.*, **9**, 163–96.

Reed, C. (1901) *The Geological History of the Rivers of East Yorkshire*, (Cambridge University Press).

Reid, C. (1887) 'On the origin of dry valleys and of coombe rock', *J. Geol. Soc. Lond.*, **43**, 364–73.

—— (1904) 'On the probable occurrence of an Eocene outlier off the Cornish coast', *Q. J. Geol. Soc. Lond.*, **60**, 113–19.

—— (1913) *Submerged Forests*, (Cambridge University Press).

Reid, E. M. and Chandler, M. E. J. (1933) *The Flora of the London Clay*, (British Museum (Natural History) London).

Rice, R. J. (1957) 'Some aspects of the glacial and post-glacial history of the Lower Goyt valley, Cheshire', *Proc. Geol. Assoc.*, **68**, 217–27.

Richards, K. S. (1981) 'Evidence of Flandrian valley alluviation in Staindale, N. Yorks. Moors', *Earth Surf. processes and Landf.*, **6**, 183–6.

—— (1982) Rivers: Form and Process in Alluvial Channels, (Methuen, London).

Richardson, D. T. (1968) 'The use of chemical analysis of cave waters as a method of water tracing and indicator of types of strata traversed', *Trans. Cave Res. Grp. G.B.*, **10**(2), 61–72.

—— (1974) 'Karst waters of the Alum Pot area', in A. C. Waltham (ed.), *Limestone and Caves of North West England*, (David and Charles, Newton Abbot), pp. 140–8.

Richmond, G. M. (1959) 'Application of stratigraphic classification and nomenclature to the Quaternary', *Bull. Am. Ass. Petrol. Geol.*, **43**, 663–75.

Riggall, J. (1980) *Lowland raised bogs survey*, Nature Conservancy Council (mscr.).

Rodgers, H. B. (1955) 'Land use in Tudor Lancashire: the evidence of the Final Concords, 1450–1558', *Trans. Inst. Brit. Geogr.*, 79–97.

—— (1962) 'The landscapes of East Lancastria', in C. F. Carter (ed.), *Manchester and its Region*, (Manchester University Press for the Brit. Ass.), pp. 1–16.

Rose, J. and J. M. Letzer (1977) 'Superimposed Drumlins', *J. Glaciol.*, **18**(8), 471–80.

Rose, L. and Vincent, P. (1985a) 'Some aspects of the morphometry of grikes: a mixture modelling approach', in M. M. Sweeting and K. S. Paterson (eds.), *New Directions in Karst*, (Geobooks, Norwich).

—— (1985b) 'The kamenitzas of the Gait Barrows N.N.R.', in M. M. Sweeting and K. J. Paterson (eds.), *New Directions in Karst*, (Geobooks, Norwich).

Round, F. E. (1957) 'The Late-glacial and Post-glacial diatom succession in the Kentmere valley deposit part I. Introduction, methods and flora', *New Phytol*, **56**, 98–126.

—— (1961) 'The diatoms of a core from Esthwaite Water', *New Phytol.*, **60**, 43–59.

Rowlands, B. M. (1971) 'Radiocarbon evidence of the age of an Irish Sea glaciation in the Vale of Clwyd', *Nature*, London, **230**, 9–11.

Ruddiman, W. F. and McIntyre, A. (1973) 'Time transgressive deglacial retreat of polar waters from the North Atlantic', *Quat. Res.*, **3**, 117–30.

Ruddiman, W. F., Sancetta, C. D. and McIntyre, A. (1976) 'North-East Atlantic paleoclimatic changes over the last 600,000 years', Geol. Soc. Amer. Mem., **145**, 111–46.

—— (1977) 'Glacial/Interglacial response rate of subpolar North Atlantic waters to climatic change: the record in ocean sediments', *Phil. Trans. R. Soc. Lond.*, B **280**, 119–42.

Ruegg, G. H. J. (1983) 'Periglacial eolian evenly laminated sandy deposits in the Late Pleistocene of N.W. Europe, a facies unrecorded in modern sedimentological handbooks', in M. E. Brookfield and T. S. Ahlbrandt (eds.), *Eolian Sediments and Processes*, (Elsevier, Amsterdam), pp. 455–82.

Rural Preservation Association (1980) *Sutton Mosses Ecological Survey*, (Rural Preservation Association, Liverpool).

Saunders, G. E. (1968a) 'A fabric analysis of the ground moraine deposits of the Lleyn Peninsula of south-west Caernarvonshire', *Geol. J.*, **6**, 105–18.

—— (1968b) 'Glaciation of possible Scottish Readvance age in North Wales', *Nature*, London, **218**, 76–8.

Schumm, S. A. and Lichty, R. W. (1965) 'Time, space and causality in Geomorphology', *Am. J. Sci.*, **263**, 110–19.

Schumm, S. A. and Shepherd, R. G. (1973) 'Valley floor morphology: evidence for sub-glacial erosion?', *Area*, **5**, 5–9.

Searle, I. W. (1973) '*An investigation into the movement of an embankment on the Potteries 'D' road at Stoke-on-Trent*', unpublished M.Sc. thesis, University of London.

Selley, R. C. (1978) *Ancient Sedimentary Environments*, (Chapman and Hall, London).

Shaw, R. P. (1984) '*Karstic sediments, residual and alluvial ore deposits of the Peak District of Derbyshire*', unpublished Ph.D. thesis, University of Leicester.

Shennan, I. (1981) '*Flandrian sea levels in the Fenland*', unpublished Ph.D. thesis, University of Durham.

—— (1982a) Problems of correlating Flandrian sea-level changes and climate', in A. F. Harding (ed.), *Climatic Changes in Later Prehistory*, (Edinburgh University Press), pp. 52–67.

—— (1982b) 'Interpretation of Flandrian sea-level data from the Fenland, England', *Proc. Geol. Assoc.*, **93**(1), 53–63.

—— (1983) 'A problem of definition in sea-level research methods', *Quat. Newsl.*, **39**, 17–19.

Shennan, I., Tooley, M. J., Davis, M. J. and Haggart, B. A. (1983) 'Analysis and interpretation of Holocene sea-level data', *Nature*, **302**, (5907), 404–6.

Shimwell, D. W. (1974) 'Sheep grazing intensity in Edale, 1692–1747, and its effect on blanket peat erosion', *Derbyshire Archaeol. J.*, **94**, 35–40.

Shortt, J. (1980) 'A brief sketch of the history of Hoghton in the county of Lancashire', (Blackburn), 6–8. (See also Price, *et al.* op. cit.)

Shotton, F. W. (1953) 'The Pleistocene deposits of the area between Coventry, Rugby and Leamington and their bearing upon the topographic development of the Midlands', *Proc. R. Soc. Lond.*, B, **237**, 209–60.

—— (1973) 'General principles governing the subdivision of the Quaternary System'. In G. F. Mitchell, L. F. Penny, F. W. Shotton and R. G. West, (eds.), *A correlation of Quaternary deposits in the British Isles*, Geol. Soc. Lond., Sp. Rep. No. 4, 1–7.

—— (1977a) 'The Devensian Stage: its development, limits and substages', *Phil. Trans. R. Soc. Lond.*, B, **280**, 107–18.

—— (ed.) (1977b) *The English Midlands*, I.N.Q.U.A., Guide, (Geo Abstracts, Norwich).

Shotton, F. W. and West, R. G. (1969) 'Stratigraphical table of the British Quaternary, *Proc. Geol. Soc. Lond.*, **1656**, 155–7.

Shotton, F. W. and Williams, R. E. G. (1971) 'Birmingham Radiocarbon dates: V', *Radiocarbon*, **13**, 141–56.

—— (1973) 'Birmingham Radiocarbon dates: VII', *Radiocarbon*, **15**, 451–68.

Shreve, R. L. (1972) 'The movement of water in glaciers', *J. Glaciol.*, **11**, 205–14.

Simmons, I. and Tooley, M. J. (1981) *The Environment in British Prehistory*, (Duckworth, London).

Simpson, I. M. (1959) 'The Pleistocene succession in the Stockport and South Manchester area', *J. Geol. Soc. Lond.*, **115**, 107–19.

Simpson, I. M. and West, R. G. (1958) 'On the stratigraphy and palaeobotany of a Late-Pleistocene organic deposit at Chelford, Cheshire', *New Phytol.*, **57**, 239–50.

Sissons, J. B. (1960a) 'Some aspects of glacial drainage channels in Britain (1)', *Scot. Geogr. Mag.*, **76**, 131–46.

—— (1960b) 'Erosion surfaces, cyclic slopes and drainage systems in Southern Scotland and Northern England', *Trans. Inst. Br. Geogr.*, **28**, 23–38.

—— (1961) 'Some aspects of glacial drainage channels in Britain', *Scot. Geogr. Mag.*, **77**, 15–36.

—— (1965) 'The Glacial Period', in J. Wreford-Watson and J. B. Sissons (eds.), *The British Isles: A Systematic Geography*, (Nelson, London), pp. 131–52.

—— (1974a) 'The Quaternary in Scotland: a review', *Scot. Geol. J.*, **10**, 311–37.

—— (1974b) 'Glacial readvances in Scotland', in C. J. Caseldine and W. A. Mitchell (eds.), *Problems of the Deglaciation of Scotland*, St. Andrews Department of Geography, Special Publication **1**, 5–15.

—— (1980) 'The Loch Lomond Advance in the Lake District, northern England', *Trans. R. Soc. Edin.*, **71**, 13–27.

—— (1981) 'Ice-dammed lakes in Glen Roy and vicinity: a summary', in J. Neale and J. P. Flenley (eds.), *The Quaternary in Britain*, (Pergamon, Oxford), pp. 174–83.

Slater, G. (1930) 'The structure of the Bride Moraine, Isle of Man', *Proc. L'pool. Geol. Soc.*, **14**, 184–96.

Smailes, A. E. (1960) *Northern England*, (Nelson, London).

Smith, A. G. (1958) 'Two lacustrine deposits in the south of the English Lake District', *New Phytol.*, **57**, 363–86.

—— (1959) 'The mires of south-western Westmorland: stratigraphy and pollen analysis', *New Phytol.*, **58**, 105–27.

Smith, A. G., Hurley, A. M. and Briden, J. C. (1981) *Phanerozoic Paleocontinental World Maps*, (University Press, Cambridge).

Smith, B. (1930) 'Borings through the glacial drifts of the northern part of the Isle of Man', *Mem. Geol. Surv. U.K.,* **3**, 14–23.

Smith, D. I. (1975) 'The problems of limestone dry valleys – implications of recent work in limestone hydrology', in R. Peel, M. Chisholm and P. Haggett (eds.), *Processes in Physical and Human Geography: Bristol Essays.* (Heinemann, London), pp. 130–147.

Smith, D. I. and Atkinson, T. C. (1977) 'Underground flow in cavernous limestones with special reference to the Malham area', *Field Studies,* **4**, 597–616.

Smith, L. P. (1976) *The Agricultural Climate of England and Wales,* (HMSO, London).

Smith, R. A. (1967) 'The deglaciation of south-west Cumberland: A re-appraisal of some features in the Eskdale and Bootle areas', *Proc. Cumb. Geol. Soc.,* **2**, 76–83.

—— (1969) 'The deglaciation of south-west Cumberland – excursion report', *Proc. Cumb. Geol. Soc.,* **2**, 119–21.

—— (1977) 'The glacial landforms and deposits of the Corney and Bootle areas', *Proc. Cumb. Geol. Soc.,* **3**, 260–4.

Soper, N. J. and Moseley, F. (1978) 'Structure', in F. Moseley (ed.), *The Geology of the Lake District,* Yorkshire Geological Society, Occ. Publ., No. 3, 45–68.

Sorby, H. C. (1869) 'Note on the excavation of the valleys in Derbyshire', *Geol. Mag.,* Decade one, **5**, 347–8.

Sparks, B. W. (1962) 'Post-glacial mollusca from Hawes Water, Lancashire, illustrating some difficulties of interpretation', *J. Conchol.,* **25**, 78–82.

Statham, I. (1977) *Earth Surface Sediment Transport,* (University Press, Oxford).

Steel, R. J. and Thompson, D. B. (1983) 'Structures and textures in Triassic braided stream conglomerates ("Bunter" Pebble Beds) in the Sherwood Sandstone Group, North Staffordshire, England', *Sedimentol,* **30**, 341–67.

Stephens, N., Creighton, J. R. and Hannon, M. A. (1975) 'The Late Pleistocene period in north-eastern Ireland: an assessment', *Irish Geogr.,* **8**, 1–23.

Stephens, N. and McCabe, A. M. (1977) 'Late Pleistocene ice movements and patterns of Late and Post-glacial shorelines on the coast of Ulster, Ireland', in C. Kidson and M. J. Tooley (eds.), *The Quaternary History of the Irish Sea,* Geol. J. Special Issue, 7, (Wiley, Chichester), 179–98.

Stevenson, I. P. and Gaunt, G. D. (1971) *Geology of the Country around Chapel-en-le-Frith,* Mem. Inst. Geol. Sci. (HMSO, London).

Straaten, L. M. J. U. van (1965) 'Coastal barrier deposits in South- and North-Holland, in particular in the area around Scheveningen and Ijmuiden', *Med. Geol. Sticht.,* N.S., **17**, 41–75.

Straw, A. and Clayton, K. M. (1979) *Eastern and Central England: The Geomorphology of the British Isles Series* (eds.), E. H. Brown and K. M. Clayton, (Methuen, London).

Straw, A. and Lewis, G. M. (1962) 'Glacial Drift in the area around Bakewell, Derbyshire', *E. Midld. Geogr.,* **3**, 72–80.

Streif, H. (1979–80) 'Cycle formation of coastal deposits and their indications of vertical sea-level changes', *Oceanis.,* **5**, 303–6.

Stuart, A. J. (1977) 'British Quaternary Vertebrates', in F. Shotton (ed.), *British Quaternary Studies: recent advances,* (Oxford University Press), pp. 69–82.

Stuiver, M. (1982) 'A high-precision calibration of the AD radiocarbon time scale', *Radiocarbon,* **24**, 1–26.

Sub Soil Surveys Ltd. (1967) '*Report of Soil Investigations at the site of the subsidence at the Cliff, Salford 7, for the City of Salford*', Sub Soil Surveys Ltd., Astley, Manchester.

Sumbler, M. (1983) 'A new look at the type Wolstonian glacial deposits of Central England', *Proc. Geol. Assoc.,* **94**(1), 23–31.

Summerfield, M. A. (1981) 'Macroscale Geomorphology', *Area,* **13**, 3–8.

Sutherland, D. G. (1981) 'The high-level marine shell beds of Scotland and the build-up of the last Scottish Ice-sheet', *Boreas,* **10**, 247–54.

Sweeting, M. M. (1950) 'Erosion cycles and limestone caverns in the Ingleborough district', *Geogr. J.,* **115**, 63–78.

—— (1964) 'Some factors in the absolute denudation of limestone terrains' *Erdkunde*, **18**(2), 92–95.

—— (1965) 'Denudation in limestone regions, Part 1 – Introduction', *Geogr. J.*, **131**, 34–7.

—— (1966) 'The weathering of Limestones, with particular reference to the Carboniferous Limestones of Northern England', in G. H. Dury (ed.) *Essays in Geomorphology*, (Heinemann, London), pp. 177–210.

—— (1972a) 'Karst of Great Britain', in M. Herak and V. T. Stringfield (eds.), *Karst*, (Elsevier, Amsterdam), pp. 417–43.

—— (1972b) *Karst Landforms*, (Macmillan, London).

—— (1974) 'Karst geomorphology in North West England', in A. C. Waltham (ed.), *The Limestones and Caves of North West England*, (David and Charles, Newton Abbot), pp. 46–78.

—— (1979) 'Karst morphology and limestone petrology', *Progress in Physical Geography*, **3**(1), 102–10.

Sweeting, M. M. and Sweeting, G. S. (1969) 'Some aspects of the Carboniferous Limestone in relation to its landforms', *Méditerranée*, **7**, 201–9.

Switsur, V. R. and Jacobi, R. M. (1975) 'Radiocarbon dates for the Pennine Mesolithic', *Nature*, **256**, 32–4.

Synge, F. M. (1952) 'Retreat stages of the last ice-sheet in the British Isles', *Irish Geogr.*, **2**, 168–71.

Synge, F. M. (1977a) 'Records of sea-levels during the late Devensian', *Phil. Trans. R. Soc. Lond.*, B, **280**, 211–28.

—— (1977b) 'The coasts of Leinster (Ireland)', in C. Kidson and M. J. Tooley (eds.), *Quaternary History of the Irish Sea*, Geol. J. Special Issue 7, (Wiley, Chichester), 199–222.

—— (1980) 'A morphological comparison of raised shorelines in Fennoscandia, Scotland and Ireland', *Geol. Fören. Stockholm, Förh. Floring.*, **102**, 235–49.

Tallis, J. H. (1964) 'Studies on southern Pennine peats. I–III', *J. Ecol.*, **52**, 323–53.

—— (1965) 'Studies on southern Pennine peats. IV. Evidence of recent erosion', *J. Ecol.*, **53**, 509–20.

—— (1969) 'The blanket bog vegetation of the Berwyn Mountains, North Wales', *J. Ecol.*, **57**, 765–87.

—— (1973a) 'Studies on southern Pennine peats. V. Direct observations on peat erosion and peat hydrology at Featherbed Moss, Derbyshire', *J. Ecol.*, **61**, 1–22.

—— (1973b) 'The terrestrialization of lake basins in north Cheshire, with special reference to the development of a "Schwingmoor" structure', *J. Ecol.*, **61**, 537–67.

—— (1981a) 'Rates of erosion, in J. Phillips, D. Yalden and J. Tallis (eds.), *Moorland Erosion Study. Phase 1 Report*, (Peak Park Joint Planning Board, Bakewell), pp. 74–83.

—— (1981b) 'Uncontrolled fires', in J. Phillips, D. Yalden and J. Tallis (eds.), *Moorland Erosion Study. Phase 1 Report*, (Peak Park Joint Planning Board, Bakewell), pp. 176–82.

—— (1985) 'Mass movement and erosion of a southern Pennine blanket peat', *J. Ecol.*, **73**.

Tallis, J. H. and Birks, H. J. B. (1965) 'The past and the present distribution of *Scheuchzeria palustris* in Europe', *J. Ecol.*, **53**, 287–98.

Tallis, J. H. and Johnson, R. H. (1980) 'The dating of landslides in Longdendale, north Derbyshire, using pollen-analytical techniques', in R. A. Cullingford, D. A. Davidson, and J. Lewin (eds.), *Timescales in Geomorphology*, (Wiley, Chichester), pp. 189–205.

Tallis, J. H. and Switsur, V. R. (1973) 'Studies on southern Pennine peats. VI. A radiocarbon-dated pollen diagram from Featherbed Moss, Derbyshire, *J. Ecol.*, **61**, 743–51.

—— (1983) 'Forest and Moorland in the south Pennine uplands in the mid-Flandrian period. I. Macro fossil evidence of the former forest cover', *J. Ecol.*, **71**, 585–600.

Tallis, J. and Yalden, D. (1983) *Moorland Restoration Project. Phase 2 Report: Re-vegetation Trials*, (Peak Park Joint Planning Board, Bakewell).

Tate, A. P. K. (1976) 'Pile driving in glacial deposits', *Quart. J. Engng. Geol.*, **9**, 280.

Taylor, B. J. (1958) 'Cemented shear planes in the Pleistocene Middle Sands of Lancashire and Cheshire, *Proc. Yorks. Geol. Soc.*, **31**(4), 359–66.

Taylor, B. J., Price, R. H. and Trotter, F. M. (1963) *Geology of the Country around Stockport and Knutsford*, Mem. Geol. Surv. G.B. (HMSO, London).

Taylor, J. A. (1983) 'The peatlands of Great Britain and Ireland', in A. J. P. Gore (ed.), *Mires: Swamp, Bog, Fen and Moor, B. Regional Studies*, (Elsevier, Amsterdam), pp. 1–46.

Taylor, R. (1975) 'The coastal salt industry at Amounderness', *Trans. Lancs. and Ches. Antiq. Soc.*, **78**, 14–21.

Taylor, T. L. (1961) *'Geomorphology of Over Wyre'*, unpublished M.A. thesis, University of Liverpool.

Temple, P. H. (1965) 'Some aspects of cirque distribution in the west-central Lake District, northern England', *Geogr. Annlr.*, **47A**, 185–93.

Ternan, J. L. (1972) 'Comments on the use of a calcium hardness variability index in the study of carbonate aquifers: with reference to the Central Pennines, England', *J. Hydrol.*, **16**, 317–21.

—— (1974) 'Some chemical and physical characteristics of five resurgences on Darnbrook Fell', in A. C. Waltham (ed.), *Limestones and Caves of North West England*, (David and Charles, Newton Abbot), pp. 115–26.

Terzaghi, K. and Peck, R. (1967) *Soil Mechanics and Engineering Practice*, (Wiley, New York).

Thomas, G. S. P. (1976) 'The Quaternary stratigraphy of the Isle of Man', *Proc. Geol. Assoc.*, **87**, 307–24.

—— (1977) 'The Quaternary of the Isle of Man', in C. Kidson and M. J. Tooley (eds.), *The Quaternary History of the Irish Sea*, Geol. J., Special Issue, 7, (Wiley, Chichester), 155–78.

—— (1984a) 'Sedimentation of a sub-aqueous esker-delta at Strabethie, Aberdeenshire', *Scot. J. Geol.*, **20**, 9–20.

—— (1984b) 'The origin of the glacio-dynamic structure of the Bride Moraine, Isle of Man', *Boreas*, **13**, 355–64.

Thomas, G. S. P. and Summers, A. J. (1983) 'The Quaternary stratigraphy between Blackwater Harbour and Tinnaberna, Co. Wexford, *J. Earth. Sci. R. Dubl. Soc.*, **5**, 121–34.

—— (1984) 'Glacio-dynamic structures from the Blackwater Formation, Co. Wexford, Ireland', *Boreas*, **13**, 5–12.

Thomas, M. F. (1974) *Tropical Geomorphology: A study of weathering and landform development in warm climates*, (Macmillan, London).

—— (1978) 'Denudation in the tropics and the interpretation of the tropical legacy in higher latitudes – a view of the British experience', in C. Embleton, D. Brunsden and D. K. C. Jones (eds.), *Geomorphology, Present Problems and Future Prospects*, (Oxford University Press), pp. 185–202.

Thompson, A. (1984) *'Long and short term channel changes in gravel bed rivers'*, unpublished Ph.D. thesis, University of Liverpool.

Thompson, D. B. and Worsley, P. (1967) 'Periods of ventifact formation in the Permo-Triassic and Quaternary of the north east Cheshire Basin', *Mercian Geol.*, **2**, 279–98.

Thorp, M. B. (1967) 'Joint patterns and landforms in the Jarawa granite massif, Northern Nigeria', in R. W. Steel and R. Lawton, (eds.), *Liverpool: Esays in Geography*, (Longman, London), 65–84.

Thorpe, P., Otlet, R. L. and Sweeting, M. M. (1980) 'Hydrological implications from 14_C profiling of U.K. tufa', Paper presented at 10th International Radiocarbon Conference, Heidelberg.

Tiddeman, R. H. (1868) 'The valleys of Lancashire', Geol. Mag. decade one, **5**, 39–40.

—— (1872) 'On the evidence for the Ice Sheet in north Lancashire, and adjacent parts of Yorkshire and Westmorland', *J. Geol. Soc. Lond.*, **28**, 471–91.

Tonks, L. H., Jones, R. C. B., Lloyd, W. and Sherlock, R. L. (1931) 'The geology of the country around Manchester and the south-east Lancashire coalfield', *Mem. Geol. Surv.*,

G.B. (HMSO, London).

Tooley, M. J. (1974) 'Sea level changes during the last 9000 years in northwest England', *Geogr. J.*, 18–42.

—— (1976) 'Flandrian sea-level changes in west Lancashire and their implications for the "Hillhouse Coastline" ', *Geol. J.*, **11**(2), 137–52.

—— (1977) in M. J. Tooley (ed.), *The Isle of Man, Lancashire Coast and Lake District*, 10th INQUA Congress 1977. Excursion Guide (Geo Abstracts, Norwich).

—— (1978a) 'Interpretation of Holocene sea-level changes', *Geol. Fören. Stockholm Förh.*, **100**(2), 203–12.

—— (1978b) *Sea-level Changes. North-west England during the Flandrian Stage*, (University Press, Oxford).

—— (1979) 'Sea-level changes during the Flandrian Stage and the implications for coastal development', in K. Suguio, T. R. Fairchild, L. Martin and J. M. Flexor (eds.), *Proc. 1978 Int. Symp. on Coastal Evolution in the Quaternary*, (Universidade de São Paulo), pp. 502–33.

—— (1980) 'Solway Lowlands, shores of Morecambe Bay and south-western Lancashire', in W. G. Jardine (ed.), *Field Guide: western Scotland and north-west England.* (INQUA. Subcommission on Shorelines of North western Europe), pp. 71–110.

—— (1982) 'Sea-level changes in northern England', *Proc. Geol. Assoc.*, **93**(1), 43–51.

—— (1984) 'Raised and buried beaches in the Durham coast', Institute of British Geographers. Excursion to the Durham coast. (Department of Geography, University of Durham).

—— (1985) 'Climate, sea-level and coastal changes', in M. J. Tooley and G. M. Sheail (eds.), *The Climatic Scene. Essays in honour of Gordon Manley*, (George Allen and Unwin, London), pp. 206–234.

—— (in press) 'The Quaternary history of Morecambe Bay', *Morecambe Bay Study Group Report*, N.E.R.C., (HMSO, London).

Tooley, M. J. and Kear, B. S. (1977) 'Mere Sands Wood (Shirdley Hill Sand)', in *The Isle of Man, Lancashire Coast and Lake District*, 10th INQUA Congress Excursion Guide. (Geo Abstract, Norwich), 9–10.

Trendall, A. F. (1962) 'The formation of apparent peneplains by a process of combine laterization and surface wash', *Z. Geomorph. N.F.*, **6**, 183–97.

Trevelyan, G. M. (1928) *History of England*, (Longmans, Green & Co., London).

Tricart, J. (1972) *Landforms of the Humid Tropics: Forest and Savanna* (Longman, London).

Troels-Smith, J. (1955) Karakterisering af løse jordarter, *Danm. Geol. Unders.* IV, 3(10), 1–73.

Trotter, F. M. (1929) 'The glaciation of eastern Edenside, Alston Block and the Carlisle Plain', *J. Geol. Soc. Lond.*, **88**, 549–607.

Trotter, F. M. and Hollingworth, S. E. (1932) 'The glacial sequence in the north of England', *Geol. Mag.*, **69**, 374–80.

Tufnell, L. (1969) 'The range of periglacial phenomena in northern England', *Biul. peryglac.*, **19**, 291–323.

Turner, J. (1981) 'The Iron Age', in I. Simmons and M. J. Tooley (eds.) *The Environment in British Prehistory*, (Duckworth, London), pp. 250–81.

Tutin, W. (1969) 'The usefulness of pollen analysis in interpretation of stratigraphic horizons, both Late-glacial and Post-glacial', *Mitt. Int. Ver. f. theor. ang. Limn.*, **17**, 154–64.

Twidale, C. R. (1976) 'On the survival of palaeoforms', *Am. J. Sci.*, **276**, 77–95.

Vincent, P. J. (1982) 'Some observations on the so-called relict karst of the Morecambe Bay Region, north-west England', *Rev. Géol. Dyn. Géogr. Phy.*, **23**, 143–50.

—— (1983) 'The dissolving landscape. Step karren', *The Geog. Mag.*, **55**, 508–10.

Vincent, P. J. and Lee, M. P. (1981) 'Some observations on the loess around Morecambe Bay, north-west England', *Proc. Yorks. Geol. Soc.*, **43**, 281–94.

—— (1982) 'Snow patches on Farleton Fell, south-east Cumbria', *Geogr. J.*, **148**, 337–42.

Walkden, G. M. (1974) 'Palaeokarstic surfaces in Upper Viséan (Carboniferous) limestones of the Derbyshire Block, England', *J. Sedim. Petrol.*, **44**, 1232–47.

Walker, D. (1955) 'Studies in the Post-glacial history of British vegetation XIV. Skelsmergh Tarn and Kentmere, Westmorland', *New Phytol.*, **54**, 222–54.

—— (1965) 'The Post-glacial period in the Langdale Fells, English Lake District', *New Phytol.*, **64**, 488–510.

—— (1966) 'The late Quaternary history of the Cumberland Lowland', *Phil. Trans. R. Soc. Lond.*, B, **251**, 1–210.

—— (1970) 'Direction and rate in some Post-glacial hydroseres', in D. Walker and R. G. West (eds.), *Studies in the Vegetational History of the British Isles*, (Cambridge University Press), pp. 117–39.

Walling, D. F. and Webb, B. W. (1981) 'Water Quality', in J. Lewin (ed.), *British Rivers*, (Allen and Unwin, London), pp. 126–69.

Wallwork, K. L. (1956) 'Subsidence in the mid-Cheshire industrial area', *Geogr. J.*, **122**, 40–53.

—— (1960) 'Some problems of subsidence and land-use in the mid-Cheshire industrial area', *Geogr. J.*, **126**, 191–9.

—— (1974) *Derelict Land*, (David and Charles, Newton Abbot).

Walsh, P. T., Boulter, M. C., Ijtaba, M. and Urbani, D. M. (1972) 'The preservation of the Neogene Brassington Formation of the southern Pennines and its bearing on the evolution of Upland Britain', *J. Geol. Soc. Lond.*, **128**, 519–59.

Walsh, P. T. and Brown, E. H. (1971) 'Solution subsidence outliers containing probable Tertiary sediments in North East Wales', *Geol. J.*, **7**(2), 299–320.

Waltham, A. C. (1970) 'Cave development in the limestone of the Ingleborough district', *Geogr. J.*, **136**, 574–85.

—— (1971) 'Shale units in the Great Scar Limestone of the Southern Askrigg Block', *Proc. Yorks. Geol. Soc.*, **38**, 285–92.

—— (1974) *Limestone and Caves of North West England*, (David and Charles, Newton Abbot).

—— (1976) Excursion report – Ingleton and Ingleborough. *Mercian Geol.*, **6**(1), 67–75.

—— (1977a) 'Cave development at the base of the limestone in Yorkshire', *Proc. 7th Int. Speleol. Congr.*, Sheffield, 1977, 421–3.

—— (1977b), 'White Scar Cave, Ingleton', *Trans. Brit. Cave, Res. Assoc.*, **4**(3), 345–53.

—— (1981) 'Origins and development of limestone caves', *Progress in Physical Geography*, **5**(2), 242–56.

Waltham, A. C. and Brook, D. B. (1980) 'The Three Counties System', *Trans. Brit. Cave Res. Assoc.*, **7**(2), 121.

Waltham, A. C., Brook, D. B., Statham, O. W. and Yeadon, T. G. (1981), 'Swinsto Hole, Kingsdale: A type example of cave development in the limestone of northern England', *Geogr. J.*, **147**(3), 350–3.

Walton, A. D. (1964) 'Meltwater channels near the Head of Trent', *N. Staffs. J. Field Studies*, **4**, 67–75.

Ward, C. (1970) 'The Ayre raised beach, Isle of Man', *Geol. J.*, **7**, 217–20.

Ward, J. C. (1870) 'On the denudation of the Lake District', *Geol. Mag.*, Decade one, **7**, 14–17.

—— (1876) *The Geology of the Northern Part of the Lake District*, Mem. Geol. Surv. England and Wales, (London).

Ward, R. C. (1981) 'River systems and river regimes', in J. Lewin (ed.), *British Rivers*, (Allen and Unwin, London), pp. 1–33.

Warren, P. T. (1971) *British Regional Geology: Northern England*, (Institute of Geological Sciences, London).

Warrington, G. (1965) 'The metalliferous mining district of Alderley Edge, Cheshire', *Mercian Geol.*, **1**, 111–29.

—— (1970) 'The stratigraphy and palaeontology of the ''Keuper'' Series of the central

Midlands of England', *J. Geol. Soc. Lond.*, **126**, 183–223.

Warrington, G., Audley-Charles, M. G., Elliott, R. E., Evans, W. B., Ivimey-Cook, H. C., Kent, P., Robinson, P. L., Shotton, F. W. and Taylor, F. M. (1980), *A Correlation of Triassic rocks in the British Isles*, Geol. Soc. Lond. Spec. Rep. 13.

Warwick, G. T. (1958) 'The characteristics and development of limestone regions in the British Isles with special reference to England and Wales', *Proc. 2nd Int. Speleol. Congr.*, Salerno, **1**(1), 79–105.

—— (1964) 'Dry valleys of the Southern Pennines, England', *Erdkunde*, **18**(2), 116–23.

—— (1971) 'Caves and the Ice Age', *Trans. Cave Res. Grp. G.B.*, **13**(2), 123–30.

—— (1975) 'The metamorphosis of karren in the north of England', *Proc. 6th Int. Speleol. Congr.*, Olomouc, 1973, Vol. 2, 435–43.

—— (1977) *The English Midlands*, 10th INQUA guide, (Geo Abstracts, Norwich).

Waters, R. S. (1957) 'Differential weathering on Oldlands', *Geogr. J.*, **123**, 503–9.

—— (1976) 'Stamp of ice on the north', *The Geog. Mag.*, **48**, 342–8.

Watson, E. (1971) 'Remains of pingos in Wales and the Isle of Man' *Geol. J.*, **7**, 381, 387.

—— (1977) 'The periglacial environment of Great Britain during the Devensian', *Phil. Trans. R. Soc. Lond.*, B, **280**, 183–98.

Wayland, E. J. (1934) *'Peneplains and some other erosional platforms'*, Bull. Geol. Surv. Uganda, Ann. Rept. 74, 366.

Wedd, C. B., Smith, B. and Wills, L. J. (1928) *The Geology of the Country around Wrexham Part II. Coal Measures and Newer Formations*. Mem. Geol. Surv. England and Wales, (HMSO, London).

West, R. G. (1960) 'The ice age', *Adv. of Sci.*, **16**, 428–40.

—— (1972) 'Relative land-sea-level changes in south-eastern England during the Pleistocene', *Phil. Trans. R. Soc. Lond.*, A. **272**, 87–98.

—— (1980a) *'The Pre-glacial Pleistocene of the Norfolk and Suffolk coasts'*, (Cambridge University Press).

—— (1980b) 'Pleistocene forest history in East Anglia', *New Phytol.*, **85**, 571–622.

West, R. G., Dickson, C. A., Catt, J. A., Weir, A. M. and Sparks, B. W. (1974) 'Late Pleistocene deposits at Bretton, Norfolk II – Devensian deposits', *Phil. Trans. R. Soc.*, B, **267**, 337–420.

Whitehead, P. F. (1977) 'A note on *Picea* in the Chelfordian interstadial organic deposit at Chelford, Cheshire', *Quat. Newsl.*, **23**, 8–10.

Whitehead, T. H., Dixon, E. E. L., Pocock, R. W., Robertson, T. and Cantril, T. C. (1927) *The Geology of the Country between Stafford and Market Drayton*, Mem. Geol. Surv. G.B. (HMSO, London).

Whittington, R. J. (1977) A Late-Glacial drainage pattern in the Kish Bank area and post-glacial sediments in the central Irish Sea', in C. Kidson and M. J. Tooley (eds.), *The Quaternary History of the Irish Sea*, Geol. J., Special Issue, 7, (Wiley, Chichester), 55–68.

Wilcock, D. N. (1967) Coarse bedload as a factor determining bed slope', Sympt. on river morphology. *Int. Assoc. Sci. Hydrol. Berne 1967*, 143–50.

—— (1971) 'Investigations in the relations between bedload transport and channel shape', *Geol. Soc. Am. Bull.*, **82**, 2159–76.

Wilkinson, I. P. and Halliwell, G. P. (1979) 'Offshore micropalaeontological biostratigraphy of southern and western Britain', *Rep. Inst. Geol. Sci.* 79/9, 65.

Wilkinson, T. P. (1971) *'The Transport of rock debris in upland streams with particular reference to Northern England'*, unpublished Ph.D. thesis, University of Newcastle-upon-Tyne.

Williams, J., Barry, R. G. and Washington, W. M. (1974) 'Simulations of the atmospheric circulation model with ice age boundary conditions', *J. Appl. Meteorol.*, **13**, 305–17.

Williams, J. K. (1978) *'The form and origin of the buried rock-head topography of part of Central Lancashire'*, unpublished M.Sc. thesis, University of Lancaster.

Williams, P. W. (1966) 'Limestone pavements', *Trans. Inst. Br. Geogr.*, **40**, 155–72.

Williams, R. B. G. (1975) 'The British climate during the last glaciation: an interpretation based on periglacial phenomena', in A. E. Wright and F. Moseley (eds.), *Ice Ages, Ancient and Modern*, Geol. J. Spec. Issue, No. 6, 95–120.

Williamson, I. A. (1956) 'A guide to the geology of the Cliviger valley, near Burnley, Lancashire', *Proc. Yorks. Geol. Soc.*, **30**, 375–406.

Wills, L. J. (1929) *The Physiographical Evolution of Britain*, (Arnold, London).

Wilson, A. C., Mathews, S. J. and Cannell, B. (1982) *The Middle Sands, a prograding sandur succession: its significance in the glacial evolution of the Wrexham–Shrewsbury region*, Rept. Inst. Geol. Sci., No. 82/1, 30–5.

Wilson, P. (1980) 'Surface textures of regolith quartz from the Southern Pennines', *Geol. J.*, **15**(2), 113–29.

—— (1981) 'Periglacial Valley-fill sediments at Edale, North Derbyshire', *E. Midld. Geogr.*, **7**, 263–71.

Wilson, P., Bateman, R. M. and Catt, J. A. (1981) 'Petrography, origin and environment of deposition of the Shirdley Hill Sand of southwest Lancashire, England', *Proc. Geol. Assoc.*, **92**(4), 211–29.

Woodland, A. W. (1968) 'Field geology and the civil engineer', *Proc. Yorks. Geol. Soc.*, **36**, 531–78.

—— (1970) 'The buried tunnel valleys of East Anglia', *Proc. Yorks. Geol. Soc.*, **37**, 521–78.

Wooldridge, S. W. (1956) *The Geographer as Scientist*, (Nelson, London).

Wooldridge, S. W. and Linton, D. L. (1939) *Structure, Surface and Drainage in South East England*, Inst. of Br. Geogr. publ. 10.

Worley, N. E. and Beck, J. S. (1976) 'Moorfurlong Mine, Bradwell, Derbyshire, and its geological evolution', *Trans. Brit. Cave Res. Assoc.*, **3**(1), 49–53.

Worsley, P. (1966) 'Some Weichselian fossil frost wedges from east Cheshire', *Mercian Geol.*, **1**, 356–7.

—— (1967a) 'Problems in naming the Pleistocene deposits of the north-east Cheshire Plain', *Mercian Geol.*, **2**, 51–5.

—— (1967b) '*Some aspects of the Quaternary evolution of the Cheshire plain*', unpublished Ph.D. thesis, University of Manchester.

—— (1970) 'The Cheshire–Shropshire Lowlands', in C. A. Lewis (ed.), *The Glaciations of Wales and Adjoining Regions*, (Longman, London), pp. 83–106..

—— (1975) 'An appraisal of the glacial Lake Lapworth concept', in A. D. M. Phillips and B. J. Turton (eds.), *Environment, Man and Economic Change*, (Longman, London), pp. 90–118.

—— (1976) 'Correlation of the last glaciation glacial maximum and the extra-glacial fluvial terraces in Britain – a case study', in D. J. Easterbrook and V. Sibrava (eds.), *Quaternary Glaciations in the Northern Hemisphere*, Report 3 International Geological Correlation Program, Project 73, 1–24. (Bellingham, Prague), 274–84.

—— (1980a) 'Reconstructing environmental conditions in the periglacial zone associated with the Last Glacial Stage', *Quat. Newsl.*, **30**, 10–11.

—— (1980b) 'Problems of radiocarbonating the Chelford Interstadial of England', in R. A. Cullingford, D. A. Davidson and J. Lewin (eds.), *Timescales in Geomorphology*, (Wiley, Chichester), 289–304.

Worsley, P., Coope, G. R., Good, T. R., Holyoak, D. T. and Robinson, J. E. (1983) 'A Pleistocene succession from beneath Chelford sands at Oakwood Quarry, Chelford, Cheshire', *Geol. J.*, **18**, 307–24.

Worthington, P. F. (1972) 'A geophysical investigation of the drift deposits in northwest Lancashire', *Geol. J.*, **8**(1), 1–16.

Wright, J. E., Hull, J. H., McQuillen, R. and Arnold, S. E. (1971) 'Irish Sea investigations 1969–70', Rep. Inst. Geol. Sci. 71/19.

Wright, W. B. (1914) *The Quaternary Ice Age*, (Macmillan, London).

Wright, W. B., Sherlock, R. L., Wray, D. A. and Tonks, L. H. (1927) *The Geology of the Rossendale Anticline*, Mem. Geol. Surv. G.B. (HMSO, London).

Wymer, J. (1981) 'The Palaeolithic', in I. Simmons and M. J. Tooley (eds.), *The Environment in British Prehistory*. (Duckworth, London), pp. 82–124.

Yates, E. M. (1955) 'Glacial meltwater spillways near Kidsgrove, Staffordshire', *Geol. Mag.*, **92**, 413–18.

–(1956) 'The Keele surface and the Upper Trent drainage', *E. Midld. Geogr.*, **1**(5), 10–22.

——(1957) '*A contribution to the geomorphology of North-West Staffordshire and adjacent parts of Cheshire*', unpublished Ph.D. thesis, University of London.

Yates, E. M. and Moseley, F. (1958) 'Glacial lakes and spillways in the vicinity of Madeley, North Staffordshire', *J. Geol. Soc. Lond.*, **113**, 409–28.

——(1967) 'A contribution to the glacial geomorphology of the Cheshire plain', *Trans. Inst. Br. Geogr.*, **42**, 107–25.

Indexes

General index

Author index

Place name index

(only specific locations cited)